P9-DWP-014

A Fireside Book
Published by Simon & Schuster
New York London Toronto Sydney Tokyo Singapore

# THE MOVIE BUSINESS BOOK

## SECOND EDITION

EDITED BY

JASON E. SQUIRE

FIRESIDE
Rockefeller Center
1230 Avenue of the Americas
New York, New York 10020



First Fireside Edition 1988
Second Fireside Edition 1992

FIRESIDE and colophon are registered trademarks
of Simon & Schuster Inc.

Designed by Quinn Hall
Manufactured in the United States of America

7 9 10 8

Library of Congress Cataloging in-Publication Data

The Movie business book / edited by Jason E. Squire.—2nd ed., 2nd Fireside ed.
p. cm.
"A Fireside book."
Includes index.
1. Motion picture industry—United States. I. Squire, Jason E., date.
PN1993.5.U6M666 1992
384'.'0973—dc20 92-20211
CIP

ISBN: 0-671-75095-X

*To my loving family*

# CONTENTS

For moviegoers and movie makers,
and to everyone who has two businesses—
their own and the movie business.

# THANK-YOUS FOR THE SECOND EDITION

One of the nicest things about returning to work on a new edition is the prospect of advancing and enhancing content.

As with the first edition, the lion's share of appreciation is extended to the contributors. They gave generously of their time, in the context of pressing job responsibilities, to share their knowledge and experiences with the reader. It's been a joyous collaboration. A bonus on a personal level was not only reconnecting with existing participants to work up new material but also seeking out new people to bring perspective to the widening focus of the movie business. All of the articles were prepared expressly for this book. Over half are new, and the balance has been carefully revised.

There are certain heroes of this edition, who responded swiftly when called upon for help: John Schulman of Warner Bros.; Ken Lemberger of TriStar; Ron Miller of the Writers Guild; Chris Huntley of Screenplay Systems; Marc Berman and Hy Hollinger of *Daily Variety;* and Edward Walters of Simon & Schuster/Fireside.

People who know their own unique contributions are Charles Bornstein, Carlynn Chapman, Annette Insdorf, Steve Kravit, Claire Ollstein, Leslie Rodier, Cathy Schulman, Ron Smith, Hal Vogel, and Walter Squire.

John Wolff and Kirsten Feist deserve special credit. John not only taught me word processing but also graciously gave me access to his advanced computer system, even on short notice. Looking back, this must have caused all manner of inconvenience, but they couldn't have been nicer.

Finally, a note of thanks to those readers who have commented on the prior edition, and an invitation to new readers that I would appreciate hearing about the content and use of this edition, which would help in my ongoing revisions.

See you at the movies. . . .

JES
West Los Angeles, California

# THANK-YOUS FOR THE FIRST EDITION

There are lots of nice people who touch the life of an enterprise such as this. Professor A. William Bluem asked me to be his coeditor on *The Movie Business* in 1970. That two-year, cross-country collaboration was a vivid education and delight. Bill died in 1974, but he lives on in his books and in the memories of his loved ones and of his thousands of Syracuse University students. If there is a guiding spirit behind this effort, it is Bill's. I miss him.

The spotlight of gratitude shines on the contributors to this text. It's their book. These gifted ladies and gentlemen, who daily practice the business on its highest levels, sensed a responsibility to join in and share their expertise and gave their time in an effort to create a legacy for the future. It's been a massive collaboration, just as movies are. Because these participants care about the spreading of understanding about the movie industry, they deserve all the credit. It was a pleasure working with them.

Columbia Pictures was generous with permission to reprint paperwork examples from the production of *The China Syndrome,* arranged through the aid of Michael Douglas, Bruce Gilbert, and Jane Fonda, coproducers, and compiled with great care and skill by Debbie Getlin. The screenplay excerpt is included through the courtesy of writers Mike Gray, T. S. Cook, and James Bridges.

Twentieth Century–Fox kindly allowed their budget top sheet and exhibition contract to be used as illustrations, and the excerpt from the *Butch Cassidy and the Sundance Kid* screenplay is used thanks to Fox and writer William Goldman.

Syd Silverman, publisher of *Variety* and *Daily Variety,* extended the courtesy of allowing excerpts from the journalism of A. D. Murphy to be included.

Many people responded generously when I called on them for help and suggestions during the preparation and editing of the manuscript: William E. Bernstein, Charles B. Bloch, Robert Breech, Leonard Chassman, Kenneth Clark, Stuart Cornfeld, Joe Davis, Stephen M. Kravit, Sidney Landau, Judd Z. Magilnick, Francis McCarthy, James R. Miller, Marc Pevers, Thomas P. Pollock, James Powers, and Steven J. Schmidt.

A warm thank-you goes to Thomas D. Selz of the law firm of Frankfurt, Garbus, Klein & Selz, New York, who rescued the ambition of this book by organizing an effort to allow me to pursue a broader, more thorough approach to the entire subject. Richard B. Heller of the same firm has also been an effective advocate. Joan Brandt of the Sterling Lord Agency in New York was instrumental in placing the completed manuscript. John G. Kirk of Prentice-Hall has provided much-appreciated enthusiasm and support for the book.

During the tough days when all this work was just beginning, my family overflowed with love and encouragement. Good friends who especially helped include Steve Alpern, Sheldon Finkelstein, Sam L. Grogg (who named the book), Robert L. Kravitz, Perry Oretzky, Basil Poledouris, and Martin Polon.

Warm thanks and love are saved for Beverly Canning Bluem, a source of inspiration.

Three teachers who care about movies and who made a difference along the way are Richard Averson, Norman O. Keim, and Howard Suber.

Much helpful typing was done by Margie Bresnahan, who always did a terrific job, usually under tight deadlines. And for providing sustenance, there was "The Apple Pan" on Pico Boulevard in Los Angeles, serving "quality forever" all hours of the day and night. Closed Mondays.

Finally, special gratitude is reserved for my brother, Walter C. Squire, who possesses an intuitive sense about the movie business and who was my sounding board from start to finish. He is always there with good counsel and criticism, and for that he deserves more credit than he knows.

JES

# PREFACE

This is an extensive revision of *The Movie Business Book,* published by Prentice-Hall in 1983 and republished by Simon & Schuster's Touchstone imprint in 1986 and Fireside imprint in 1988. The lineage descends from the pioneering text *The Movie Business: American Film Industry Practice,* published by Hastings House in 1972. It was my privilege to work with Professor A. William Bluem as his coeditor on that book.

We began in 1970, and it was pretty discouraging at first. Early efforts to enlist knowledgeable industry members were unsuccessful. One famous lawyer explained that he was unwilling to share "industry secrets" and suggested that that was the reason why there was no book on the movie business around at the time. But there were enough forthright professionals who understood the need for such a work, and their involvement and expertise made *The Movie Business* an industry standard, the primary text in the field, adopted at colleges all over the country.

The warm reception of that first work coincided with the growth and recognition of film study on campuses. Institutions rich in courses in movie production, history and aesthetics began to broaden their curricula to introduce the business of motion pictures as an interdisciplinary study often coordinated with law and business schools as part of a well-rounded treatment of mass entertainment and popular culture.

*The Movie Business Book* came out in a favorable climate as well. Curiosity about the behind-the-scenes money workings of the business was on the rise, and books that documented movie-industry excess, such as Steven Bach's *Final Cut* about *Heaven's Gate* and David McClintick's *Indecent Exposure* about the Columbia-Begelman scandal, became best-sellers. TV viewers watching *Entertainment Tonight* and other programs became conversant in the latest weekend grosses and wanted to know more. The book was adopted as a text by professors who recognized its structure as a one-semester survey course. By 1990 a growing international readership of students and professionals began inquiring about a revision. This second edition is in response to those requests.

To cover the detail and scope of this book properly, the idea was to go to the source: industry specialists with specific expertise. If there is one point of view imposed upon the reader, it is that of the producer, who, more than anyone, must harness the variety of elements involved in making and marketing a picture.

A note on style: The contributors will often use terms that are interchangeable in industry jargon. Therefore a movie is also a "picture" or a "film," and a studio is also a "financier-distributor" or a "major."

The sequence of chapters is organized to cover the life span of a movie, with every section building upon the preceding one. After an orientation in section I, subsequent chapters follow the process of writing, financing, dealmaking and production, through distribution to theatres, and subsequent exposures. The format breaks down as follows:

I. *The Creators.* An orientation, this section focuses on the most visible of the moviemakers. The producer, represented here by David Puttnam, often chooses the story, does the hiring, is responsible for the money spent, and is the chief executive on a picture, staying with it longer than anyone else. The director, personified by Sydney Pollack, is the captain on the set, working closely with the producer on decisions that precede and follow production. The director is usually the single creator most closely identified with the finished picture. The range of creators includes a comic genius, Mel Brooks; a sensitive writer-director, Joan Micklin Silver; and a filmmaker who does it all himself, Henry Jaglom.

II. *The Property.* Following the overview of section I, we return to the beginning of a movie's life, the word. The work of the writer, discussed by William Goldman, is represented in the marketplace by a literary agent, such as Lee G. Rosenberg (one of the founders of Triad Artists), whose job it is to sell the work so that it becomes a property, in a process usually involving submission to a buyer, perhaps through the story editor, Eleanor Breese. Such a purchase might increase the property's value in published form, as described by Roberta Kent of STE Representation Ltd. and Joel Gotler of Metropolitan Talent Agency, Beverly Hills.

III. *The Money.* Money spent for services rendered is the basis for almost all activity in the business. In this section the practices of the movie marketplace are examined, as buyers and sellers trade off. The trends of protecting investments and issues of increasing costs are examined by Peter J. Dekom of the Los Angeles law firm of Bloom, Dekom and Hergott. The diversity of types of financing, detailed by Norman H. Garey, include venture-capital strategies, described by Sam L. Grogg (based on his experience as a principal of FilmDallas), and widen to an overview from entertainment-industry analyst Harold L. Vogel of Merrill Lynch in New York.

IV. *The Management.* The collective expertise responsible for decision making is covered next. Movie-executive philosophy is described from the buck-stops-here vantage point of Mike Medavoy, chairman of TriStar Pictures; is interpreted by industry consultant Richard Lederer; and is reflected upon by producer David V. Picker, who has served as president of United Artists, Paramount, and Columbia Pictures.

V. *The Deal.* A fundamental truth of the business is that all major transactions are structured to be set down in written agreements. These agreements are key reference documents. The basic elements of every deal are: what is the property or service; who is providing it to whom; what is the time frame; and what is the compensation. A chief architect for many deals is the entertainment lawyer, represented herein by Norman H. Garey, who negotiates the picture deal on behalf of his producer-client with the financier-distributor's business-affairs chief executive, such as Stephen M. Kravit, executive vice president of Kings Road Entertainment. The producer would probably have to secure overbudget protection, as described by Norman G. Rudman of the Los Angeles law firm of Slaff, Mosk & Rudman, and, in pursuit of specific performers and other creative people, will have countless dealings with talent agents, as detailed by producer Barry Weitz.

VI. *The Shooting.* All of the complex preparation and military-style logistical planning is played out during principal photography. Compliance with the budget and shooting schedule is the daily responsibility of the line producer, personified by Paul Maslansky, the nuts-and-bolts executive on the set. This frees the director, represented here by James Bridges, to create with the other artists the necessary illusions amid the constant pressures of time and money.

VII. *The Selling.* Getting the picture on film is only half the battle. The entire enterprise is in financial peril unless the audience can be seduced to pay for experiencing it. The critical relationship between theatrical distribution and exhibition is analyzed by longtime *Daily Variety* financial editor A. D. Murphy. Positioning the paying audience to the first point of sale—the theatre—is the role of motion picture marketing, detailed by Robert G. Friedman, president of Warner Bros. worldwide theatrical advertising and publicity.

VIII. *The Distributors*. The distributors are the vital link to the paying public. Part salesmen and part magicians, theatrical-distribution executives carry on a love-hate relationship with exhibitors to get the product into theatres, while weighing the outlay of advertising dollars with the marketing executives against box-office revenue. Distribution companies range from the studio financier-distributor exemplified by D. Barry Reardon, president of Warner Bros. Distributing in Burbank, to the independents, represented by Ira Deutchman, president of Fine Line Features, a division of New Line Cinema in New York. The highly lucrative area of home-video distribution is explained by Richard B. "Reg" Childs, president and COO of Nelson Entertainment in Beverly Hills.

IX. *The Exhibitors*. The theatrical exhibitor, or theatre owner, is the first retailer in the movie business and has the earliest direct contact with the customer. At that initial point of sale the exhibitor handles the cash, as covered by A. Alan Friedberg, chairman of Loews Theatres, based in Secaucus, New Jersey. At once the consummate showman and pragmatic businessman, the theatre owner is bound together with the distributor by their need for each other. Similar business problems are shared by large theatre chains, as described by Stanley H. Durwood, chairman and CEO of AMC Entertainment in Kansas City, and Gregory S. Rutkowski, vice president, West Operations, of AMC Theatres in Los Angeles; and by independent exhibitors such as Robert Laemmle of Laemmle Theatres in Los Angeles. The next retailer of a motion picture is the home-video store, whose total revenues outperform theatrical, as detailed by Van Wallach, senior editor of *Video Store* magazine, based in New York.

X. *The Audience*. All of the energy and money described up to now has been applied to create a product and an atmosphere around it that will attract a paying audience. With all of the distractions and competition for the leisure dollar, the moviegoer will make a conscious decision to see a specific movie in a theatre or watch one at home. The demographic makeup of American moviegoers and their viewing patterns is detailed in the statistics chapter, along with figures that constitute a global report card for American movies and country-by-country total grosses from the top-ten export markets. In the United States, choosing which movie to see may be influenced by its rating, part of an industry support system described by Jack Valenti, president of the Motion Picture Association of America based in

Washington, D.C. A satisfying moviegoing experience is lik
spark related purchases, especially for children, of the toys and
products created through merchandising deals, as outlined by
ford Blum, president of the Imagination Factory, based in
Angeles.

XI. *International.* This is a global business, as proven by the sta
that approximately half of an average studio picture's revenue is
overseas. This growing market is analyzed country-by-country
detailed overview covering delivery systems and sources of reven
Peter J. Dekom. Anyone becoming involved in coproductions or
seas picture-making will come across overseas tax incentives and
ernment subsidies, as reviewed, market-by-market, by Nigel Sin
of the Beverly Hills law firm of Sinclair, Tenenbaum & Co.,
Steven Gerse, senior counsel, Walt Disney Pictures in Burbank
example of an independent movie company positioned as an inte
tional player is offered by Barbara Boyle, president and cofound
Sovereign Pictures in Los Angeles.

XII. *The Future.* In a world where technological changes occu
breakneck pace, the movie business has been an adapter more tha
innovator. Every generation has seen new technical marvels that h
enhanced the way we view and relate to movies: Television brou
movies into the home; home video freed the consumer from the
anny of passive television, making the viewer an active programm
Each technological marvel was first ignored by the studios, then
braced as a new, substantial source of income. With high-definit
television beckoning, the cycle continues. Futurist Martin Polon
Polon Research International, based in Newton, Massachusetts,
plains the impact of technology on production, on distribution,
theatres and in the home.

This work sets out to shed light on the entire spectrum of feat
movie-making, from the creation, through production, distribut
and exhibition to paying audiences at box offices and beyond, throu
out their exploitable lives. The idea is to cover this territory from
firsthand vantage points of specialists on the industry front lines, a
service to moviegoers and moviemakers.

From this flow other purposes: to provide the movie audience w
exposure to the pressures and priorities of business decisions th
affect, on multiple levels, the final product unspooling at the lo
theatre and at home; to acquaint people working in any aspect
motion pictures, home video or other delivery system with how th
jobs relate to the entire motion picture process; to remove misconce
tions or resistance about the money side of movies and make t
concepts involved clearer and more accessible; and to act as a reco

itudes and styles of doing business that may take
the passage of time.

s a basic understanding of how the movie busi-
with a deeper fascination for the process and
he journey along the way, these goals will have

# INTRODUCTION

*by Jason E. Squire*

This book is all about the business side of movies. Spawned a century ago, the American motion picture matured, in one generation, into a mixture of art and commerce. It continues to have a profound and worldwide effect on behavior, culture, politics and economics.

At its simplest the feature film is the arranging of light images to win hearts in dark rooms. At its most complex it is a massive venture of commerce, a vast creative enterprise requiring the logistical discipline of the military, the financial foreshadowing of the Federal Reserve and the psychological tolerance of the clergy—all harnessed in private hands on behalf of the telling of a story.

In the commercial movie industry the idea is to make movies that attract customers in one form or another who cumulatively pay enough money for the privilege so that all costs are recouped, with enough left over to satisfy investors and make more movies. The profit motive is at work here, but the formula that attracts audiences is as elusive as can be.

## MOVIES AS PRODUCT

A motion picture is an extremely perishable commodity. It lives and has value as long as it is on people's minds and in their frame of reference. The value of a movie to its owners changes as the amount of public access to it increases and as it ages. A successful movie can remain in theatrical release for six months or more, and a failure can be gone from theatre screens after two weekends. But movies can have rejuvenated value each time they are sold in a library of product to a delivery system or format, reaching a new audience in a given territory.

The life cycle of a motion picture in the United States consists of tiers. Theatrical exhibition is the primary one, the first tier, and marketing in subsequent formats will often be keyed to commercial success in theatres. The first weekends of a theatrical run will call for an after-overhead formula of some 90% of box-office gross returned to the distributor, while later weeks, if the picture has staying power ("legs"), will find the exhibitor receiving a growing share of revenue, as the product ages.

The second tier in a picture's life, perhaps six months after theatrical release, is when it is available in video stores as an over-the-counter videocassette sale or rental item and has virtually disappeared from neighborhood theatres. The distributor's share is calculated on a percentage basis, and the product has made a significant jump from the theatre into the home.

The picture ages further before it is available to the millions of subscribers of home pay-TV services, the third tier. Revenue to the distributor from this market is based on a formula geared to subscriber fees to the pay-TV or pay-per-view service.

Another time frame (or "window") generally passes before the picture can be shown on television, either local or network, the fourth tier of exploitation. At this point the movie can be seen via a variety of permutations: on an established TV network; on an ad-hoc national network of stations created to show a library of, say, one studio's recent films; or in a group of other films, sold in a TV syndication package to specific local stations for a certain number of runs over a period of years. The picture has decreased in value because of its prior exposure, and revenue is turned over to the owners based on negotiated payments triggered by each run.

Pictures originating as made-for-network (broadcast, cable, or pay) television movies are subsequently sold via domestic television syndication as further product for local TV station libraries and are sometimes released theatrically in certain other countries.

Overseas markets follow the same general pattern, as dictated by the territorial owners of distribution rights and the sophistication of home technology available.

Naturally there are some pictures that will alter this pattern of release, shifting sequence because of specific deals or perhaps returning to certain formats as revivals because of their uniqueness, depending on what the market will bear. In the United States *Rocky Horror Picture Show* pioneered the concept of weekend midnight showings and became quite lucrative; and a version of *A Christmas Carol* can always be found on a local TV station around that holiday.

## THE MOVIE BUSINESS

Like any business, the motion picture business exists to make money. That's the only way more movies can be produced and the ongoing network of production, distribution, and exhibition can be fed. Comparisons can be loosely made to other industries: production encompasses research, development, and manufacturing; distribution can be compared to wholesaling; and exhibition to retailing. But there the comparison ends because the public's demand and use of entertainment products such as motion pictures are unlike the demand and use of any other product. For example, in no other business is a single product fully created, at an investment of tens of millions of dollars, with no real assurance that the public will buy it. In no other business does the public "use" the product and then take away with them (as Samuel Marx observed in *Mayer and Thalberg*) merely the memory of it. In the truest sense it's an industry based on dreams, and the service that is rendered is entertainment, leaving (at its best) an afterglow of emotions and recollections.

A note about "business." The very word puts some people off. When applied to movies, it once conjured up conflicts pitting West Coast against East Coast, the creative community against the business community, art against commerce. That's all changed. The motion picture professional of any specialty learns early to mix the creative and business sense, out of self-defense.

What's also changed is that "the movie business" or "industry" embraces a wider definition of movie-related (or inspired) product and after-theatrical delivery systems. Since its beginnings, reference to the industry meant licensing to theatres only; later on, it included television showings and venues such as colleges and airplanes. Today home video is the single most lucrative market in the movie business, with total revenue exceeding that of theatrical exhibition. Added to that are other formats, so that movie companies are also in the related businesses of satellite, laser, and pay-per-view licensing (and any other way that movies may be exposed to a consumer for a price), along with movie theatres, book publishing, theme parks, music, merchandising and real estate.

In earlier days, as the industry grew, the studios were vertically integrated production/distribution/exhibition factories. Then, by the late 1940s, the U.S. Justice Department concluded that this structure was anticompetitive, and forced divorcement of one element in order to enhance competition. The studios chose to lose exhibition. Forty years later legal overseers deemed that the climate had changed, and permitted certain studios to buy or invest in theatre chains. Today this return to vertical integration is further demonstrated in typical studio divisions such as home video, television syndication, music and merchandising.

The movie industry defies strict analysis from a traditional business point of view. Any profiling of it points to certain concepts not characteristic of other industries, concepts that can prevail only in an industry whose product is creative. In this regard it is important to know how the money is arranged for, spent, protected and returned. This book is an effort to organize, present and examine all that.

What are the hallmarks of the movie *business*? That movies are a collaborative medium; every picture-making experience is different; financing movies is an enormous crapshoot; there are no hard-and-fast rules; many of the essential decisions and choices spring from intuitive leaps; most successful practitioners possess a personal mix of creative and business sense; creative and business decisions often rely on relationships and personalities; an entire investment is made before anyone knows if the product is marketable; and as far as profits are concerned, the sky's the limit.

These elements focus on a cyclical industry, difficult to chart, rooted in creativity and intuition, involving narcissism and greed, capable of dismal losses and euphoric profits, engaged in an ongoing seduction of the paying audience. At the initial retail level there is the significant act of motivating people to leave their homes to go to see (or rent or buy) a specific movie. The single universally acknowledged marketing tool for this product—favorable audience word-of-mouth—can perhaps be influenced, but it cannot be controlled. As distribution executive Peter Myers once noted, it is a supremely democratic form of entertainment, where customers vote for one movie over another by simply putting down hard cash.

The high risk inherent in the business points to why conservative capital has historically shied away from motion picture investment, although control of motion picture companies has always been attractive to a broad spectrum of players. Industry observers insist that "it's not a business" as a shorthand way of explaining its unpredictable elements, while movies continue to attract creative entrepreneurs and artists who are devoted to the production and dissemination of this most valuable and influential national resource and export.

At its core the industry can be reduced to competing for the work of creators in a limited talent pool that shifts each year based on the strength of specific works and audience acceptance. But the list of producers, directors, writers, actors and other practitioners who have earned commercial success and an industry following is quite exclusive. Therefore energies are intensely spent daily to compete for this limited talent pool and to create new entries into it.

What has happened in the movie business to warrant such intensity? There have been far-reaching economic changes during each of the last few decades. Financial stakes are higher than ever, a trend that can be traced back to (among other forces) the shift to network television advertising in the 1970s and the resulting national-release pattern. That decade also witnessed a redefining of box-office potential, with certain break-out successes underscoring the increased value of revenue from merchandising, book-publishing and recorded-music sales.

The 1980s were the boom years of home video, viedocassette recorders and cable, and their attendant revenue streams (which were used as cushions for higher budgets), as well as the vertical growth of those entertainment companies that purchased theatre chains and developed home-video and other divisions. The decade also witnessed an escalation in costs for both production and marketing.

The 1990s are seeing movie companies taking a global view. International export markets are developing increasing shares of revenue through wider access to customers via delivery-system advances. At the same time a surge of acquisitions has resulted in the majority of American studio ownership resting in overseas hands. The most influential of all these changes can be broken out in chronological order: network TV advertising, *Star Wars,* home video, the eighties cost spirals and the global industry.

## NETWORK TV ADVERTISING

Throughout the 1960s motion pictures were released regionally and advertised with a mixture of newspaper and local television campaigns. Then, a distribution company named Sunn Classics began experimenting with massive, national network television advertising for the regional release of family pictures. That move to saturation national TV advertising is an indirect but significant reason for the escalating costs of today's movies.

When the studios recognized the cost-efficiency of network television, that expenditure forced the simultaneous blanket national release of certain pictures. Over the years, with the advent of theatre multiplexes, print runs escalated from 500 to 1,000 to 2,000 or more for the most commercial product to maximize audience access, generate heat and make the most amount of money in the shortest period of time (thereby reducing interest costs on capital needed for the production and marketing of the product). With more wide releases, the die was cast, and advertising and print costs would continue to rise.

The massive studio releases of 2,000 or more prints leave less room for independent distributors. However, there are a number of active niche-marketing distributors who specialize in smaller-scaled economics and release pictures with a more hands-on approach than the studios. These companies struggle to carve out release patterns and, when one of their pictures breaks out in popularity, the revenue can be enormous, as with any studio picture. In today's movie business, nonstudio filmmakers and distributors work hard at making pictures and at securing and maintaining screens in the face of intense studio competition, and their story is also represented in the pages of this book.

## *STAR WARS*

*Star Wars* rewrote the economics of the movie business. This landmark picture, released by Twentieth Century Fox in 1977, not only broke through the upper limits of traditional grossing potential via repeat business but also redefined worldwide income in book, record and merchandising sales. When the dust cleared, brand-new adjunct businesses had been established that changed the profile of movie economics by generating substantial revenue in other media. With *Star Wars* product merchandising went through the roof. The success of a variety of movie-related books firmly solidified the publishing offshoot of movie tie-ins; and soundtrack albums sales came into their own, proving that an orchestral score could shoot to the top of the charts, much like pop scores of the day, such as *Saturday Night Fever. E.T., Batman, Ghost, Home Alone, Terminator* 2 and others would follow, but at the time, there was nothing, nothing like the economic impact of *Star Wars.*

## HOME VIDEO

Once upon a time you couldn't own movies, and if you were lucky enough to see them on TV, you couldn't decide when. This was consistent with the long-held industry sensitivity over product ownership. Movies were licensed, never sold, until the "home-video revolution." That term has become a cliché, but the impact of home video on the movie business was nothing short of revolutionary. It has expanded the movie-watching aesthetic, added a new household appliance—the VCR—to popular culture and changed television viewing from passive to active. In less than ten years home video grew from an infant market to a source of revenue that exceeds theatrical in size. And true to history, the studios at first did not embrace this new technology. They waited on the sidelines and watched as young upstart companies began reaping enormous sums before testing the waters themselves.

Today every studio has a home-video division, and every household with a video camera is an instant studio. Since filmmaking is a natural form of expression, anyone can learn its aesthetics by simply shooting video and reviewing the results. When instinctive corrections are made in the next try, the language of film is being adopted. On household screens, personal home movies have become interchangeable with major motion pictures. Relatives and friends are seen on the same screen as movie stars.

Feature films have never been more accessible. Home video has changed people's moviegoing habits and increased the numbers of movies we see. The question is no longer whether to see a movie but whether to leave the house to see a movie. And that choice will continue to be influenced by other seductive alternatives both inside and outside the home in future competition for the leisure dollar.

Along the way there will be losers. For instance, network television viewing has declined severely in the face of more choices through cable, music, pay and interactive TV. It is a measure of the movie business's resilience as a primary leisure-time activity that its attendance hasn't declined in the face of growing alternatives. But it hasn't increased either. Movie attendance has remained stable (some say stagnant) over the last fifteen years.

## THE EIGHTIES COST SPIRALS

Throughout the 1980s there evolved a dual cost spiral in the upward trajectory of production budgets and marketing costs. Increased production costs were being rationalized to allow for the shooting time needed for the kind of complicated physical production (including special effects) that audiences were responding to and, among other increases, to pay the higher salaries negotiated for certain key artists by their agents, who had taken on a proprietary role in motion picture deal-making. Increased marketing costs were being rationalized in that they were further benefiting markets that followed after theatrical release, especially home video.

The other, more seductive culprit was home video itself, whose ever-increasing revenue was being factored into production deals, promising to offset a significant percentage of negative cost. This provided a false sense of security to movie executives who were approving higher budgets in light of a seemingly ever-expanding cushion. But by the end of the eighties home video values had stopped growing, having stabilized into a mature market. In contrast studio picture budgets continued to rise, increasing 185% between 1980 and 1990, while marketing (prints and advertising) costs rose 169% in the same ten-year period.

By 1990 high-risk, high-reward picture-making was supposedly warranted in light of the potential global return from theatrical, home video and other formats. Meanwhile agents continued to achieve more money for their clients, while the audience hungered for ever more impressive filmmaking. Once certain levels of artist compensation were reached (gross multiples instead of net, for example), there was no going back. Today the mainstream studio product, whose mass audience expects "tentpoles" of entertainment during the summer and winter holidays, continues to escalate in cost, making the stakes higher than they have ever been. Added is the pressure of a home-video market that has seen the evaporation of "B picture" revenue, and must rely chiefly upon mainstream "A product." The result is more stress on the industry as the cost of movie-making has grown incrementally.

Shooting days are more costly; advertising is more costly; mistakes are more costly. It's more of a crapshoot than ever, at a table restricted to high rollers. Strategies to reduce costs inevitably follow (some are expressed within the pages of this book), but the short-term solution has found certain studios either restructuring, merging or being purchased by larger international parents to achieve financial security.

## THE GLOBAL INDUSTRY

As a measure of how global the industry has become, one need only recognize that more than half of the American studios have overseas owners. Columbia and TriStar are owned by the Sony Corporation of Japan; MCA/Universal is owned by the Matsushita Electric Industrial Company of Japan; the parent company of Twentieth Century Fox is News Corporation of Australia. (As of this writing, MGM-Pathé is in the hands of an overseas bank, Credit Lyonnais of France.) These parent companies come from successes in other businesses: Sony and Matsushita, known in the United States as hardware manufacturers, and News Corp., known for newspaper and magazine publishing. The remaining American parents, the Walt Disney Company, Paramount Communications and Time Warner, are similarly active in nonmovie, entertainment-related businesses. Because of the erratic nature of the feature film business, each studio has had to achieve some form of internal stability from more cash-predictable enterprises, such as theme parks (Disney and MCA Universal), cable TV and publishing (Disney, MCA Universal, Paramount and Time Warner).

This kind of diversity is increasingly essential for any major player involved in movies, to offset potentially long periods of cash-flow uncertainty: the between-hits periods that are the downside of the movie business. Some parents entered the business to provide synergy for their other products while offering this all-important cash flow. For instance, Sony and Matsushita purchased studios to assure software entertainment for their hardware consumer electronics. And after its acquisition by News Corp., Twentieth Century Fox bought television stations, successfully establishing what amounts to a fourth national TV network, to the surprise of critics. Today the company makes theatrical and TV movies, and both types have their TV debut on the Fox family of stations.

Actually the movie business has always been an international business, just never so dominated by the English language as it is today. The pantomime of silent films was a universal vocabulary easily exported with title cards conforming to the local language. The coming of sound created dialogue barriers that, along with issues of politics and economics, curtailed the growing film industries until after World

War II. Then a flourishing of artistic expression in countries including those in Western Europe and Japan—coupled with tourism and the natural curiosity that goes with it—created a cultural appetite that led to the business of importing and exporting motion pictures and later television programming.

In the United States the studios fully financed pictures and distributed them worldwide. But in the 1960s and '70s, in a new global approach, entrepreneurs such as Joseph E. Levine and Dino DeLaurentiis raised money by combining financing from different countries, in exchange for distribution rights divided among those countries. Today, as American studios have shifted emphasis—they are essentially banker-wholesalers now more than picture-makers—product is also being generated by new independent entrepreneurs who bargain with territorial investors and distributors (studios, satellite delivery systems, etc.) to carve out the best possible deal.

The international entertainment marketplace extends far beyond theatrical exhibition to new commercial video formats found in worldwide homes, offering a mix of indigenous and English-language programming. For program suppliers the trend toward privatization of television in the European Community constitutes new sources of potential revenue, although certain restrictions apply. The EC market represents 100-million more customers than in the United States. Burgeoning delivery systems into the home are growing along with the entrenched VCR market, so that estimates project that VCRs are in nearly 50% of EC homes, cable TV in nearly 40% and satellite delivery in 25% of EC households.

Overseas many countries support movie industries whose product divides into two types: one that is domestic, that simply doesn't travel, and another that is suitable for export because of its content and style. This second type is made available through sales organizations to a growing global market in a variety of formats. Studios and entrepreneurs alike are positioning themselves and joining in configurations to reach wider movie audiences, and the global market is growing. Japan stands out among the rest as the largest export market for American films, so that it is logical that Japanese companies have been investing in American studios and production companies.

The system of buying and selling movies for import and export is based on supply and demand. Variables such as currency fluctuation, government regulations, frozen funds, local taste, censorship and pirating come into play. The highlights of this commerce are film festivals and selling markets worldwide.

Although the issues of market potential, investment regulations, film content and audience taste vary around the world, the basic business skills applied to commercial movie-making remain the same. In developing a screenplay, action may be rewritten to conform to budget; in deal-making, all parties must be protected; in securing funds, ingenuity often prevails; in production, the creative mix is influenced by pressures of time and money; in distribution, imaginative selling is required; and in exhibition, the anxiety of anticipating public response is universal. Pictures are a gamble, and businessmen work to apply formulas and intuition to reduce odds. Because the product requires the spending of huge amounts of money, it is up to lawyers, accountants and agents to construct some framework of responsibility and recourse to support the creators and financiers.

All these elements are basic to the movie-making ordeal and know no boundaries. The business practices and deal-making procedures that govern most of these transactions are based on American models. And these practices and procedures constitute the heart of *The Movie Business Book,* Second Edition.

I
THE
CREATORS

# THE PRODUCER

*by* ***DAVID PUTTNAM,*** *who in the mid-1970s had gained a reputation as champion of new directing talent by backing the first pictures of Michael Apted,* Stardust; *Alan Parker,* Bugsy Malone; *Ridley Scott,* The Duellists; *and Adrian Lyne, with* Foxes. *Subsequently he worked with directors including Alan Parker again,* Midnight Express; *Hugh Hudson,* Chariots of Fire *(which won both the British and American Academy Awards for Best Picture of 1981); Bill Forsyth,* Local Hero; *and Roland Joffe on both* The Killing Fields *and* The Mission. *In 1986 Mr. Puttnam was asked to become chairman and chief executive officer of Columbia Pictures, a post he relinquished 15 months later to return to producing. His recent pictures include working with director Michael Caton-Jones on* Memphis Belle *and Istvan Szabo on* Meeting Venus. *In addition to his many civic and academic honors, Mr. Puttnam serves as president of the Council for the Protection of Rural England and chairman of the National Film and Television School. He is based at Pinewood Studios, outside London.*

*The producer has one absolutely crucial week on a movie; it may even come down to three days: the Wednesday, Thursday and Friday of the second week [of shooting]. . . .*

There are almost as many ways of functioning as a producer as there are producers. Sensible producers devise a working system that maximizes their personal strengths—personality, knowledge, talent—and minimizes their weaknesses, bringing in people to compensate for those weaknesses. It's dangerous, in fact (and not particularly helpful), to mandate a way of producing pictures or to try and be someone else's type of producer. I admonish the reader to take the following observations as mere advice and to strike out on one's own to develop a personal style of producing.

In many respects I am something of a throwback. I started out in the late fifties and sixties working in an advertising agency where, as an account executive, I was not allowed to have much input into the creative process. Luckily I was able to reconstruct the job in such a way as to give myself a fair amount of say as to who wrote and designed the ads I worked on. Coming to movies from this experience, I had the confidence to apply the same skepticism and reconstruct the role of the producer in a way that offered the type of creative satisfaction that I was seeking.

The first film I produced, *Melody,* fell together more by luck than by judgment. Although I was more of a passenger, I made sure to pay close attention to the process, and didn't make too much of a fool of myself. The second film (*The Pied Piper*) was just the opposite, a catastrophe. I learned a great deal from my mistakes and completely revised my thinking. In those days a producer's job was simply packaging. I decided that if I was going to have ghastly experiences like this second picture, I'd like them to be my fault, instead of trying to ride herd on a situation over which I had no real control. As a result on the third film, *That'll Be the Day,* I set the parameters around which I've worked ever since: very hands-on, very involved, with semiautobiographical story elements that I can specifically relate to. Now let me tour through the picture-making process and explain how I apply these parameters to the job of producing.

With few exceptions I generally either conceive of my stories or find them in newspapers. *Chariots of Fire, Local Hero* and *The Killing Fields* all had their origins in newspaper reporting. It's impossible to explain why one story intuitively imposes itself on your imagination rather than another—a bit like trying to justify falling in love with one woman in a room in which many would seem equally attractive. This is organic and should stay that way!

The casting of screenwriters to fit material stems from reading vast numbers of screenplays, becoming familiar with writers and their craft and knowing the tone and content you want. Monte Merrick was chosen for *Memphis Belle,* a movie with ten major roles, because I had read a screenplay of his wherein he juggled several characters, giving them all quite specific identities.

In my development process I normally pay for an option to buy the chosen material, say an original screenplay, making sure to have the right to extend the option period for a smaller additional sum of money, whether or not the extension money is applied against the purchase price. Because I work slowly, my option period is longer than most, normally covering two years, extendable for a third year.

In terms of cutoffs in a screenplay deal I am quite singular; I stay with a writer over many, many drafts. Over the years I've observed that the batting average of writers coming in and rewriting is low. This puzzled me until I went to Columbia. There it was clear that a studio executive has an average of forty minutes for a script conference and is under great pressure to report progress at weekly production meetings. The easy way to apparent "action" is to decide to replace a writer in the course of that script conference rather than taking the time necessary for the complex, tedious, difficult task of properly developing a screenplay from one draft to the next. Unless a writer feels "written out," I try to stick with the same one, since that person probably understands the story problems and solutions better than anyone and, I would hope, remains eager to implement them; it's the devil you know as opposed to the devil you don't know—and probably can't afford!

A fundamental task of a producer is to make the project as risk-averse as possible. My advice is to bring in bargains, relatively inexpensive pictures on or under budget, since you can't mandate the success of a movie. Here is where the American system is forgiving and the European system is not. In the American system, if you deliver a *good* movie in a timely and responsible manner but the picture fails at the box office, you are not regarded as a failure. With the next project you are remembered as the person who delivered the last picture properly. In Europe, no matter how well you produce a film, if it's not a critical or commercial success, there is the smell of failure about *you*, and the next project may suffer.

The key to remaining stable is being prepared to think the unthinkable, being prepared to "walk away from this picture," which can be an expression of either power or sacrifice. I've done it a few times and saved myself an awful lot of grief, most notably on *Greystoke*. During preproduction I suddenly realized I didn't know *how* to produce the film; it was too complex. I didn't believe it could be done for the amount that Warner Bros. wanted to spend, and I didn't want to be on the sharp end of their disappointment. I walked away and saved myself a year or more of deep pain.

Once a project is green-lighted, there are certain dangers to watch for. One is being pushed into a fixed release date, which is lethal for a producer. This got to be known as "shooting a release date" and doesn't happen very often (though there are well-publicized examples). Once a producer is trapped into this, all of the normal controls that exist in the making of a movie can go out the window. Another danger is building the film's existence around one important cast name. Once this occurs, you are to a very great extent at that person's mercy. There are several films that never got beyond the development stage because of the rewriting demands of major actors who were attached too early on.

Another danger is the pressure to deliver a budget before locations have been adequately scouted. This is always counterproductive, since a cheaper way of making the movie results from a full assessment of all the possible locations with the production designer and director. Money wisely spent in preproduction is an *investment*. Spending an extra $100,000 in preproduction can save $1-million during shooting. Sometimes it's difficult for studios to understand this, since preproduction expenses seem like high-risk money (in that there's still a chance the picture can be canceled), but their attitude and their professionalism in this respect is steadily improving.

My advice in budgeting is: Don't lie to yourself. The temptation is there, especially if a picture won't go forward if it's budgeted over a certain figure. When you begin to trim a budget to reach a financier's rigid number, you start questioning items, which *can* lead to disaster. Do we really need a unit doctor? Do we need a standby ambulance? No, until something goes wrong, just as with insurance. On location for *Memphis Belle* there was a plane crash, and three fire engines raced to the scene in seconds. At that moment I realized that if anyone had asked during preproduction whether we needed three fire engines standing by, the answer probably would have been to cut back and save money. After that I wrote to Warners suggesting a firm corporate policy that, notwithstanding budgetary pressures, certain areas should *never* be cut, including medical, security and fire. In this way if someone is irresponsible enough to cut back in those areas and there is an accident, a company can honestly point to this rule in their terms of business and protect themselves.

In scheduling the same rule applies: Don't lie to yourself. Also, keep aside some sequences that you know you could complete the movie without and place them at the end of the schedule. In other words, don't find yourself shooting sequences in week 2 that, if push came to shove, you could drop. Keep them tucked into the end of the schedule so that they can be deleted if need be. They can also act as a valuable spur to a tired director!

On *Chariots of Fire* I made a scheduling decision that turned out wonderfully well in hindsight. There was a rather tenuous financing arrangement involving two companies who were funding it not out of any particular enthusiasm but because of other, more complex commitments. I wanted to impress them, so I scheduled a rather big production sequence for the first week. Sure enough, in week 2, when problems began to arise, I had something wonderful to show them. There was a sense of relief on the part of the financiers, who later generously agreed to invest additional money.

This leads to another bit of advice: While it can be important to shoot a complicated production sequence in the first week to get it behind you, tuck it into Thursday or Friday. It takes at least seventy-two hours for a crew to get to know each other. It may appear as if everyone is functioning properly that first Monday morning, but the real work won't emerge until Wednesday. You've got to be emotionally and, if possible, financially prepared literally to throw away the first two days of shooting.

The producer has one absolutely crucial week on a movie; it may even come down to three days: the Wednesday, Thursday and Friday of the second week. All the most fundamental decisions will have to be made within that time. Whatever is going wrong must either be changed then or you will be stuck with those problems for the duration of the production.

Two examples involve the relationship with the director and the working rhythm of the film, both of which become established very quickly. As producer you should already have done your homework and learned about the director's style and the other artists' abilities by asking those who have worked with them in the past. Your choices should have been based on this *and* on intuitive judgment. Certainly problems during that precious first week of shooting will ring alarm bells in your head, since they may well be contrary to those early assumptions. If need be, this is the time to have a confrontation with the director. There is a week's rushes to look at, enough material to judge the competency of the director as expressed through the visual style and the performances. If changes are required, there is this narrow window of opportunity; now's the time. If the working rhythm indicates, when charted out through the schedule, that you will be two weeks over, it must be adjusted, with the director's cooperation, by the end of that second week, or you *are* destined to go over.

Once a film has started shooting, the pivotal relationship is between the producer and the director. The best "in-production" relationship I ever had was with Roland Joffe on *The Killing Fields*. Because it was an enormously complicated film, we would dine together almost every night, talking over that day's problems and the next day's challenges, and that was very productive. There was never any equivocation. When we had crises, he totally understood them; when things were working well, he knew why. The key is trust. If the director suspects the producer is likely to lie about having more money (or less resources) available than there actually is, the basis of the relationship will be undermined quickly. It is demoralizing for a director to think the producer has a hidden agenda, just as it is demoralizing for a producer to believe the studio has another agenda altogether.

Let me emphasize that once a film is green-lighted, the key is absolute trust. A producer must have the confidence to be able to say to the director, "This is not working," and the director must have the confidence to respond, "How am I going to dig my way out of this?" Then you work together to dig your way out. There *will* be crises; there's never been a movie made without crises. It's the ability to deal with those crises that is the making of a good or a bad producer. Once a film starts, producing is, for the most part, crisis management; you have no other real function. In most other respects you're in the way!

During shooting I try to time my appearances so that they have some relevance. I try to be there at the beginning of the day (if you're there, why isn't everyone else?), during lunch and at the end of the day. It is very good for the crew to see the producer lunching with the director and other artists, for it communicates a sense of continuity, of family. I always reserve the right to decide what time the movie wraps, rather than hand that over to the director. If that is lost, you have effectively lost budgetary control of the film. If the director decides how long to shoot, the director controls the budget. There are exceptions. On a recent shoot in Hungary we took out a completion bond with a local studio, so I didn't have to look at my watch once. Whether we wanted to wrap at six or at nine, we had a fixed cost; any overage was their problem.

The core working group from the producer's perspective during shooting is the production manager, first assistant director and the production accountant. They're your SWAT team, who help sort out your problems. With their input you bring a series of sensibly thought-out options to the director, and the director decides among those options. As an example, on *The Duellists* we were behind from day 1 due to weather. I would meet with Ridley Scott every night and discuss options to get back on schedule, such as condensing a sequence and moving on after lunch or cutting the size of a crowd, and he would decide.

Let's consider the issue of going over budget. As noted earlier, the first line of defense is to recognize the pattern early (again, by the beginning of week 2) and confer with the director in order to correct it by reworking the schedule or your shooting system in order to catch up. But what if the picture is clearly better off for going over one week? In that case, instead of trying to hold to the schedule, you must pay for the overage by imposing savings from other below-the-line categories. Either decision is made by working closely with your SWAT team and presenting the alternatives to the director. It's important not to become ideological about it, not to see it as a war between producer and director, because like it or not, you're in it together.

In the unlikely event that the director does not respond constructively in the face of such an overage, you have a battle on your hands. It also means you've made a flawed judgment in selecting that director. One of the things I try to do in preproduction is to feel out that type of issue, to create one or two minicrises to see how the director reacts. During the location reconnaissance ("reccie"), if option A is closed down, how flexible is the director in exploring option B? What you don't want is a "pussycat" director—"Everything is fine, whatever you say"—because that kind of lack of conviction will end up on the screen. The ideal director is someone who is decisive and flexible: firm in what he or she wants; flexible in understanding the nature of the producer's problems and in helping solve or at least address them.

The value of choosing to work on location is realism. Another important element, especially in exotic locations, is the ability to photograph indigenous people rather than shipping in extras. Since travel time is wasted time, another concern is how close the crew lives to the location. If travel time is an hour and a half, that means three hours are wasted each day, and that's both expensive and stupid; every fourth day is effectively wasted.

Currency is another issue on distant locations, and my advice is to buy currency forward. Once you *know* what the picture is going to cost, buy the currency for the entire location schedule; don't be a victim of fluctuation. As an example, assume the pound moves against the dollar from $1.72 to $1.99 over several weeks. If the movie is budgeted in dollars but you are spending pounds, the cost of your dollars has escalated by some 15%, a heavy unforeseen overage. To avoid this, convert your financing into the local currency as soon as possible. (If a studio balks, reach agreement with them such that they can keep any benefit from fluctuation but that they protect you from any losses.) Don't play the currency market—it's not your job, and you've got a movie to produce.

Logistics are extremely important, since shooting on location is parallel to a military operation, calling for moving great numbers of people and equipment, keeping lines of communication open, seeing that everyone is well fed and maintaining flexibility so that plans can be changed at short notice.

As producer you have the absolute obligation to keep the crew happy, well fed and not exploited in terms of working ridiculous hours, which may become counterproductive, making them tired and irritable. At the end of week 2 it's a good idea to throw a little unit party, which brings people together and consolidates relationships. One should be reasonably quick to criticize but even faster to praise. I advocate a "slush fund" in a budget of perhaps 1% of the budget from which you can buy small things such as birthday presents or a round of drinks or award bonuses where deserved, all of which, in addition to a couple of unit parties, generate good morale and a sense of well-being.

Outsiders wonder why so many people are needed on location. In my experience there are in reality seldom more people than are necessary, especially in support staff, who can always fill in where needed. The most expensive element in shooting a film is *time*. The addition of an extra person can often fill a gap so that every given hour in the day is properly used. Don't scrimp with your key personnel. Beware of bargains. If proper planning with an experienced director of photography avoids *one* day's error, he has paid for himself.

During production it's important to prepare a weekly progress report reflecting the budget changes, specifying over- and under-budget shifts from the previous week. I force myself to spend half a day a week on this. In every area where the change is significant I write a detailed explanation. This is not only for the record; I have found that writing about these problems makes me address them head-on and search for solutions in the coming week. I'm a great believer in putting thoughts on paper.

If you've prepared properly and done your numbers right, the relationship with the financier during production should be amiable. Stay in contact through phone calls every week or so, reporting progress. I adamantly resist sending rushes back to the studio; they have every chance of being misinterpreted. The *only* people who can really understand the rushes are the director and the editor and, to a limited degree, the producer. Rather, after a few weeks I send them a series of selected takes, to offer a flavor of the movie. On *Midnight Express* the rushes for two out of the first three days were unusable. Had Columbia seen that material, they might have panicked. Luckily we were on location in Malta and were able to cover by rescheduling and get back on track. Crews are great at covering for each other!

Editing normally begins *during* shooting, and your relationship with the editor is most important. I've done three-quarters of my films with the same two editors and trust them implicitly. The director must come to trust the editor's judgment as well. But let me emphasize that the director-editor rapport must be such that the editor tells the director the *truth*. If the rushes aren't working, the editor is the first person who knows it and must be comfortable enough in the relationship to convey that to the director. Bluntly, the editor must not be someone who looks to the director to remain employed. On location we tend only to see rushes on video, so the daily call to the editor at home base to assess the quality of the rushes is vital. On *Mahler* I asked Ken Russell if he wanted to look at the rushes. He asked if we had any room in the schedule to reshoot, and I said no. Then he said, "What's the point in looking at the rushes?"

I try to bring the composer on very early, if possible at the screenplay level. The three essential creative contributions on a movie exist on two tiers. On the first tier is the director, writer and composer, followed by the trio of production designer, editor and cameraman. What you see on the screen is an amalgam of their work, and the producer's job is to ride herd on them. In the case of *Chariots of Fire* it's hard to say what the movie would have been like without Vangelis's brilliant score. Since I happen to enjoy music, my research into finding composers is a constant delight. On *The Killing Fields* we decided that, visually, many sequences would end with machinery wiping the human element off screen. We sought a cacophony of sound, and I thought of Mike Oldfield's *Tubular Bells*. He was keen to do the film and came on board before shooting started.

Now let us assume we have completed principal photography on schedule and we turn to the final phase, postproduction. It's important to do postproduction close to home, which in my case is in England. This way you are familiar with all the elements and can maintain control. I've worked with the same dubbing mixer for twenty years, and there is a mutual trust and a shorthand that develops. If there is something he cannot achieve, I sincerely believe it can't be done.

During the latter stage of postproduction there is the question of when to preview. We tend to do a rough, four-day mix to give a sense of the sound track, including some music, for the previews. Only after the previews will we do the full-blown sound mix, finely detailed.

Previews are immensely valuable. It's the first time you learn how your preconceptions fit with those of the actual audience. First, the film will *feel* long. Individual cuts, scenes, moments and rhythms you've grown to like will have to be adjusted, because you can feel the audience wanting the film to move on. Next, you forget how smart the audience is, and find that the opening reel is far too expository; you just don't need that body of exposition. I've always tried to leave some money in the budget for reshooting a couple of days of odds and ends, useful inserts learned about in the previews. This "shooting to the cut" is remarkably economical (with an almost 1:1 shooting ratio) and can solve enormous problems. Not every studio will agree to this, but it's extremely valuable and often comes down to a measure of their enthusiasm for the picture.

What's the ideal preview audience? It depends on the picture. We previewed *Memphis Belle* to very broad audiences and took our chances. On *Meeting Venus* we previewed to preselected audiences of people who had seen films like *Dangerous Liaisons* or *Reversal of Fortune*. Before I preview for the studio in Los Angeles, I pre-preview in the U.K. to discover obvious problems and make adjustments in advance.

There are two types of preview: for production and for distribution. The production preview points up the strengths and weaknesses of the movie and allows us to make improvements before the final mix. Then you lock the film and move to the marketing and distribution issues. For distribution previews we show essentially a finished movie and are addressing marketing decisions and market positioning.

Conferring with marketing executives is an ongoing process. After the film is locked in, the key relationship for me becomes Rob Friedman, head of marketing at Warners (see article by Robert G. Friedman, p. 291). It is up to him and his colleagues how to position the film, and up to Joel Wayne, who cuts the trailers and does the print ads, with enormous responsibility resting on Barry Reardon, who handles the release pattern and theatre dates (see article by D. Barry Reardon, p. 309). I trust these executives totally and prefer to bow to their expertise. If we disagree, I would on balance rather have them misjudge a movie and remain confident in me as a colleague than win some hysterical disagreement over a specific movie and destroy the relationships. Also, since I am not prepared to accept interference in *my* area, I must in fairness take a similar view and respect what they do.

As far as release patterns, *Memphis Belle* was an interesting fluke. It was originally set to be released in the U.S. in August 1990, but that was delayed by the Persian Gulf War. Since we couldn't change the U.K. release date (we were tied to the fiftieth anniversary of the Battle of Britain), the picture opened there first, and five weeks later in the U.S. This changed the pattern for British journalists, since they always decide how much space to give the opening of a movie based on its success in America. We all had to do a lot of improvising, and we added to the advertising budget. The picture opened strongly in the U.K., and this gave everybody confidence.

Approaching the release date, marketing expenses are committed, theatres are committed and tension is enormous. You die a little, because there's not much you can do to affect the outcome! If a picture opens stronger than expected, rejoice and chase it with resources to maximize it. At this point an extra $1-million or so must be available to drive it over the top. If a picture opens poorly and it is in broad release, there's *nothing* you can do. If it is in platform release, there may be ways to rescue it if reviews are good and exit surveys are positive. Perhaps the movie can be repositioned through recutting the trailer and rethinking the advertising campaign. Jeffrey Katzenberg at Disney is particularly admirable in that on at least two occasions he has gone back on a second week, spending more than the opening week, to reposition a movie and try to rescue it. The opening weekend for *Memphis Belle* in the U.S. was good but not great. Since we knew women loved it, in the U.K. and Japan we sold it as a fairly romantic film. In the U.S. we emphasized it as more of an action film, trying to attract young men. In hindsight we probably should have kept after the audience of women that we knew were responsive—you live and learn.

Once a picture is released, it takes on a life of its own as it proceeds through the varied release patterns and viewing formats to the increasing movie audience around the world, making, one sincerely hopes, some kind of positive impact.

Meanwhile my ambition remains to go on exploring the creative and commercial potential of movie-making, especially in the area of European coproduction, and working with Warners to try to produce a truly *great* movie that defies the tyrannies of language and nationality!

# THE DIRECTOR

*by* ***SYDNEY POLLACK,*** *who has earned distinction as a director with such theatrical feature credits as* Havana, Out of Africa *(winner of seven Oscars, including Best Picture, Best Director and Best Screenplay),* Tootsie *(New York Film Critics Award),* Absence of Malice, The Electric Horseman, Bobby Deerfield, Three Days of the Condor, The Yakuza, The Way We Were, Jeremiah Johnson, They Shoot Horses, Don't They?, Castle Keep, The Scalphunters, This Property Is Condemned *and* The Slender Thread. *His fifteen pictures have received 43 Academy Award nominations (four for Best Picture), and eight appear on* Variety's *list of "All Time Rental Champs." Mr. Pollack has also received the Golden Globe, the National Society of Film Critics' Award, the NATO Director of the Year Award and prizes at the Moscow, Taormina, Brussels, Belgrade and San Sebastian film festivals.*

*Every time the director says, "Let's* try *it this way," and not "Let's* do *it this way," money is being spent at enormous rates. . . .*

For me the first stage in developing a film is, obviously, finding and working on the story. Source material can be in almost any form, from prepublication galleys of a novel to a treatment (or a long synopsis of a story), to firsthand discussions with writers on original ideas, to magazine or newspaper articles. Development is simply bringing the project from a raw, nonproduceable state to a produceable state. In practice, what it means to me is "kneading" the material like dough, through plain hard work with the screenwriter, until it has characters and a "world" that I feel enthusiastic about and comfortable with.

Some writers like to have lots of meetings, get the director's input, then go off alone to complete an entire draft screenplay. This is a gamble that sometimes leads to success and sometimes doesn't. If the writer is off on an unwanted story tangent, the problem will usually be compounded throughout the screenplay. Other writers like to give a director four to five pages at a time, discuss them, and rewrite as necessary. A middle ground might have the director and writer meeting several times before the writer goes off and completes half a screenplay, fifty to sixty pages. Then they might meet again and discuss any problems or new ideas growing out of those pages. Another method might find the writer and director writing together in a room, arguing about characters and dialogue. This is the closest form of collaboration and depends on a strong relationship between the two. Naturally, whatever choice is made originates with the personalities and work habits of those involved. Since I am not a writer, I selfishly look for a kind of alter ego who can come closest to creating a work that accurately depicts what I visualize or care about. In the best of circumstances a shorthand is developed, a mutual intuition.

All creative work is some form of a controlled free-association process, which doesn't evolve 100% consciously. For me, working with the writer means page-by-page analysis (sometimes to the exasperation of the writer), covering details ranging from something as specific as dialogue rhythm to something as general as the central idea—a theme, spine, armature—to be used as a guide. It is useful to identify and state problems, through the posing of questions. Sometimes you just talk, or "hang out." Nothing is ever really wasted. It may appear to be wasteful to go for a walk and have some coffee together, but it is all part of the process. If you can clearly articulate a story problem, for example, then it remains on your mind, consciously or unconsciously, always percolating. That evolves into specific work when the writer turns to rewriting. A final technique for me is to sit with the writer as I type through the screenplay myself. This way I absorb it and also begin staging the scenes in my head. Sometimes I find holes or story problems during typing that are not otherwise apparent.

As far as reading material, I sometimes use readers for the initial screening because of the sheer bulk of submissions. Often I've chosen material that has no objective merit, only subjective merit. In those cases the value of the project becomes clear only after it is developed to conform to a subjective vision. On the other hand, there have been many projects that I have turned down but that other directors have made into wonderful movies.

It is impossible to end any discussion of material selection today without referring to agency packaging. Although I frankly don't like the idea of being part of a package, I would not have made *Tootsie* otherwise. The script came to me with Dustin Hoffman attached from our mutual agent, Michael Ovitz of Creative Artists. Although I responded to the idea of the story, there were problems with the screenplay, so I initially passed. Michael persuaded me to develop the script to see if I might find an approach that was satisfying. This worked, happily, but I would never have been involved without his perseverance.

Agents have become very powerful. In order to be competitive, they have had to invent new ways of agenting. In earlier times most agents functioned as deal-makers and morale boosters, and it became a matter of choosing a compatible personality. But agents began to define new tests of their agentry. One way is to involve a director earlier than someone else, even though he or she has not been offered that script. This is sometimes an uncomfortable situation to be put in: agents sometimes submit photocopies of scripts that they are not actually supposed to be reading. The triumph for the agent may be to boast access to the director. Another "emblem" of agenting can involve information, whereby one agent demonstrates better connections than another. The newer, tougher breed of agents define their merit by a new set of standards, having to do with being packagers, politicians, "intelligence" gatherers and information traders, to ingratiate themselves and be viewed as distinctive, in order to compete.

Whether or not some of this behavior is arguably unprincipled, it represents a change in the function of agents and how they are perceived in the business.

A result of their new power is that agencies today share with studios much of the decision-making process as to what movies get made. That process was once exclusively studio-driven. The question is, Has this resulted in better films? I don't think so. Do these pictures make more money? Without question, yes. Is that the primary reason for the existence of Hollywood? Unequivocally, yes. In this way the agencies are helping accomplish what the system is designed to do.

Economics is the most inhibiting factor for a mainstream director making a film. Increments of overbudget twenty years ago were minuscule compared with those of today. When you start moving in fractional increments of $40-million, it's a lot different than $10-million. Most creative people, with the exception of actors and directors, create alone and in silence. With actors and directors it's a little like taking your clothes off in public. There are hundreds of technicians; if we are on a street, there are horns honking and the clock is ticking away. Every time the director says, "Let's *try* it this way," and not, "Let's *do* it this way," money is being spent at enormous rates. The average Hollywood studio out on location spends between $85,000 and $135,000 a day. The high range often applies to period pictures, calling for rentals of vintage cars and wardrobe for extras whose costumes must be fitted and sewn. Because the numbers are so high, one is far more conscious of dollar pressure when making choices during shooting. Is take 7 really better than take 1? How much better? $5,000 better? $10,000 better? No one can decide that except the director, who must try to be intransigent while hoping that every decision is making the picture better and therefore not wasteful.

Once the studio green-lights the picture, preproduction begins. On *Tootsie,* this ran three and a half months; on *Out of Africa* it was a year. What accounts for this spread? *Tootsie* was a modern-dress picture shot in the studio and on location in New York. For *Out of Africa* farmland had to be planted, locations had to be scouted, a period reproduction of a town had to be built, clothes designed for 1913 had to be made, extras had to be found.

In preproduction the main focus is on finalizing the script, budgeting, scheduling, casting and location scouting. As a director who also produces, I use a line producer to help me make economically sound decisions, with an eye to what the creative consequences will be. That person signs the checks, advises on union rules and negotiates the deals with most of the below-the-line personnel, such as location managers, production designers, prop manufacturers and costume designers, among others.

The person functioning as line producer for me does the same job that is known variously as production manager, unit production manager or associate producer, and the actual credit designation has to do with experience as well as the hierarchy and division of labor on a specific picture. There are directors who turn all of the producing decisions over to one person, who is justifiably termed the producer. My own feeling is that the person who is designated "producer" should be the person who also has a direct say in creative areas, including selecting the director, casting and heads of departments.

The first thing usually asked a director is "How much do you think it will cost?" Sometimes it is difficult to know in the beginning, but I make an educated guess. Then I work with the production manager/line producer to make a *breakdown,* a scene-by-scene itemization of what will be required: How many days will it take? Should a set be rented or built? Will we be on location or in a studio? How many extras? How much will living expenses be? How large a crew? How much action will there be? Where will we need multiple cameras? (For a specimen of a breakdown sheet and other production paperwork, see article by James Bridges, p. 254.) Sometimes, in trying to be conscientious, you might decide to shoot a complex sequence in one night. The production manager might say no, it involves 800 extras, and after they are dressed, you won't be able to start shooting until three hours after their call, it needs two nights. For every segment of location shooting it is important to save interior sequences as cover sets, to turn to in case of bad weather. It's just common sense.

Casting impacts budget. If a high-end star is committed, the production is saddled with special charges on top of the star salary, such as hiring the star's hairdressers and makeup artist, secretary and security. Each requires another hotel room on location, another airfare ticket, another per diem, another driver (at teamster driver rates), another car rental, plus fringes.

The *shooting schedule* is derived from the breakdown pages, the scheduling of production days in a certain order. Scheduling is a compromise between the aesthetic advantage of shooting in sequence versus the practical, economic needs dictated by shooting out of sequence. Pictures are rarely shot in continuity. The goal is to strike a balance. Obviously in a relationship story one tries not to shoot the emotional breakup scene before the characters have met. For *The Way We Were* we shot act 1 first, which was the college sequence, but within the act, we shot out of order. Then we had to mix act 2 (New York) and act 3 (California) together, because we built the New York interiors in Burbank. When we moved to New York after shooting the college scenes, we shot only exteriors, including the end of the picture, and then had to skip the emotional interior sequences pending our return to the West Coast.

Let's assume we have gone through the entire script in such detail and come down to 60 days of shooting. Since we have discussed the amount of manpower and equipment needed in front of and behind the camera for each shooting segment, we can arrive at a rough budget figure covering per-day costs. Add to that items that are fixed costs, such as site rentals, set building and insurance, along with postproduction costs, for a below-the-line budget of, say, $15-million. This would not include the above-the-line costs, which cover the salaries of the creative people: writers, producers, director and actors. The studio may then say, "We can't spend that much money." So we start to cut. Instead of carrying a crane and the additional men required to operate it for ten weeks, one might agree to group all the crane shots into a two-week period. We forget the 400 extras in scene 24 and do it with 200 well-spaced extras. But nothing saves money on a film as much as cutting days off the schedule; other changes become relatively minor. Eventually, with more discussion back-and-forth with the financing studio, there is a shooting schedule and matching budget that is approved by everyone.

The choice of crew is next and is obviously extremely important. Not only do their various creative and mechanical abilities contribute to the final film, but every moment they save is an extra moment that can be spent creatively. Every director researches the background of proposed crew members religiously. What pictures have they done? What is their personality? How fast are they? Do they get on well with other crew members?

The early stages of production depend upon the nature of the individual picture. If it requires an enormous amount of set construction, for example, the production designer is one of the first people hired. He may have a practical problem: to build a town that will take two months to construct and we are three months away from shooting.

I start thinking about a cinematographer from the first day I come on a picture. We may sit down together two months before shooting, just to talk. He may be finishing another picture, but we will get together for a couple of evenings, run some pictures together, talk about an effect, a concept, a way of working, *f*-stops, light level, color level, focus. These aspects of photography are all part of any filmmaker's vocabulary. It's possible to shoot the same scene twice, doing the same thing with the same actors; yet if they are photographed differently, the mood and emotional effect will obviously change.

*They Shoot Horses, Don't They?* posed a common problem. Since color film tends to glamorize and enrich, the challenge was how to make it look unglamorous. Phil Lathrop, the cinematographer, and I finally did two things: Since we were shooting the entire picture on one set, I needed the freedom to shoot 360 degrees for variety. That meant no standing lights; otherwise, they would be photographed. Lighting people from above, from over their heads, is usually ugly, casting harsh shadows, which happened to be perfect for this story. Next, we experimented with force-developing the film, that is, underexposing the film and keeping it in the developer a few minutes longer, making it grainier, milkier and taking some of the glamorous texture off it.

Location scouting starts with the creative and ends up with the practical—the terrible compromise that characterizes filmmaking. First you look for the physical requirements, and for a certain kind of light and positioning in relation to the sun. If the plan calls for filming all day, you usually try to avoid shooting in the direction of sunrise or sunset. Otherwise, every couple of hours the location will be lit differently, and it will be difficult to match shots. This is an economic issue. Instead shoot against the light, or with the light behind you, for the minimum amount of change. Shooting toward south or north, shots can be matched over a longer period of time than shooting east or west. One would consult weather charts for the area for the specific time the shooting is planned, as opposed to when the scouting is done. You would also look for enough area for the crew to park. It's often misleading to scout on holidays. The idea is to anticipate the exact conditions of shooting. Inquire as to possible distractions. Will there be planes overhead? Traffic nearby? A different season? Again, common sense. I try to minimize looping, so it's important to make sure the location will be as quiet as possible. There is a world of difference between the immediacy of the emotional sound of a live performance compared with the artificial sound duplicated months later in a looping studio.

During principal photography I take great pains with performances. I try to rehearse only the day of shooting, to find a certain freshness. Other directors spend detailed rehearsal time prior to shooting. Sometimes I imagine a role differently from the way the actor sees it. If interpretations differ, there are only three options: Be persuasive and articulate enough to convince the actor; be open to modifying your approach if the actor is persuasive; or fire the actor. I've had to make the third choice only once.

While shooting, the director is closest (in no order of importance) to the cinematographer, the production designer, the first assistant director, the editor and often the costume designer. Then, on a more technical level, the sound people, script supervisor, prop people and others. Every director works differently. I look to the script supervisor for thorough notes for the editor, not for shot coverage. Generally I don't storyboard, except on a large-scale sequence involving lots of extras and expensive equipment. Sometimes I will draw stick figures for the crew, including a shot list, to visualize where we are going. Much of this is like diet: Moderation is the answer. A shot list can help organize thoughts but should never blind a director to spontaneous opportunities; on the set you always try to see what's unfolding before your eyes.

There are two philosophies about production, and both work. The one I subscribe to involves making the film at the very moment you are making it. In the other, the shooting is the photographic record of something that has been preplanned and rehearsed; Hitchcock worked that way. For me the film gets made at the moment it is happening. I try to be very alert to what is in front of me at that moment that isn't planned, that is new, but I don't hold that out as a preferable way to work for anyone else.

During production weather is often the most unpredictable factor. On *Out of Africa* there was an evening call for 5,000 extras and it began to rain. The water soaked and ruined the hand-made costumes, and everyone had to be sent home. If it were an intimate scene, we could have rewritten it for the rain. But in this case we had to take the loss on the costumes and shoot an alternate sequence while they were made again. This is an example of the kind of crisis situation a director sometimes faces in managing the financial impact of location emergencies. The bigger the production gets, the greater the potential for catastrophe. And problems can come from the most unexpected sources. On *Havana* a cargo container with all our wardrobe was stolen at the airport by someone who thought it would be valuable to them. Even governments can take advantage. In Kenya during *Out of Africa* the government imposed a customs duty on every roll of film coming into the country, even though the film was obviously not remaining there. The film didn't qualify as a duty item under their customs rules, but that didn't stop them. On *Havana* we were careful to complete shooting in the Dominican Republic two weeks before elections, for fear of violence.

There's no magic to running a set. It is a matter of choosing professional craftsmen with compatible personalities. (Nobody likes a screamer or hysterics.) You try to strike a balance between being relatively comfortable and relatively tense. That edge of tension is helpful, but it has to be balanced with a kind of confidence and relaxation so that people trust you, want to listen to you and believe you know what you're doing (even though it may not always be true). One mistake a new director can make out of insecurity is to "show them who's boss." Crew members see through that right away.

Surprisingly, one of the most important aspects of making a movie is staying in shape, keeping your stamina up for the sheer physical rigors of shooting. People underestimate the workload: up at 5:30 A.M., in bed at 11:30 P.M. six days a week, over 40–70 days. (*Out of Africa* shot for over 100 days.) Everyone has his or her own technique to stay fit. I try to exercise every morning, follow a diet during the week loaded with carbohydrates and vegetables, very little protein. On the weekends this can loosen up.

There are all kinds of problems that lead to falling behind schedule, and solutions involve common sense and ingenuity. If it starts to rain, you might try to include that in the scene; if an actor is sick, try to place his material in another scene; if there is a stop date with an actor, try to shoot all his scenes first. Sometimes a complicated sequence scheduled for two days may be redesigned as one shot, perhaps an elaborate camera move completed in half a day. Another way might be to shoot with very long lenses, too far away to see the actors' lips move. Shoot the scene, record the dialogue five minutes later, and two pages can sometimes be shot in an hour. These are extremes, but they work. Sometimes they're even preferable to your first ideas.

This business has its own etiquette for problems. A new director will surely get calls and visits from management. I got them every day when I was starting out. With an established director, management may be worried, but they communicate it in a more subtle way. Perhaps they make suggestions, in a spirit of cooperation, such as asking you to reduce the size of a sequence, or to change locations in the interest of speed, or to move a driving scene to a walking scene. In looking for cost-saving changes, I would usually call the writer, who might come up with solutions as well.

I view the dailies each night and discuss with the editor the takes I like best. We might use a piece of take 1 and another piece from take 2 and so on. The editor starts assembling the film from these designated takes as soon as possible. I view this *rough assembly* about two or three weeks after filming. This is always demoralizing for me because it never looks like what I've imagined. Then the fine cutting begins. Working with two or even three Kem editing machines to use the maximum number of heads, I usually put the edited version on one head and the outtakes of the same scene on the other heads, to compare them. Then, with the editor, we begin to re-mark and re-discuss each scene. We work through the film, scene by scene, in this way, which takes about six weeks to two months. I don't try to shape the movie at this point, but try to get each scene to play as well as possible. This concerns details like who to favor in a scene, whose face to be on for a given line, how much time to allow between reactions, whether to start a scene on a close-up of an actor, on an object, on a two-shot or on a master. Each decision has its aesthetic consequences.

Once I am satisfied that this is the best version I can achieve for each scene, we start considering the shape and rhythm of the overall movie: This scene is too slow; that scene should be cut in half; that one's not needed; this should be moved up front; we need a new scene here that we don't have; some voice-over is necessary; *that* point is never going to be clear; the audience won't understand what we meant in scene 6 when it was set up in scene 2, so we have to do some looping to put a line off-screen to set up scene 6 in scene 2. At the same time I usually start laying temporary music to the cut picture, which helps me test the emotional result.

Then postsynchronization (or *looping*) occurs, which involves bringing the actors back to an audio studio, perhaps to add certain dialogue for clarity or rewrite purposes, or maybe for voice-overs. This process is speeded up with ADR, computer-assisted automatic dialogue replacement.

Next is a *spotting session* with the composer, who was probably chosen before shooting. We watch the movie, discussing in great detail the texture of the music and where it should be placed. What sort of emotion is needed? Should there be a solo instrument with a lot of air around it? In tempo or not? With rhythm? A thick sound? A transparent one? Wistful? It gets very specific. We might decide, "Start the music when she turns her head there and continue to the point where he walks out the door." That calls for a musical cue that is exactly 130 feet 6 frames in length. The music editor creates a master log of all these choices and times. Then the composer goes off to write the score. Sometimes the composer might play some key melodies early on for me, before they are fully orchestrated.

Then several things are happening at once: The titles are laid out and opticals are ordered (such as dissolves, fades and superimpositions); the mixers are choosing sound effects and processing and cleaning up dialogue for the final mixing of the picture; and the theatre trailer is probably being prepared, along with approaches for the print advertising campaign.

All of the postproduction elements require close attention, and most point up the importance of sound. The sound editors and I discuss such things as street sounds; sounds offstage; background ambience in any given scene; the atmosphere of a night at the campfire; the echo of the animals on the plains. These details infuse scenes with reality and mood—a kind of emotional underpinning. There is a premix period of about two weeks, for the primary sound engineer (the dialogue engineer) to conform the dialogue tracks electronically and equalize them. Microphones are not like human ears; they hear things in "patterns." Depending on the angle of the microphone during shooting, the equalization and sound of a voice might be completely different from shot to shot. If part of a scene was shot in the morning and part was shot in the afternoon, and the mike was not in precisely the same position, the sound will be different and must be conformed. There can be different background noise on a track in the third line of a speech than on the first line of the speech. If a close-up was shot later than the master, the cut-in from the master to the close-up might be jarring aurally when this background noise changes. The background can be "cleaned out." But when a sound frequency is cleaned out in the background, the same frequency is also cleaned out in the voice. To avoid this, there are devices, filters, that shut off automatically when the voice starts and turn on when the voice ends.

Around the same time there are the *scoring sessions,* which I always find very exciting. Days are spent with the composer and often with a full symphony orchestra, who perform the musical cues to picture on a soundstage. Assisting the composer, who usually conducts, there might be an orchestrator. Some composers do their own orchestrations, some don't. There is also the contractor, who handles the business aspects of the sessions. Issues of performance are discussed, and evaluations are made that sometimes lead to instantaneous rewrites of a particular cue.

At the final mix the three sound sources—dialogue, music and effects—are carefully blended with the picture in stereo (left, middle, right) and interwoven to fill out the emotional content of each reel, a process that moves at about a reel a day (a two-hour picture runs twelve reels). In this process there are decisions on equalization and balances: At what level of consciousness should we hear the wind, versus the music, versus the dialogue, versus the seagulls, versus the soldiers marching in the distance, versus a woman singing, drunk, in an alleyway? The musical mix from scoring can also be modified in the final mix. And since the picture is usually in stereo, one is also conscious about assigning information to the left, middle, right and/or surround channels.

It's possible radically to alter what an audience feels by tiny variations from literal reality in the sound track: when a sound starts and when it fades; whether it cuts off or overlaps slightly into the next scene; whether it previews the coming scene by beginning just a touch before the cut. Quite often I'll make a straight cut visually but soften it by cross-fading the sound behind it rather than cutting from one sound to another.

Reverb can be a valuable tool to make a sound dissolve. It can be thin and transparent; you can hear through it, much the way you see through an image dissolve. On a mixing board, where one knob controls the source and one controls reverb, pushing the reverb and pulling back on the source at the same time can create a sound that is increasingly transparent and less opaque, becoming ethereal. This can be a useful effect. There was a section in *Out of Africa,* for instance, where I needed Karen (Meryl Streep) to realize that Denys (Robert Redford) was dead. He was hundreds of miles away, but she needed suddenly to *sense* it while she's just standing there. We communicated this with sound, suddenly chopping off the source sound and pushing the reverb all the way so that there was an odd, eerie echo. The sound track can guide the audience often as much as the visual.

Sometimes, in reaction to an artificial emptiness in the background sound of a scene, I wonder how to fill it. A woman's laugh? Two people whispering? Would construction sounds add to the scene? One of my favorite background tracks is the sound of someone practicing a piano, scales in the distance. It's a very evocative sound—moody.

Next is working with the cinematographer at the laboratory in timing the print, correcting the density and the color-balance. Density refers to lightness and darkness, and color-balance is concerned with choosing the warmer (toward red) and/or colder (toward blue) tones. Does the scene require a warm or a cold feeling? More red? More blue? Balancing is a subjective art. Often flesh tones in a given scene are the best guide. Since everybody sees differently, this is also subjective. What's too warm to my eye may be cold to someone else's. We look at a *mute answer print*. It has no sound because until now we have been working with magnetic sound on strips of film separate from the picture. The merger of sound and picture will happen in the printing process.

Meanwhile the sound has been transferred to an optical negative and sent to the lab. A bunch of squiggly black lines become squiggly white lines on the positive, which will be "read" on a projector and translated into sound. The laboratory takes all of the technical data from viewing the mute answer print, joins the sound negative with it and strikes the first final-viewing answer print. The response to all of the postproduction questions should be found in the answer print. From there all the other prints are made.

Next there is the film's transfer to video. This can be done in two ways: either from an approved positive color print or from the cut negative. Tests are made from each to decide which should be the source. In the video transfer it's possible to make up for all kinds of imaging mistakes that cannot be corrected on film.

Marketing, which accelerates with a completed film, is a mysterious process. (See article by Robert G. Friedman, p. 291.) The tension in marketing is often found between the need for the marketing team to persuade as large an audience as possible to attend on opening weekend and the wish the director has for the marketing campaign to be honest as to what they are going to see. Marketing executives are always straddling this conundrum. Since many pictures are either made or broken on their opening weekend, there is enormous pressure to bring people into theatres that weekend at all costs. A primary tool is the trailer, which isolates the dramatic high points while perhaps slightly misrepresenting the film as a result. This forces a film into a category—love story, action, suspense, mystery—which may not be a proper fit. If the picture is misrepresented in its campaign and the wrong audience shows up, they will not like the movie, generating bad word-of-mouth. How does marketing represent a picture truthfully and still get the maximum number of people to attend? That can be a dilemma.

Sometimes I travel around the United States and sometimes other countries to help sell the picture. The major publicity tours can be divided into sectors. Heading east, the major markets are England, Belgium, France, Germany, Italy, Spain, Sweden, Holland and Hungary. To the west there are Australia and Japan. In the south there are Mexico, Brazil and Argentina.

As far as preparing for the overseas markets, it is nice to see that most European countries are leaning toward subtitles instead of dubbing. This is a relief for directors, since dubbing a picture can sound dreadful. In dubbing, different actors are used, and this loses subtleties of performance and ambience. I try to minimize this by working with translators and subtitlers, using dialogue transcriptions, and by approving the voices of actors for other languages. For a measure of how potent the global market has become, *Out of Africa* did 65% of its business outside the United States. It was thought of as an American picture since it was made by an American company, with two American stars (even though one of them plays a Dane), and is about English characters. Yet it made most of its money overseas.

# MY MOVIES: THE COLLISION OF ART AND MONEY

*by* ***MEL BROOKS,*** *who was one of the writers on Sid Caesar's classic television program* Your Show of Shows. *Then he turned to movies, where he won an Oscar for writing and directing a short subject,* The Critic. *He won another Oscar for writing* The Producers, *also the first feature he directed, and went on to write and direct* The Twelve Chairs, *and cowrite and direct* Blazing Saddles *and* Young Frankenstein. *Brooks managed to add more hyphens to his credits for subsequent pictures* Silent Movie *(actor-director-cowriter),* High Anxiety, History of the World, Part I, Spaceballs, *and* Life Stinks *(producer-actor-director-writer). His production company, Brooksfilms, has produced such pictures as* The Elephant Man, My Favorite Year, Frances, The Fly, *and* The Vagrant. *In addition to industry awards, he has received France's Legion of Honor, Order of Arts and Letters.*

***What's the toughest thing about making film? Putting in the little holes. . . .***

I'm primarily an observer of life who formalizes his observations by writing them down. Some of us have no need to tell anybody else about those observations. I happen to have a need to pronounce myself. I started my career as a drummer; I'm sorry I stopped because it still is the best and the loudest way of calling attention to myself.

I began as a writer and I'm still basically a writer. I directed a summer stock company when I was a little boy in Red Bank, New Jersey; I also directed some theatre in the borscht belt and was a drummer-comic there. I'd been writing *Your Show of Shows* and then *Caesar's Hour* on television for years when I decided that it was time for me to leave and do something else with my life. Maybe become a housewife.

When I moved away from Sid Caesar and went into the real world of writing television specials, it was very difficult when the director or producer would say "Thank you'" and that was it; I had no control over the material. When I wrote my first movie, *The Producers,* I decided that I would also direct it to protect my vision.

*The Producers* was first written as a novel. It talked too much, so I made it into a play, which ended up with too many locations, so I turned it into a film. It was right out of my own life experience; I once worked with a man who did make serious love to very old ladies late at night on an old leather office couch. They would give him blank checks, and he would produce phony plays. I can't mention his name because he would go to jail. Just for the old ladies alone he would go to jail. I wrote my heart out, and the movie, I think, is one of my best, though it was not very commercial.

I directed *The Producers* because I didn't want anybody interfering with my words. It was a very difficult course for me to chart because I didn't realize that movies are so expensive.

The story of *The Producers* is a very interesting one. I wrote the script together with help from my assistant, Alfa-Betty Olsen, and every studio said, "Please, no. Try not to come back here." An agent, Barry Levinson (no relation to the filmmaker), knew producer Sidney Glazier. Barry set a meeting (in those days you didn't "take" a meeting) with me, and Sidney, who read it, shook my hand and said, "It's the funniest thing I've ever heard. We're going to make a movie out of this." He went to Joseph E. Levine (Embassy Pictures at that time) and said to Joe, "I will raise half a million dollars if you put up half a million dollars. You distribute the film; it will cost a million dollars to make." Joe Levine said yes but "Brooks can't direct." I said no.

Levine wanted a real director. At a luncheon meeting I ate very nicely; I didn't want to make any mistakes, nothing dropped out of my mouth. I didn't eat bread and butter because I didn't know whether you should cut the bread or break it. Meanwhile Joe ate everything; I had nothing to worry about. At the end of the meal Levine turned to me casually and said, "Hey, kid, you think you can direct a picture?" I said, "Sure." He said, "Swear to God?" I said, "Yes." He said, "O.K. Go ahead," and shook my hand. He was impressed with me; he thought I was nice and cute and funny. He hadn't read the script, but he liked the idea. Like the old Hollywood producers who never read anything, he liked to hear things. "Tell it to me," they'd say. "He goes to this place called Shangri-La." "Yeah?" "And there are these people, they look young, but they're really old." "Sounds good, sounds good."

Sidney Glazier raised half a million dollars from a company called Universal Marion Corporation, and Joe Levine and Embassy put up half a million. I cast the picture with Zero Mostel and Gene Wilder, and we shot it in eight weeks; the budget was $946,000.

When we finished the picture, we took it to the Lane Theatre in Philadelphia for a sneak preview around Christmas 1967. Nobody came. Twelve hundred seats. Eleven people showed up. The movie was over, and I got on a slow train back to New York. Then, since we had shown the picture to some critics earlier in Philadelphia, the reviews came out. They were horrible. Joe Levine was going to shelve it but Sidney Glazier prevailed upon him to wait and to start a little campaign of sneak previews in New York. We opened at the Fine Arts Theatre in Manhattan. Word of mouth had spread; there was a line around the block; you couldn't get in. Levine opened the picture slowly elsewhere, handling it carefully. Eventually it became a moneymaker, though it took four years to get its money back. Levine had risked almost a million dollars. He put up $500,000 for the negative and about $400,000 for prints, advertising and openings. It took a lot of perseverance and work to make that picture become profitable.

I learned a little about movie financing from *The Producers* because I had a participation in the movie, and it was important for me to learn why I wasn't getting any money for my participation. It's impossible for a profit participant to make any money on a movie unless it's a gigantic hit because overhead and interest are always being charged to the film.

My second picture was *The Twelve Chairs*. For it I actually went to Yugoslavia and learned there are two basic costs in a movie: *below-the-line,* which refers to materials and the technical aspects of the film, such as personnel, set construction, wires, lights, cameras, and transportation; and *above-the-line,* the more creative aspects of the film, such as the property itself, writers, producer, director, stars, and principals. Extras are (no offense) below-the-line.

When I finished my next picture, *Blazing Saddles,* I screened it for the Warner Bros. executives. It was quiet. Not even the world-famous bean scene got a laugh. I turned very pale. Thank God it was dark. John Calley, who was in charge of production at the time, and Ted Ashley, who was running Warner Bros., were very nice to me and said, "It's crazy. We like it. Forgive the fact that there was no laughter. These people are studying their various jobs in connection with the picture, and they don't know what to make of it." I said to myself, "It's a failure."

Mike Hertzberg, the producer, was alone with me after they had all cleared out of that screening room, and suddenly he's on the telephone saying, "Yes, yes, screening room twelve, eight o'clock. Be there; invite people." I said, "What are you doing?" He said, "We're having a screening of the same movie. Tonight. All the secretaries at Warner Bros. I'm getting 200 people to see it." Eight o'clock comes, 200 plain humans are packed into this room. They're very quiet and polite. Frankie Laine sings the title song. The whip cracks start; laughter begins. We go to the railroad segment. Lyle, the cruel overseer, says to Cleavon Little, "How about a little good old nigger work song?" The audience gets a little chilled. Cleavon Little and the other guys working on the railroad begin to sing, "I Get No Kick From Champagne." People leave their chairs in ecstasy, float upside down, and the laughter never stops from that moment on.

We next had a sneak preview in Westwood, and that, too, was successful. It was then that the studio executives screened *Blazing Saddles* again and changed their minds. The picture opened to mixed reviews. My films have never gotten unanimously good reviews; I hope they never will. "Everything Mel Brooks has learned about films," said *The New York Times,* "seems to have been forgotten in this mess called *Blazing Saddles.*" When my next picture, *Young Frankenstein,* was released, one critic said, "Where was the great anarchic beauty of *Blazing Saddles*?" It takes them a while to like my films.

When *Blazing Saddles* opened, it got fairly bad reviews and was an instant hit in New York. But Warner Bros. opened it in about 500 theatres in one day across the country. It did well only in 15 big cities. In every other place, like Lubbock, Amarillo, Pittsfield, and Des

Moines, it died. The picture didn't get the word of mouth it needed, so it closed. The Warners advertising executive, Dick Lederer, was my guardian angel. (For an article by Mr. Lederer, see p. 179.) He said, "I think we should spend $3-million to advertise it. Pull it out of all these little cities, open it in the summer when it's gained a reputation, and spend $3-million to support it." There was a mixed vote in the room; the deciding votes belonged to Ted Ashley and John Calley. They said, "O.K., we go with you. Let's spend $3-million, and we'll try and see what we can do." To their credit, *Blazing Saddles* opened wide in June to tremendous business around the country. It's done over $80-million in rentals worldwide in 1974 dollars.

In my experience directing comedies, I've found that timing for laughs is critical. How much space should you leave on screen for laughter before you go on to the next sound on the sound track? Once you've shown the picture to an audience, you and the editor decide. You say, "Look, they're laughing pretty heavily here. Can I have some more frames before we cut from that shot?" That helps. There are ways to find the proper rhythm of jokes and laughter. The Marx Brothers actually went on the road with some of the comedy sequences from their films to test the timing on live audiences.

After a while I can judge within a few seconds either way just how much laughter we can get. Sometimes I'm dead wrong. In *Silent Movie* there was a sequence that no one will ever see; it's on the cutting-room floor. The sequence is called "Lobsters in New York." It starts with a shot of a neon sign that reads "Chez Lobster." The camera drops down to restaurant doors and pulls back. The doors open, the camera goes inside, and we see greeting us a huge well-dressed lobster with claws and tails; around the camera come two other very well-dressed lobsters in evening clothes. The maitre d' lobster leads them to a waiter lobster in a white jacket, who leads them to a table. They order, then follow the waiter lobster to a huge tank. In the tank little people are swimming around. We thought this was hysterical. The lobsters choose some people, pick them up squirming around, and the sequence ends. Every time we saw this sequence, we were on the floor laughing. When we showed it to an audience of secretaries—the first audience to see any of my films—they did not laugh at all at "Lobsters in New York." They stared at each other. Not one snicker. Finally we got some embarrassed sounds and yawns. We threw out the entire sequence as a result. That was one of the surprises that comedy screenwriters get from time to time.

When I did *Silent Movie,* I used a Sony videotape camera that showed us exactly what we'd just shot when we played it back. I also consulted on the set with the three writers who had written the picture with me. The same team later wrote *High Anxiety* with me, and I wanted them around to be harsh critics. Film is such a collaborative process that unless you're writing a very personal story, it's helpful to write with another person. You become a mini-audience right there and then; multiple judgments enter into what you're doing, and that's important early on.

Writers! Do not discuss embryonic ideas. Incipient ideas are your own and nobody else's. When you have coffee, don't talk to other people in the business about your ideas until they are fully written and registered with the Writers Guild. You will not get help; you'll get envy and you'll get stealing. Your ideas are private. Not only will an idea get stolen, but you will let the vapor of creation escape when you tell it. Talk to yourself through the paper; write it down. It's a good exercise, and sometimes it makes money for you.

Comedy is a rough form to sell to studios and independent producers. It's the most mercurial cinematic item. Every once in a while a good comedy comes along. *Sibling Rivalry,* Carl Reiner's film, is a terrific comedy, but nobody saw it because the studio didn't support it enough. Then there is the "Friday-night phenomenon." If it doesn't do terrific business on opening night, marketing monies are soon scarce. Studios rarely take a chance on comedy. When I made my studio deal at Twentieth Century Fox, I didn't know *Blazing Saddles* was going to go through the roof. I just wanted security, a place to work and an assurance that I would make at least three movies. I still respect Fox for taking a chance on me when I was relatively unknown and for giving me the opportunity to make my three movies for them. I don't need studio front money now; all I really need is a studio's distribution expertise and muscle.

When I presented the idea of *Silent Movie* to Fox, their mouths dropped open. They were very shocked, but they didn't want to turn me down because of my track record. That's "The Green Awning Syndrome." After Mike Nichols made *The Graduate,* I said, "If Mike Nichols went to Joe Levine and said, 'I want to do *The Green Awning,*' the answer would be, 'The what?' '*The Green Awning.*' 'What is it?' 'It's a movie about a green awning.' 'Does any famous star walk under the green awning?' 'No, all unknowns.' 'Are there any naked women near the green awning?' 'No, no naked women.' 'Are people talking and eating sandwiches and scrambled eggs on outdoor tables under the green awning?' 'No, it's just a green awning.' 'Panavision?' 'No, just a green awning. It doesn't move.' 'How long would it be?' 'Two hours.' 'For two hours nothing but a green awning on the screen? No talking, no dialogue, nothing? All right, we'll do it.' " That's the Green Awning Syndrome. When I said, "Silent Movie," in 1976, they said, "Sounds interesting." I knew in their hearts they were saying, "Oh, God! How can we say no without hurting his feelings, without losing him?" I explained later that there might be some great movie stars in it.

As it happened, I worked with five stars in *Silent Movie*: Liza Minnelli, Paul Newman, Anne Bancroft, Jimmy Caan, and Burt Reynolds. I chose them because I knew they were pleasures to work with. Fox was more amenable to the idea of *Silent Movie* when I told them there would be stars in it. "No dialogue in 1976? That's a toughie." They were very brave to make that picture. I wasn't frightened at all until the first dailies. I said, "What the hell are we doing? I can't hear anybody! This is crazy. You've done it, Brooks, this is it. Sanity has finally caught up with you." But it all worked out, and the picture was a huge success.

I wanted to keep the writing team on *Silent Movie* together, but they were ready to go off in different directions when the picture was over. Casually at lunch one day I said, "What about a movie called *High Anxiety*?" "What's that?" "It's Hitchcock." "Oh, yeah?" We discussed it and then wrote it in sixteen weeks. We knew that we had to have the Psychoneurotic Institute for the Very, Very Nervous, and I knew that I would be Professor Richard Harpo Thorndyke. Six years I was in analysis, and I wanted to get even with them. *High Anxiety* is a tribute to Alfred Hitchcock; there were many scenes from his films in it, and stylistically it's very much like him. Through that picture I could talk about psychoanalysis and psychiatry, which I care about a lot. I can't make fun of anything I don't care about. The budget of *High Anxiety* was $4.3-million; I brought it in for $3.4-million.

In order to direct comedy, the first step is to understand the script thoroughly: how it translates into sound and action and how the characters relate to each other. Once it is understood, you get a picture in your mind that is always altered by the specific gifts of the actors and actresses. My secret is very simple. Early on I have three or four casual readings of the entire movie script around a big table with all the principal players. Questions come up. I will say things like, "Gee, that's a strange approach to that line; I heard it differently. But I like it." An actor will invariably say, "What do you mean? How did you hear it?" I'll say, "Well, forget it." He'll say, "No, no, please." In that way I sneak in a line reading and he thinks he discovered it himself.

I try to keep the atmosphere on the set buoyant and relaxed because there's a lot of tension in making a movie. The cinematographer is always worried about the lights. If he's outdoors, he's a maniac because a cloud might go by. The actors are always worried about getting the scene right, and often when they've done it perfectly, they will ask for one more take because they're not sure. First, there is the artistic pressure of capturing life itself through that one-eyed monster; second, there is the budget, in which every dollar is a second that is ticking away. The pressure is especially great when you're doing a big scene with a lot of extras.

For the writer-director there is often a conflict when you'd like to do something lavish in a scene, but you're not sure whether it's worth the cost. For example, at the same time that I'm writing a big grand-salon scene in *History of the World, Part I* for the French Revolution sequence, I think of the budget and how much it will cost to reconstruct the Palace of Versailles. How am I going to get those verisimilitudinous qualities . . . which is very difficult to say and to get. In this case I spoke to my friend Albert Whitlock, who paints the greatest mattes in the world. He can paint on glass and take a few live characters and paint Versailles around them. If I were to try to construct Versailles, the picture would cost $100-million.

On my pictures since *History of the World, Part I,* Brooksfilms has managed to retain all or most of the overseas rights. Emile Buyse, who ran Fox's international operation for years, was stolen by me to run Brooksfilms' overseas sales, and he is the architect behind this. Generally, a domestic distributor puts up roughly 60%–70% of the negative cost, and the balance is financed through overseas sales. (I take a reduced fee for my multiple services so that the picture is not saddled with huge above-the-line costs.) Recently Brooksfilms has taken Canal Plus of France as its international partner in the following arrangement: Canal Plus pays for 40% of the negative cost for designated Brooksfilms, in exchange for being a 50% partner in the net profits those films derive from overseas in all markets and formats.

I direct a film to protect the writing. I produce a film to have total business control as well as creative control over the film's future. Little by little, in defense of the initial vision, I've learned to put on other hats. In the movie business it's important to understand the nature of money and how to sail through those terrible white waters with reefs and sharks, where art and money meet. For example, there are little nuances to watch for in contracts with a domestic distributor. They might throw in a clause that is easily overlooked, stating that revenue from airline play on any and all airlines serviced in the United States is the property of the domestic distributor. Seems innocuous, but it can be dangerous, costing hundreds of thousands of dollars. All monies derived from airlines that are based abroad and fly foreign flags should be constituted as *international revenue* to keep it from falling into the domestic distributor's pocket.

Another business issue is whether to go union or nonunion. The answer is, sometimes it's cheaper to go union. Mistakes that nonunion crews can make can be incredibly costly: loading the film incorrectly, not checking the gate for scratches or dirt and focusing improperly can all be disastrous. Here are some guidelines on keeping costs down: Rehearse as many days as possible before shooting, never rehearse on shooting days and curtail the number of shooting days.

Transportation is costly. Don't move your cast and crew from a studio to location more than once. Schedule all your location shooting together. On *Life Stinks* we shot on location for the first seven weeks and finished up the last five weeks in the studio. In writing the film I was very conscious of trying to keep most of the action shot on location happening during the day. Night shooting is very difficult, very tiring and in the end very costly. God is very nice, giving the filmmaker a lovely sun to light the world with. Use it in good health.

Generally, in production, I take as much time as each film needs. Now, I take even more time in the distribution of the picture. What does the one-sheet look like? Where is the picture opening? Is there a good sound system? Is the theatre equipped with Dolby? Newspaper advertising is archaic and very costly; I would rather spend money on a good television campaign.

Since I'm very proud of my image abroad as a filmmaker, I'll travel to help sell my films. It's possible to get a lot of coverage in European newspapers simply by going there and doing interviews. I won't do that in the States because I don't want the same high level of exposure here. My pictures do very well in Sweden, Germany, Italy and France; in England they do pretty well, but not as well as I'd like because of the high cost of advertising there. China is an exciting market. If everybody in China went to see my movies and paid one penny to get in, I'd be rich. Besides, I love Chinese food.

My advice as to the best way to break into the business is to write. Or be a big movie star. If Tom Cruise wants to direct a picture, they'll let him. Very few of us are going to become big movie stars, but we can write if we apply ourselves. To me the vapor of human existence is best captured in film; it's a great molding of all the primary creative arts. If a writer is talented, his talent can open the door to directing. The director, in the end, is the real author of a movie.

It's very important for a creative personality—writer, director or actor—to have good advisors. I have a lawyer, Alan U. Schwartz (May The Schwartz Be With You), who has been a friend for thirty-five years. I also have a good business manager-accountant, Robert Goldberg, who protects me from studio accounting. Studio accounting should be a Busby Berkeley musical. When do they stop taking money? They take overhead on interest and interest on overhead. If a picture costs a million dollars to make, it's a third more just because of studio accounting procedures. If it's $15-million to honestly and actually produce, it will cost $26-million for the same picture to be done at a major studio. For their part, they're risking a lot of money, which is how they justify it.

As for the future of the business, I see bigger profits on fewer pictures. The home-entertainment technology scares me more than anything because I want an audience to laugh at my movies. I want people to sit in a dark theatre, let the silver screen bathe them with images and have them laugh as a group. It's thrilling to hear a lot of people laughing together. But with the direction of current technology, it seems we'll have tiny little groups at home, or sometimes even one skinny person watching a big fat Mel Brooks movie. You can't get a lot of laughs that way. I wasn't born to make one thin person laugh; I was born to make a lot of fat and skinny people sit in the dark and laugh together.

What is the toughest thing about making film? Putting in the little holes. The sprocket holes are the hardest thing to make. Everything else is easy, but all night you have to sit with that little puncher and make the holes on the side of the film. You could faint from that work. The rest is easy: the script is easy, the acting is easy, the directing is a breeze . . . but the sprockets will tear your heart out.

# THE WRITER-DIRECTOR

*by JOAN MICKLIN SILVER, who writes and directs theatrical films, television movies and for the theatre. Her pictures include* Hester Street *(which earned her a Writers Guild nomination for best screenplay);* Between the Lines; Chilly Scenes of Winter *for United Artists;* Crossing Delancey *for Warner Bros.;* Loverboy *for TriStar; and* Big Girls Don't Cry . . . They Get Even *for New Line. Her television films include* Bernice Bobs Her Hair *for PBS;* Finnegan Begin Again *for HBO; and* Parole Board, *also for HBO. On stage, Ms. Silver has directed many plays and musicals, including* Album, Maybe I'm Doing it Wrong *(from the songs of Randy Newman) and* A . . . My Name is Alice, *co-conceived with Julianne Boyd. Born in Omaha, Nebraska, she lives in New York City with her husband, Raphael.*

*The advice I have for women in film is to toughen up and leave your sensitivities at the back door. . . .*

I began my career as a screenwriter. After writing several screenplays, I finally sold one to a studio, which put on a second screenwriter with whom I shared screen credit. That screenwriter worked very closely with the director on his particular vision of the material. The result was that the finished film didn't represent my conception of the picture. I learned quickly that the only way I would have half a chance to get what I wanted up on the screen was to direct it myself.

I approached the educational film company for whom I'd already written some scenarios and asked if they would let me *direct* a short film for them as well as write it, and they did. It was a scripted educational film with actors that described the immigrant experience to high school students. The story dealt with a Polish Catholic family that comes to America in 1907; the main character was a twelve-year-old boy. After directing two more short films for that company, I very much wanted to direct something longer. I remember meeting a television producer who was extremely fond of one of my short films. He was doing an hour show, 46-minutes of material (minus the commercials), and told me he could not in good conscience hire me because, although I had done 30-minute projects, I hadn't demonstrated the ability to do 46-minute ones. Well, I remember running out on Sixth Avenue, screaming, and realizing that opportunities were going to be extremely rare for me as a woman who wanted to direct.

My husband, Ray, was aware of my troubles, and he said if I could write a low-budget film I wanted to direct, he would try to raise the money for it. That became *Hester Street*. The technical expertise necessary to direct a feature was learned right there on the job. One advantage on that first picture was the support of a terrific crew, such as the cameraman, Ken Van Sickle, who was unusually sympathetic, patient and willing to take me through the ropes. There were also two friends working for the same educational film company who recommended crew and provided equipment rental lists and other guidance.

In the screenwriting, *Hester Street* was deliberately designed as a low-budget film. There was a small cast, a limited number of exterior production sequences and a minimum number of locations. The screenplay was based on a novella written in the 1890s by Abraham Cahan.

During production I was faced with a number of handicaps that stemmed from the limited budget of $350,000 and a 35-day shooting schedule. For example, there's a scene where the husband goes to meet his wife and child on Ellis Island. We didn't have the money to re-create it as it was done in *Godfather II,* but we could create a part and make it seem like the whole. The dramatic high point of that sequence was the separation between those coming to meet people and the new arrivals waiting to be recognized. By limiting the action to an area around a wire fence separating the two groups, we captured the drama that suggested what was going on beyond camera range. Another example of this technique can be found in a television film I directed, *Bernice Bobs Her Hair*. The story begins with a dance at a country club, but we could not afford enough extras to really mount a dance. My solution was to have the action take place everywhere but the dance floor: in the ladies' dressing room, on the veranda outside and so on. We hear the music and have a sense of the dance going on, but never actually see it; still we lose none of the drama.

When the time came to re-create the *Hester Street* exterior, we had scheduled only four and a half days. By some miracle, it didn't rain. This was the big physical production sequence; a lot of money went into dressing that large exterior location, which was actually a section of Morton Street in Greenwich Village. With the help of an excellent production designer, Stuart Wurtzel, we redressed that street, removing signs and streetlights and adding pushcarts and other props. The people who lived on Morton Street were extraordinarily helpful. For example, for the last scene of the picture the camera had to be placed on a fire escape for an overhead shot of the wife, boarder and child walking down the street talking. The fire escape we wanted for our camera placement was outside the apartment of a student, and we had to leave the windows open because of the cables. He sat inside with his overcoat on all that time studying.

One thing that afflicts low-budget filmmakers is a nagging worry about money. During shooting, time literally *is* money. On *Hester Street* the location of a particular scene hadn't been properly secured, and it took two hours for us to gain access. Even while we were shooting the scene, I was thinking of the loss of those two hours and how to make up for it.

Another problem of low-budget filming is the level of technical expertise in the crew. A less-experienced crew requires more rehearsal time, particularly if there is a complicated camera move. The problem is not only the time passing but also the need for extra crew rehearsals, which can tire the actors unnecessarily. (Stand-ins are a luxury not available on low-budget films.)

Although I'm a writer, I'm not so fiercely attached to my dialogue. Some directors who don't write feel a little more in awe of the screenplay, a greater responsibility to it. But since I'm also the writer, I don't feel any compunction about cutting or changing the script—and I always cut it. Being a writer-director is a happy combination on the set, because there are times when something unforeseen happens dramatically, and I'm able to do a new scene on the spot.

*Hester Street* was my first directed feature. As one goes along in directing, more confidence is developed. I remember once we did a master of an important scene that included a lot of camera moves; it was the scene in which the wife makes a marriage proposal to the boarder. I thought the master was glorious, but since it was such an important scene, I covered it with two-shots, singles, etc. Of course, when I looked at the rushes, the master was great and I used it. If I had been more experienced, I'd have known that I didn't need further coverage. I felt the master was beautiful; I was just afraid to trust that first reaction.

*Hester Street* enjoyed wonderful critical acclaim and had the advantage of being distributed in 1975, when movie production by the studios was lower than usual, so securing theatres wasn't terribly difficult. All of the distribution and advertising was Ray's job. He nurtured the release and publicity of the picture very carefully and handcrafted the distribution pattern into a strong commercial success.

After *Hester Street* some of the studios wanted me to bring them my next project, "anything you want to do." So I brought them *Between the Lines,* and they said, "Well, not that one." But that was the picture I wanted to make. Ray said we could either continue trying to raise studio financing or take our share of the profits from *Hester Street* and make the movie ourselves. We might have spent years trying to raise all the financing, which would have further delayed the project; so, we decided to finance it internally.

*Between the Lines* was screenwriter Fred Barron's original project. Although I supervised his rewrites, it was basically his script. We had an extremely spirited cast, and there was a lot of improvisational work done during the 44-day schedule. The picture was budgeted at a little under $800,000 in 1976. Although there was more money than on *Hester Street,* budget limitations were still a problem. It was a New York–based union production, though the action of the story takes place in Boston. I would have preferred shooting the whole picture in Boston, but on location everybody's hotel bills and per diems must be added to the budget. Therefore we shot only two weeks in Boston and built the interior of the newspaper set in a New York studio. Shooting went faster in the studio, though, since there were not a lot of distractions, such as noise problems.

In 1978 I read that three young actors had purchased Ann Beattie's book *Chilly Scenes of Winter,* which I'd known and loved and actually looked into acquiring at one time. I said to Ray, "Somebody bought my book. Do you think I ought to get in touch with them?" He said, "Yes." I contacted them; they were looking for a writer-director, and we got together. We submitted the project to Twentieth Century Fox, which put up the development money for the screenplay. Then United Artists bought the project from Fox, and it was made as a UA picture with a new title, *Head Over Heels.*

On *Head Over Heels* the budget was $2.2-million—small by Hollywood standards in 1979—on a 42-day shooting schedule. Our actors cost more than on the prior picture, and we were paying for a big Hollywood crew, but we benefited from their expertise. For example, the first assistant cameraman did a remarkable job following focus, just by eyeballing it. There was one scene in which the main character is at a candy stand going back and forth in front of the camera. In each take the actor did it differently, yet each take was in sharp focus. The assistant cameraman was maintaining focus, and he wasn't even looking through the camera eyepiece; the camera operator does that. That level of expertise is *de rigueur* in Hollywood.

*Crossing Delancey* cost $5.5-million on a schedule of 40 days in 1988 and was one of those happy projects where everything worked out well. The picture was structured as a negative pickup, and I had final cut. Production went smoothly, the weather was good, the studio—Warners—was cooperative and the cast was superb. But it also posed challenges. For instance, the exteriors were particularly difficult. We were able to capture a lot of the street vibrancy, shooting on Essex Street on Manhattan's Lower East Side. But New Yorkers are very blasé about having crews around, and some people thought nothing of walking right into the shot; we just had to work around that. One unusual problem occurred in casting. For the day the circumcision scene was scheduled, we needed an eight-day-old baby who hadn't been circumcised. Credit goes to Judy Klaman, the extras casting person who had been tracking several babies. Sure enough, a mother gave birth just in time and allowed us to use her baby in the scene.

Another casting challenge arose on *Big Girls Don't Cry . . . They Get Even,* released by New Line Cinema. There were eleven speaking roles for children, including the thirteen-year-old girl who played the main character. The picture was shot in 42 days in 1991 and budgeted at $7-million. We had terrible weather problems shooting at Mammoth Lakes, a mountain resort in northern California. Instead of the Indian summer the area had enjoyed for several years, our shooting was interrupted by snow, rain, sleet, and hail. The cast members were very good sports about this, even though the screenplay called for frolicking in the water. In the rushes the clapper-holder would be wearing a down jacket and heavy scarf against the cold, while the cast would be performing in bikinis and bathing suits.

Naturally I am always looking for good stories. What is interesting is the kind of changes that occur in the main characters presented in movies. Until recently buddy pictures about men or relationships based around men had been the norm. But now there are more successful pictures wherein women are equally as interesting as men, as in *The Silence of the Lambs,* or are the leads, as in *Thelma and Louise*.

When I function as writer-director, my role changes through the course of a picture. At first the screenwriter dominates, since it's from the screenplay that one budgets and finds locations; the screenplay is the blueprint of the film. Sometimes I write things into the screenplay that are not strictly necessary but are helpful to readers. For example, I am usually quite deliberate about descriptions or specifying how an actor says a line. Also, sometimes a scene contains more dialogue than I intend to use when I direct, because dialogue makes the scene more comprehensible to the reader. As I write, I'm very careful to include the information that the production manager, assistant directors, cameraman, prop man, set designer and others need to plan their jobs. Through the screenplay I'm leading these creative people, giving them guidance for their respective planning from the printed page. Once we're rehearsing, I pare away at the script and work with the actors, who find new ideas and make suggestions. Then the screenplay begins to shift and take on the life that the actors bring to it. This may require dialogue changes.

Once I'm directing, I think about the movie in a different way. I'm always considering the editing, how one scene ends and another begins, for example. During shooting, I'm cutting in my head and thinking about getting in and out of scenes. Also, on the set I'm always conferring with the cameraman, not only on camera placement and lighting but on cutting as well.

As an actor's director, I'd rather use a slightly less beautiful shot if it contains a better performance. In one shot in *Head Over Heels* the main character and his stepfather walk out of a hospital, stop at a certain point and talk. The actors kept rehearsing and missing their marks, moving to where they were not lit. I wanted to tell them, but I didn't want to break the dramatic momentum they were building. The cameraman, Bobby Byrne, offered to relight the situation so that the actors were still well lit even if they missed their marks. Because of the cameraman's understanding, the actors did not have to be interrupted.

There are always business decisions that are part of the creative process in filmmaking. For example, good actors want to try scenes over. But the director must decide whether new ideas or extra takes are worth it and whether there is time and money to try them. The luxury that a higher budget brings to a shoot is the luxury of time. Time can be badly used; directors can become self-indulgent. But, for the most part, those directors who have more money and therefore more time have a chance to make a better film.

The editing process is fascinating, and the relationship between the editor and the director is usually a close one. I'm delighted with and eager to try editors' suggestions in the cutting room. Films have so many components, and the shifting and layering of these components during editing and sound mixing is a really interesting process.

As a woman I've had to overcome certain obstacles in order to direct features. Happily, I'm married to somebody who wanted to help me overcome them, who thought I was being unjustly denied opportunities and that what I needed to do was simply to make a feature. When I started working, I looked to people like Barbara Loden, who made *Wanda,* and Shirley Clarke (*Jason, The Cool World*). Today I see the industry easing up; a number of women now have projects in development at different studios, though the attitude of that television producer is still around. There is a higher consciousness generally; there are more female production executives and assistant directors, more women on production crews. The only place where there seems to be a lack of women is in the camera department. Not too many camerawomen have opportunities to work on feature films, particularly union films.

The advice I have for women in film is to toughen up and leave your sensitivities at the back door. There will be a lot of rejection, which goes with the territory. Writing and making short films is a good way to break into the industry; those who persist will probably prevail. That's the advice I'd probably give to a young man who wants to be a filmmaker too.

# THE INDEPENDENT FILMMAKER

*by* ***HENRY JAGLOM,*** *whose films as writer-director include* A Safe Place, Tracks, Sitting Ducks, Can She Bake a Cherry Pie?, Always (But Not Forever), New Year's Day, Eating, Venice/Venice *and* Lucky Ducks. *He began his career as an actor and was featured in films directed by Jack Nicholson, Dennis Hopper and Orson Welles (each of whom he subsequently directed in return). Mr. Jaglom is unique among filmmakers in that he produces his pictures financed through presales and now distributes them through his own company, Rainbow Releasing, a subsidiary of his Los Angeles–based International Rainbow Pictures.*

***I had come upon a formula that has lasted to this day: If I could make a picture for $1-million, I could aim at 10% or 15% of the audience and show a profit. . . .***

While it has been said that I "do it all myself" and this is rare in American filmmaking, I see it as the business-person part of me taking care of the artist. A creative person can't just stop with the work. You have to learn how to get your work seen, how to force your vision against all the pressures that try to stop you. I've seen many wonderful, creative people trampled on because they have not known how to take care of the business aspects of themselves. Orson Welles was the prime example of this.

Early in my career it was clear that there was no established business mechanism to support the kind of films I wanted to make. I came out to Los Angeles in the late 1960s as an actor and began peddling my own scripts. After the success of *Easy Rider,* Columbia wanted to reward people who had worked on that movie (I had worked on the editing), which allowed producer Bert Schneider to let me direct my screenplay *A Safe Place* (1971), starring Tuesday Weld, Orson Welles and Jack Nicholson. Bert Schneider gave me creative control and final cut on that movie, which spoiled me for life. Columbia could not be bothered with it; they had *Nicholas and Alexandra.* My picture was a small poetic art film, made for a big company, and didn't do any business. Although I was offered other studio pictures based on a gift they thought I had directing actors and creating screen reality, I would no longer have control. They wouldn't give me final cut, no matter what else they offered.

I had to find another way to make the kind of intimate, small-scale human picture that appealed to me. It took me five years to finance my second film, *Tracks,* starring Dennis Hopper. I had wanted to make a film about the Vietnam War, but at the time, the war was still going on, and the only thing less popular than making a film about the war was wanting Dennis Hopper to star in it. It would never have been made had it not been for an economically corrupt system that existed at that time involving tax shelters.

*Tracks* was originally financed in 1973–74. My partner, Howard Zuker, raised $1-million through dentists and doctors investing $25,000 to $50,000 apiece, each of whom could write off seven or eight times that amount under the tax shelter laws of the time. The investors were not interested in picture content; they were only interested in the write-off. In fact, when Columbia refused to release Bert Schneider and Peter Davis's brilliant documentary about the Vietnam War, *Hearts and Minds,* we took that initial $1-million raised by Zuker and used it to get *Hearts and Minds* away from Columbia and released by Warner Bros., and started raising money all over again for *Tracks,* which is why the film didn't come out until 1976.

At the time, another roadblock was that there were no small distribution companies, which were later to proliferate. There were only majors, and none of them wanted to release *Tracks;* it remains the only picture of mine that never had a real theatrical release. After playing a couple of weeks in a few cities, it was gone.

Luckily, since *A Safe Place* was an extraordinary success in Europe (playing many years in Paris) and *Tracks* performed strongly as well (sharing Italy's Donatello award with Woody Allen's *Manhattan*), I was able to finance my third film in Europe. (At the same time, tax shelter financing was ending in the United States.) Then it was slowly dawning on the major distributors that there was money to be made releasing smaller, artistic pictures, so they launched specialty divisions: Triumph at Columbia, Orion Classics at Orion, Universal Classics. Other companies began emerging to release independent pictures just in time for my third film, *Sitting Ducks* (1980), which happened to be quite commercially successful. It took ten years since my first picture to get my third picture released. *Sitting Ducks* was distributed by United Film Distribution, an arm of United Artists Theatres (no relation to United Artists Corp.).

Now I had come upon a formula that has lasted to this day: If I could make a picture for $1-million, I could aim at 10% or 15% of the audience and show a profit. At the same time, videocassettes and cable technologies offered new sources of income for films. And there were now these independent distributors, including Vestron, New World, New Line, Goldwyn, Atlantic, Island, Alive, Castle Hill, Skouras and later Miramax, who realized that if a picture cost $1-million and made $5-million, everyone would be very happy. If that same picture were made by a studio for $10–12-million and made $5-million, it would be a disaster. These smaller distributors were doing well handling smaller-scale quality pictures such as mine, with less advertising costs than the majors, films aimed at intelligent, adult audiences who wanted to see serious work about human relationships.

The question became how do I, from an economic standpoint, devise a system to finance films for that minority audience? First, I found that Europe was an ideal source of financing. For example, a German distributor comes to me with a contract offering $200,000 for their territory for each film I make. They will give me $1.5-million for my next six films; the films are called "Jaglom #8, 9, 10, 11, 12 and 13." They don't ask for titles or casting, but only that the films be in the English language, 35mm, color, no less than eighty minutes, no longer than two and a half hours, and "signed by Jaglom," as they put it. With another $150,000 from England, $100,000 from Italy, $75,000 from Scandinavia, for example, and similar amounts from other European territories, I can easily achieve my $1-million, still retain all rights in the United States and, most importantly, keep my ownership of the negative. Since then, happily, my pictures have increased in value, and new formats such as home video, cable and satellite have added to the size of distributors' advances, and even created many markets where formerly there were none, or only one.

Since I own the negative in the United States, I can then find the most sympathetic small distributor who will release my pictures properly, aimed at that 10%–15% of the movie audience who we know will positively respond to my films. *Can She Bake a Cherry Pie?* (1983) was released by Castle Hill; *Always* (1985) was released by the Samuel Goldwyn Company (for an advance of $1-million); and *Someone to Love* (1987) was codistributed by Castle Hill and my own company.

When I saw how it all worked, I decided I could release my own pictures, so I began Rainbow Releasing as a subsidiary of our production company, International Rainbow Pictures, and we handled *New Year's Day* (1989) and *Eating* (1990) through this division. There is really no mystery as to how it works. For a small, artistic "quality" picture in America, there are two or three theatres in each city to choose from. That's all. You show them the picture; they make an offer; you discuss advances and terms and how many weeks for the length of the run, and do this for each city. Once a theatre is chosen, we allocate a print, trailers, posters and press kits, all self-generated. This way we control the quality of the print, what photos are sent, the advertising campaign and the one-sheets. I am so hands-on that I even draw the artwork title for most of my movies, which has become a kind of signature, I guess.

My release strategy has the first opening in Los Angeles, where I have a strong following, usually in a Laemmle theatre. (See article by Robert Laemmle, p. 359.) In the sixteenth week in Los Angeles I frequently do better than in the fourth week, thanks to word-of-mouth. In Boston my pictures can run for close to a year. In each city my audiences tend to build through word-of-mouth, so the picture must be able to hold in the theatre for this to occur. Obviously I must find sympathetic exhibitors willing to allow the time. On *Eating* I found a small New York theatre in a wonderful location in Greenwich Village willing to hold the picture for many months. Once word-of-mouth catches on, I can spend less on advertising, which in any case can never compete with the big company's full-page ads and TV campaigns. To my delight my audience is growing with each picture. The result is I tend to play in fifty to seventy-five cities theatrically and make $4–5-million worldwide, from all sources.

Let me underscore the importance of finding theatres willing to have my pictures settle in for a period of time. After all, I make movies for an audience that I hope will feel less isolated, less crazy, because they see something that touches upon some truths about their own lives. A film of mine is not for everybody, but those people who want it should be able to find it. My job is to find ways to make it available to them.

Overseas this is equally important. After *Eating* was shown at Deauville, I was able to make a deal with the best French distributor, MK2, which greatly improves my position in that territory, and of course I made sure to visit Paris to do press to support the release of the film there. It became a smash hit! Finding that ideal distributor in each overseas territory is an important step, just as finding a proper theatre in each American city is. With each move I am building relationships with distributors and exhibitors, supportive journalists and growing audiences to position my next film and also to benefit my prior pictures. Further, as new technologies emerge in growing territories, this means more revenue. Since I own the negatives, I continually make new deals for old films as opportunities arise. For example, I just sold *Tracks* to an Asian country for a satellite delivery system that didn't exist when the picture first came out. *Sitting Ducks* has even played Mozambique!

My system is working well. Part of the reason is that I don't have ego about spending a lot of money. It's the other way around; my ego is in striving for quality for *little* money. As of this writing, I am simultaneously editing my ninth (*Venice / Venice*) and tenth (*Lucky Ducks*) pictures, am in preproduction on my eleventh (*Happy Ending*) and am basically financed through my fifteenth. I am always juggling three or four movies and working on the release of one or two others, supporting them through their various stages throughout the world.

Orson Welles once said to me at lunch, "The enemy of art is the absence of limitations." Economically and creatively that's the most important advice you can be given. You have limitations; you don't have $1-million to blow up that bridge, so you have to create something else on film to produce the same effect. Instead of having money to hire hundreds of extras, you have to sneak a cameraman in a wheelchair through the streets of New York City and steal the shot, which gives you a look of much greater reality. With economic limitations you are forced to create art. As an example, on *Can She Bake a Cherry Pie?* I could not afford to rent a restaurant. So I asked the owner if I could put a few chairs outside, as if it were an outdoor café. The scene was shot, and there was all this wonderful, natural background activity *inside* the restaurant. Afterward the owner kept the outside area as a new feature of his restaurant. "The absence of limitations is the enemy of art." For those of us who have the pretension to claim to try to be artists in this business so many love to call their "industry," money limitations are not a problem, they are a spur to be creative.

Another part of the formula to keep costs low is that I pay all my actors union scale. If they happen to be stars whose names help sell the picture, such as Orson, Jack Nicholson, Dennis Hopper, Sally Kellerman or Karen Black, I give them a percentage of the profits, which in my case is *actual* profits, as opposed to the fancy bookkeeping "profit participation" of studio pictures.

To give you a sense of my working process, I get my actors together and work with them in creating their characters and *encourage* the dialogue out of them, rather than predetermining it, using *their* language, *their* memories, *their* personalities to fit into *my* overview. Movies, to me (as opposed to plays or novels), exist when they are shot, so you cannot predetermine what the look, nuance or emotion of a scene will ultimately be. You create it, and *then* it exists, not the other way around.

Afterward I go into my editing room and "write" the movie with what the actors have given me on film. My most scripted movie was my first, *A Safe Place,* which was also the most poetic and difficult to make. Every movie since has had less of a written script and been developed more in my head and on my feet. I always go into a project with a strange sense of knowing, in my mind, what the whole movie is supposed to feel like. Orson Welles said the difference between me and other filmmakers is that others first write their story and try to find their theme *in* that story; I start with my theme—the end of a marriage, the loneliness of a generation, trying to move on in life, eating disorders—and *then* work at that, figuring that my story will emerge. I don't think of myself as writing and directing a movie so much as *creating* a movie, and I love the process of trying to take these pieces, like a jigsaw puzzle in my brain that keeps changing shape, and putting them together. I thought this was a very original way to work until I started reading about the great filmmakers of the silent period and found it's how many of them made pictures. Chaplin made up a lot of his work as he went along.

I will generally shoot for a few weeks, then edit for many, many months, then go back to shoot some extra days. It is like working on a painting; how do I know I'm finished until I finish? I don't want to predetermine it; I want it to become real, authentic, alive, and follow its own needs. Because of this fluidity, shooting can sometimes overlap. One day I found myself shooting sequences for three movies: I had been editing *New Year's Day* and *Eating* while shooting *Venice/Venice,* when I decided to pick up some bathroom scenes for *Eating* and also needed an opening and closing shot for *New Year's Day*. I had my crew intact for *Venice/Venice,* so all I had to do was gather together a few of the actors from the other two movies and change the furniture around a bit. "The enemy of art is the absence of limitations." When you're forced to create, you create. And it frees you.

Dividing my energies among four movies at a time is exhilarating. Returning to the editing of one movie after concentrating on another sheds new light on the first one. It's all part of the same organic process. And if I happen to take a year to edit a movie, there is no economic pressure, no money on loan from banks, nobody worrying, as long as I turn in a picture each year or so to my exhibitors around the United States and my distributors in Europe. And each year the audience continues to grow, which is very exciting.

Videocassettes have been a boon to independent filmmakers. Not only do they provide a source of revenue, but they are a great leveler. In video stores my pictures are next to multimillion-dollar studio pictures, for rent at exactly the same price. Since my economics are so low, I exert my energies upon reaching the literate movie audience, without regard to demographics or popular opinion, giving me the freedom to make films exactly the way I want to. I'm enormously lucky in that regard.

For home video I used to sell to the highest bidder, taking the best deal I could get. Then I found that Paramount handled films with great care and sensitivity, and now they release all of my films on home video. They give me a generous advance, which is more than enough to cover prints and advertising on the theatrical release. This increases the value of the film when it goes to home video, since impressions from one market carry over into the next. I have also made some overseas-territory deals with Paramount and have sales representatives handling my films for various other international markets.

The amazing thing about all this is that we all end up making quite a bit of money, although that was never the driving force. With a budget of $1-million (and my movies shouldn't be made for more) plus $500,000 for advertising and promotion, $5-million in revenue worldwide means that we make a $3.5-million profit, far more than I ever thought a movie of mine could pull it. So the irony of this modest filmmaking formula is that it is surprisingly profitable, and I have complete control over my pictures, with none of the bureaucratic, creative or financial headaches that my counterparts who work for studios seem always to be complaining about.

When making films at this level, there must be multiple ways to create attention, to make up for the lack of millions of dollars in advertising. For example, we are now involved in book publishing and merchandising; there are book versions of the movies (published by Samuel French) and we sell T-shirts in theatres. Who knows what's next?

To sell my films in overseas territories I attend the Cannes Film Festival, Deauville, the Venice Film Festival, Toronto, wherever I'm invited. I screen my movie, discuss terms with interested distributors and make my deals. My main market is Western Europe; I have no audience so far in Asia and very little in South America, but it will come. Recently Eastern European countries from the former Soviet bloc are becoming interested. Because my movies are very verbal, dubbing them doesn't work that well; my audience has to be literate and comfortable reading subtitles. I do believe though, that it's just a matter of time before that 10%–15% audience share that is mine in the United States (and higher in Western Europe) will slowly find my movies in the emerging free nations of Eastern Europe and eventually Asia and Africa as well.

My advice to new filmmakers is quite simple: *Do not accept anyone's word that something is impossible*; it's the limitation of the person telling you that makes them say it can't be done. Do not acknowledge that there's a wrong way to do something just because somebody else says so; they just don't know how to do it yet. To make a movie, you need as much money as you've got, not a penny more. If it's just $20,000 you can raise, take a video camera and go make a movie. Someone will see it and you'll be on your way. Don't talk about it; do it. This is a great time for independent filmmakers because of all the new formats that need to be fed. The business today is analogous to book publishing, with studios releasing the equivalent of mass-market paperbacks and independents making the "hardcovers" for a smaller audience. Thanks to home video, they are all side-by-side on the video-store shelf. There are no rules to any of this; anyone following rules is falling into traps laid by people with limited imaginations.

II
THE
PROPERTY

# THE SCREENWRITER

*by* ***WILLIAM GOLDMAN,*** *a distinguished novelist and also a screenwriter who has won Academy Awards for his original screenplay* Butch Cassidy and the Sundance Kid *and for his adaptation of* All the President's Men. *He has adapted his novels* Marathon Man, Magic *and* The Princess Bride *for the screen and has written the screenplays for* Harper, The Hot Rock, A Bridge Too Far *and* Misery. *Among his other novels are* The Temple of Gold, Soldier in the Rain, Boys and Girls Together, Tinsel, Control *and* The Color of Light. *His book* Hype and Glory *deals with his experiences as a judge at the Cannes Film Festival and the Miss America Contest.*

*What movies get made reflect the executive mentality; what movies are successful reflect the audience. . . .*

Writers have always been secondary in Hollywood. But ask *any* director and he will tell you he is only as good as his screenplay. There is no picture without a script. When you read that a producer announces a new $25-million picture from a novel he has bought, that's nonsense. No one knows what a film will cost until there is a screenplay. There is no film; there is no anything at all in this world until there is a screenplay. A screenplay is gold.

Hollywood is constantly shifting. It's a whole new and unpredictable ball game. Now one can write anything. Since no one knows any more what will or won't go, almost anything has a chance of getting made. Now it seems possible for a writer to say what he wants through film and make a living at it.

One of the things that no one tells an eager author in college is that if he writes a novel, the chances are that he won't get it published. And if he does get it published, he might make a thousand dollars or maybe even two. It takes years and years to become an established fiction writer, and one can hardly support a family that way. There aren't more than a handful of writers who can actually make a living out of hardcover fiction writing. Film writing, on the other hand, not only pays, it overpays. And it is a way for one to exercise his craft and still feed his children—both critical aspects of a writer's life.

There is more interest in screenwriters today than ever before because of money. People are beginning to wonder why screenwriters get so much money since the star makes up his part and the director has all the visual concepts. The answer is that it all starts with the word: the screenplay. The reason that the director gets all the publicity is because he's the most visible person during shooting, which is the only time the press is allowed around a picture. They are not present during preshooting, when the writer, producer and director are working on the script or are assembling the cast with the help of the casting director. No one is present postshooting, when the editors and composer are working their magic. And though the press may be on the set during a day's shooting, they're not around the night before, when that day's schedule is mapped out. At this critical session, the production designer will say, "We must have the door here," and the cinematographer will say, "Well, if you move the door here, I can give you this shot coming in, which will scare everybody," and the director will agree or disagree, or he won't know. He's just one of many people going down the river on this boat, hoping they get past the rapids.

Movies are a group endeavor. There is a group of six or eight technicians who are essential to the collaborative process: the writer, director, cinematographer, cutter, production designer, producer, production manager and sometimes the composer. As for writers, we are more essential than the public gives us credit for, but no more essential than the other technicians. But our visibility is low because few of us go out on publicity junkets. Basically we are very dull people.

Some authors start out, no doubt, knowing they want to write screenplays. I am basically a novelist, and I fell into screenplay writing rather by misinterpretation. It happened at a time when I was in the middle of a monstrous novel called *Boys and Girls Together*. I was hung up in the thing, and to try to unstick myself, I wrote a ten-day book called *No Way to Treat a Lady,* which was published under another name. It is a short book with 50 or 60 chapters. Cliff Robertson got hold of it and thought it was a screen treatment rather than a novel. At the time he had a short story called *Flowers for Algernon,* which eventually became *Charly*. He asked me to do the screenplay, but when he saw the results, he promptly fired me, hired a new writer and went on to win the Oscar for Best Actor.

The whole sequence of events did prompt me to learn more about screenwriting. I bought the only book available—called *How to Write a Screenplay,* or some such title—and discovered that screenplays are unreadable. The style is impossible and must be dispensed with. It always has those big capital-letter things that say, "EXT. JOHN'S HOUSE DAY." I realized that I cannot write this way. Instead I use run-on sentences. I use the phrase *cut to* the way I use *said* in a novel—strictly for rhythm. And I am perfectly willing to let one sentence fill a whole page. Here's an example from the ending of *Butch Cassidy and the Sundance Kid:*

CUT TO:

BUTCH

streaking, diving again, then up, and the bullets landing around him aren't even close as --

CUT TO:

SUNDANCE

whirling and spinning, continuing to fire and --

CUT TO:

SEVERAL POLICEMEN

dropping for safety behind the wall and --

CUT TO:

BUTCH

really moving now, dodging, diving, up again and --

CUT TO:

SUNDANCE

flinging away one gun, grabbing another from his holster, continuing to turn and fire and --

CUT TO:

TWO POLICEMEN

falling wounded to the ground and --

CUT TO:

BUTCH

letting out a notch, then launching into another dive forward and --

CUT TO:

SUNDANCE

whirling, but you never know which way he's going to spin and --

CUT TO:

THE HEAD POLICEMAN

cursing, forced to drop for safety behind the wall and --

CUT TO:

BUTCH

racing to the mules, and then he is there, grabbing at the near mule for ammunition and --

CUT TO:

SUNDANCE

throwing the second gun away, reaching into his holster for another, continuing to spin and fire and --

CUT TO:

BUTCH

He has the ammunition now and --

CUT TO:

ANOTHER POLICEMAN

screaming as he falls and --

CUT TO:

BUTCH

his arms loaded, tearing away from the mules and they're still not even coming close to him as they fire and the mules are behind him now as he runs and cuts and cuts again, going full out and --

CUT TO:

THE HEAD POLICEMAN

cursing incoherently at what is happening and --

CUT TO:

SUNDANCE

whirling faster than ever and --

CUT TO:

BUTCH

dodging and cutting and as a pattern of bullets rips into his body he somersaults and lies there, pouring blood and --

CUT TO:

SUNDANCE

running toward him and --

CUT TO:

ALL THE POLICEMEN

rising up behind the wall now, firing, and --

CUT TO:

SUNDANCE

as he falls.



In this sequence I've used the proper form, but I never want to let the reader's eye go—it's all one sentence.

A writer needs to find his own style, something he is comfortable with. For example, I use tons of camera directions, all for rhythm. It often upsets the directors, who shoot the scenes the way they want them anyway. But it *looks* like a screenplay, and yet it is *readable*. The standard form cannot be read by man or beast.

Anyone wanting to be a screenwriter should write a screenplay—not an outline or a screen treatment or a novel that then has to be adapted. A studio can have over a million dollars tied up in a property between the time it is purchased as a novel and the time a script is ready. And this is aside from subsequent production costs. If an author writes a screenplay, it is already there to be seen and judged. The company can say right off, "Yeah, we'll shoot it," or "No, we won't." If it sells, it pays the bills. And besides this essential aspect, it is a legitimate and honorable kind of piece to write.

Background reading and research can be important for a writer. For one thing, sometimes he just stumbles upon something that really grabs him and that he knows he wants to do something with. It was way back in 1958 or '59 that I first came across the material about Butch Cassidy and was moved by it and knew I wanted someday to write a movie about it. I continued researching the subject off and on for ten years, finding things to read that added background and depth. There is a lot available on Cassidy but almost nothing on Longbaugh (Sundance). Larry Turman, a good friend, who produced *The Graduate,* was very important in helping me to structure it.

Since I am basically a novelist, it never occurred to me to ask for advance money on "spec" based on an outline that I might sell to someone. I just wrote as if I were writing a novel. This is an unusual occurrence, at least for a Class A picture. The professional screenwriter doesn't usually just write an original screenplay and then look for a market. If he makes his living as a screenwriter, what he probably does is "buckshot" it. That is, he writes 10 outlines and circulates them, hoping that one of the 10 clicks and someone gives him money for it. He then writes the full screenplay with financial backing.

I wrote the first draft of *Butch Cassidy and the Sundance Kid* in 1965 and showed it to a few people, none of whom was interested. I rewrote it, really changing very little, and suddenly, for whatever reason, everyone went mad for it. Five out of the seven sources in Hollywood who could buy a screenplay were after it. It was this unexpected competition—not my particular skill with the rewrite—that sent the price so high.

Authors who write in various other forms of fiction and nonfiction besides screenplays often have two agents, one on each coast. The one on the West Coast handles the film material, while the New York agent handles all of the other manuscripts. My Hollywood agent at the time, the marvelous Evarts Ziegler, handled all negotiations for *Butch Cassidy.* My only contact with the deal was that he called me in New York every day to keep me posted on the bidding and warned me to stand by the phone to get his call when the bidding was over. It was up to me to give the final okay. The screenplay was finally bought by Fox.

No doubt many authors write a film imagining a certain actor in a specific role. Right from the beginning I had Paul Newman in mind. Actually as I wrote the picture originally, I saw Paul Newman and Jack Lemmon in the main roles. Jack Lemmon had just done a movie called *Cowboy,* and I thought he would do a fine Butch Cassidy. Paul Newman had done a movie about Billy the Kid, and I saw him as the Sundance Kid. As the years went on, Lemmon disappeared from my mind, but Newman agreed that he would play the Sundance Kid. Then, when George Hill (who was eventually signed as director) read the script, he mistakenly assumed that Newman was going to play Butch. When that happened, Newman, who wasn't really eager to play Sundance, was delighted to change roles. Then the long search began for the actor who would play the Sundance Kid. Every star in Hollywood was up for it. There were arguments about certain choices. Under such circumstances an author doesn't have very much power. Long ago Hollywood decided that the way to keep people quiet is to overpay them. An author paid all that money should go home and count it and be content. I was in there arguing, and so were others who had more influence, notably Newman and Hill. We finally won the battle, and Robert Redford, who in those days was not nearly so well known as some of the other candidates, got the part.

I was really fortunate. Overall I happened to be delighted with *Butch Cassidy.* In many ways it is better than what I wrote; in many ways it isn't; and in many ways it's different. My script was much darker and, I think, would not have been so successful. And most of the credit for its coming off so well I give to George Roy Hill, the director.

*Butch Cassidy* is an example of an original screenplay. I've also adapted my own novels (*Marathon Man, Magic, The Princess Bride*) and books written by others (*All the President's Men, A Bridge Too Far*). The hardest thing to write is an original because it's creative; the easiest thing is an adaptation of somebody else's. On a straight adaptation, I don't have to deal with the anguish of the original writer. But when I'm adapting my own work, I think, "That was so hard to write, I'd like to keep it." I'm not as ruthless as I should be. The Faulkner phrase "You must kill all your darlings" is basically true.

For example, one scene in *Marathon Man* that I cared about was the run. The hero runs along and fantasizes that legendary runners come alongside him and get him out of a scrape. In the first draft screenplay, I wrote it as a fantasy, as in the book. John Schlesinger, the director, said, "I can't shoot this; it's a literary conceit and it won't play." When a director says, "I don't know how to make that play," it's best to change it, rather than risking that his uncertainty will show through in the film.

The assignment on *A Bridge Too Far* was unique because it was financed not by a studio but by one man, Joseph E. Levine. In order for the story to be told properly and for it to be faithful to Cornelius Ryan's book, one had to have a lot of stars. The use of stars would help the audience organize the several parallel stories to be told. This affected the writing of the screenplay since, in a scene of two characters talking, if the scene legitimately belonged to character A but character B was cast with a star, I would flip the scene so that it would favor character B.

Films that become successful tend to reinforce our expectations; films that are not as successful but are equally competent, tell audiences things they don't want to know. For example, everyone in this country thinks of *A Bridge Too Far* as being a commercial failure. In fact, it was a giant success in Japan and Great Britain and did very well all over the world, except for the United States. It told Americans something they didn't want to know: Battles and wars can be lost. All over the world people have lost wars and know that kind of suffering and agony firsthand; we don't.

On *All the President's Men*, if there was a contribution in the screenplay that was valid, it was deciding to end the film in the middle of the book, on a less-than-triumphant note for Woodward and Bernstein. Instead of having them get saved by the cavalry at the end, the idea was to have the audience apply what they knew and fill in the ultimate victory. No one knew at the time that *President's Men* would become a very successful film. People were saying, "Haven't we had enough of Watergate?" Nobody knows what's going to work. Hollywood is based on a search for past magic. "Redford and Newman worked twice (*Butch Cassidy* and *The Sting*); if we could only get Redford and Newman in a picture we'd all be rich." The reality is that nobody knows. One can guess that a movie about some robots in the future will work, and that George Lucas will handle it well, but Universal didn't think so. They passed on *Star Wars* when they had *American Graffiti*.

How close is a writer allowed to the actual production? To a degree the answer lies in how big a writer he is. The bigger the name, the more likely he is to have a say about the details of production. Generally, the answer is that the writer gets as close to the production as his director allows. The production is really the director's baby. If he has faith in the author's judgment, the director will be more willing to tolerate his presence during filming. If the director doesn't want him, there is nothing the writer can do about it.

An author is blessed if he has a director who is interested in working closely with him as he prepares for production. The time when the author is most essential is in the story conferences with the director prior to filming. It is during these very crucial days that he tells the director over and over again exactly what he meant. Talking it all out in minute detail with the director can clarify the content and ensure the director's chance of a clean and accurate interpretation. It is during these conferences that scenes are cut, added or otherwise modified. In *Butch Cassidy*, for example, the screenplay was changed, but never basically. Certain scenes were cut; the musical numbers were added, but the thing that makes the movie work—the basic relationship established between the two men—was left essentially unchanged.

In one specific instance I had written an atrocious scene, the opening scene of Robert Redford and the card game. Everyone said, "Get rid of it! It stinks!" And I kept saying, "I know it stinks, but it's the best I can do." And all the time I was going through that pressure, George Hill kept saying, "You're not going to change it!" George knew how to make it play. He took the scene and put it in sepia, which gave it an old look. And he had what is probably the longest close-up in modern film history on Bob Redford. It's about ninety seconds of solid Redford, and the scene really plays. He gets a tremendous tension out of it. This is a striking example of how a good director can take even a rotten scene and make it work.

I went out to Hollywood in June 1968. George Hill was already there. For about ninety days, he and I met every day, spending most of each day talking about every aspect of the script and coming up with ideas for it. These meetings lasted until mid-September and included a two-week rehearsal period prior to actual filming. Until the filming began, I was involved in many decisions that were made, but the final work necessarily was that of the director. I returned in the middle of production for one week of shooting at the studio between location work in Utah and Colorado and in Mexico. On this visit I saw four or five hours of rushes that George had shot in Utah and Colorado and gave him my reactions. That basically was my contact with the production of the film.

My own feeling is that I don't want to be around on a film I have written. There are times when an author can be helpful. In *Butch Cassidy*, for example, there were a couple of scenes misdone. Had I been around, I could have said, "Oh, no, no, no, no—I meant this." You see, they were actually miswritten, and I didn't realize it until I saw them on film. They are not, incidentally, in the final film. Had I been around, I could have said, "I miswrote that. Don't play what I wrote; play it this way."

Generally, however, I don't like to be around for two reasons. First of all, because I am the screenwriter, nobody really wants me around. If a line is misspoken with the proper emotion or spoken properly without the proper emotion, there can be problems. The writer thinks the actors are ruining his lines, and the actors resent the author's presence. And similar tensions can arise between director and author over interpretation. Second, although there is nothing more exciting than your first day on a movie-star-laden set seeing all your dreams come true, by the second day you are bored with it. By the third day everything is so technical that you are ready to scream, "Let me out of here!" The idea of standing around for seventy-two days of shooting, bothering people and saying such insignificant things as "The line is 'There's the fireplace,' not 'Where's the fireplace,' " is madness. Since the author just doesn't know when he might be really helpful, he might as well stay away and avoid the agony for himself and everyone else.

In *Adventures in the Screen Trade* I wrote that nobody knows anything in the movie business because no one can predict popular taste. In 1990 *Teenage Mutant Ninja Turtles* had one of the biggest openings in the history of movies. The distributor was an independent company, New Line. All the studios turned it down. I would have done the same thing. But studio executives were scratching their heads, wondering "How did we miss on that one?" The fact is they don't know, since they're in this blizzard, and the snow won't ever stop coming down on them. It's a fascinating business to watch from a distance, and I'm glad I never had to be a studio executive.

What movies get made reflect the executive mentality; what movies are successful reflect the audience. I have no idea what they will like; I try to write a screenplay that I will like, and I pray. If I want to continue working in pictures, it's essential first that my screenplay gets made and second that it gets made properly. After all, the business pays attention only to writers who write movies that are commercially viable. But, beyond it all, nobody really knows which films will be big. There are no sure-fire commercial ideas anymore. And there are no unbreakable rules. Classically, westerns have villains. *Butch Cassidy,* however, the most successful western ever made, has no tangible villain, no confrontation in the usual sense. Perhaps the success for the movie with kids is in the concept of the "super posse," a force that follows them and makes them do terrible things that they cannot control.

My advice to screenwriters starting out is hustle, pester, embarrass yourselves to get to any contact in the business you can, and move to Los Angeles, because that's where the business is. Also, you must be able to handle rejection.

If screenwriting were the only kind of writing I did, however, I think I would find it desperately frustrating. When I write a novel, I take it to my editor. He says, "This stinks and I want you to change it." If I agree with him, I say, "Okay," and I change it. If I don't agree with him, I can say, "Good-bye." It is my baby and I can fight to the death. I can either not get it published at all or get it published as I want elsewhere. At least it is *my* fight to make if I choose. In films an author doesn't have that right. In films he must assume the director or producer will be ultimately responsible for what the finished product is and whether it works or not. And, of course, there is no guarantee that he will get a director or producer who will listen to him.

One thing that really pleases me about movies today is that advertising and publicity and critical reviews don't mean anything anymore. *Butch Cassidy* opened in New York to pretty bad notices but tremendous business. Happily, the reviews are totally unimportant on a film. No one except maybe the critic's mother is going to go to a film or stay away from a film because the critic says it's good or bad. Movie audiences will not be lectured to. It is a golden time.

# THE LITERARY AGENT

*by* ***LEE G. ROSENBERG,*** *one of the founders of Triad Artists, Inc., a full-service talent agency based in Los Angeles. Educated at the Choate School and Harvard College, Mr. Rosenberg held production positions in films and television before cofounding the literary agency Adams, Ray & Rosenberg, which thrived from 1963 to 1984. In 1984 Mr. Rosenberg founded Triad Artists, one of the largest talent agencies, which serves a wide range of clients in entertainment markets around the world.*

***. . . in the negotiation of subsequent covering documentation there is a give-and-take that usually takes from the writer. . . .***

Writers are born, not trained. Talent is genetic. A writer may have talent, but what one learns from instruction or by examining screenplays or movies is craft, the form in which talent expresses itself. Durable success is dependent upon this duality.

In the entertainment industry a writer's career combines creativity and business. The business life of a writer is protected by the literary agent. Literary agents are employed by writers to seek out and formalize employment within the context of wider career guidance.

To begin a study of the agent-client symbiosis, let us examine a writer's initial problem: finding and engaging an agent.

This is often difficult for the unknown writer. First, simply saying one is a writer doesn't make it so. Second, once a writer has something to offer—perhaps a completed screenplay—he must attempt to interest an agent. Consider the direct phone call, for example. The impression of personality and intellect the writer might make on an agent willing to take the call is possibly more important than coming to the office and saying, "Here's my script." Motivating an agent to take calls like these may depend upon whether there has been prior contact about the writer from someone the agent knows and trusts or whether he has received a letter from a writer that sounds reasonable, intelligent and perhaps reflects the writer's ability to communicate an original point of view. The inventive applicant may attract the interest of even a busy agent, who might then agree to listen to him. Compounding the issue is that many agencies, ours included, do not accept unsolicited material unless it comes from a source whose judgment is respected.

However the writer manages it, after the initial connection he then must put a screenplay on the agent's desk. Sitting and conversing accomplishes little; the agent must read the script. And this points up the greatest problem of all: finding time for reading. If an agent is conscientious and successful, there is a prodigious amount of reading to do, from at least three sources: his own coterie of clients; submissions from the marketplace presented to his clients for assignment; and writing samples from potential new clients, on the recommendation of others in the industry. Following the submission of a writing sample the writer must be very patient.

If the agent is impressed, he may call the writer to discuss the work and arrange a meeting to talk things over. Considering this meeting a first date that may lead to marriage, the agent projects his own personal chemistry while seeking to get acquainted with the author as a person. As in any marriage, a couple never knows whether they have made the right choice until they have lived together for a few years. At this first encounter the parties begin to assess each other. The agent will react to the material as a sample of the writer's work and as an index of his ability to handle the visual medium. The writer will be seeking a sense of confidence that this agent is the person to guide his career.

If it's good, the writer's work will be circulated among other agents within the agency. In an enterprise calling for so subjective a judgment on the part of the reader, there is added insurance of success if the agents ultimately shouldering the work of representation are all enthusiastic about the writer. For the alliance to succeed, there must be mutual trust and genuine enthusiasm between agency and client.

For new clients at Triad we have a signing committee that analyzes how the potential newcomer fits into the overall client mix. A talented beginner will require infinitely more time than an established writer, so the signing decision is also evaluated in terms of time versus the point in the career at which the agency is entering the life of a client. Periodically our agents gather at retreats to consider whether the balance of our client list is appropriate. In the end we all understand that the lifeblood of the agency—and of the industry—is new talent.

I keep an eye on new clients to make sure they are being placed in the marketplace with early assignments in proportion to their talent. Recently I faced an interesting dilemma. A junior agent had come across a writer who had just graduated from an eastern college and had written a spec (speculative) script for a television series, and then was hired on to that series as a sort of intern. The writing, as a measure of promise, was excellent, but promise implies a period of time before its value is realized. So there was considerable internal debate as to whether the agency should make the investment in time to nurture this talent. The debate was intricate, subjective and abstract. Discussion ranged over various components as well as qualities of the writing: dialogue, structure, characterization, craft, and so on. Finally the agents decided to take on the responsibility.

This writer was then signed to standard agency agreements of two years' duration. The Writers Guild of America provides that the first-term writer, one who signs with an agency for the first time, may terminate the agreement after eighteen months upon written notice.

We do not employ the writer; the writer employs us. If the agency loses enthusiasm and therefore effectiveness for a client in the marketplace, the agent must so advise the client. This can be a terrible dilemma, which must be resolved with candor and courage.

If the client seeks to terminate but has an unexpired contract, that creates an issue. Usually a substantial amount of time, effort and overhead has been invested in behalf of that client. The agent's inclination is to resist giving up the client's contractual future without recouping that investment.

However, poor conditions may exist in the agency-client relationship that can be improved. Perhaps there has been inadequate communication between the writer and the agent, causing the writer to feel his interests have been neglected. His complaint may clarify things and initiate satisfactory correction. In this case the contract allows the agency legally to hold the client while the parties have an opportunity to reassess their attitudes and creates an involuntary cooling-off period. If the parties still can't resolve their differences, the client has the right to dismiss the agency and hire another, with the understanding that his present agent may not wish to work out a commission settlement arrangement with the new agent. Of course the agency may release the client completely from any continuing obligations.

The agency commission is 10% of all gross monies received by the writer, whether from profit participation or from payment made for rights and/or services. All monies go directly to the agency. We usually deduct the commission and issue the client's check within three to five days.

Once the agency contract is signed, the agent begins the process of seeking work for the client. Generally the writer has either written a spec script that may lend itself to packaging or auction or is seeking employment as a writer-for-hire. The acquisition of rights and services is governed by the Writers Guild of America Theatrical and Television Basic Agreement. (Inquiries as to content should be addressed to the Writers Guild of America West, 8955 Beverly Boulevard, Los Angeles, CA 90048; phone 310-550-1000; or Writers Guild of America East, 555 West 57th Street, New York, NY 10019, phone 212-245-6180.) Some contractual terms covering sale of a spec script, as well as contractual terms of a script written for hire (including performance of specific services such as drafts, rewrites, polishes) will enjoy a congruency, since they both fall under the WGA Basic Agreement. In the case of a writer making his first sale, I strongly recommend that terms of the WGA Basic Agreement be "included by reference" in the deal.

Let's examine selling strategy for a spec script. First we determine whether there is a duplication in the marketplace. If another project exists that is similar, it reduces the value of ours. When three competing studios each developed Robin Hood screenplays with directors, the first to attract a major star—Kevin Costner—knocked the others out of competition. Another went on to be made for television, and the third dissipated.

Second we determine whether there is a studio with a special relationship with a producer, director or star who is particularly suitable for the screenplay. This judgment is often intuitive, based on the joint experience of agents covering the tastes of potential buyers.

As the market plan falls into place, it includes a list of buyers targeted for submission, when to submit the project, predisposing elements of the project, and how the chess game is likely to unfold. Perhaps a marketing device will be added. With one screenplay, *The Ticking Man,* alarm clocks were sent out with the script, set to go off at a particular hour. This created a stir among buyers, who got themselves mutually excited about the script.

As the agent moves in to close a deal, he must nail down as many specific details as possible. If he is dealing with people he knows who are knowledgeable and have studio bureaucracies backing them, he is usually safe in making verbal commitments. With anyone else, however, he will be wise, before proceeding, to insist that a memorandum agreement be drawn up and, in extreme cases, that funds be put in escrow. Proving damages in a broken verbal agreement is excessively difficult and costly. Our agency generally executes a seven-page *deal letter* outlining substantive points including price, general rights acquired, and certain other stipulations that may be unique in a given case. Details are set down so that a lawyer can actually draft a contract from our letter. Once the deal letter has been initialed by the parties concerned, our agents record it in synopsis form in an internal *deal memo,* including the client's name, commission, the buyer's corporate designation, the starting date of services, and many other items distilled from the deal letter. After agreeing upon the deal, there is an administrative period while contracts are drawn up and payments are made. The agent can help to keep lines of communication open between writer and buyer and to work out problems that might arise from misunderstandings or differences of opinion.

In a typical deal the following points are discussed between the agent representing the writer-seller and, say, the business affairs executive representing a producer-buyer.

First is the duration of the initial option period, and of subsequent option periods, which could be of any length and of any number.

As to monies paid for the option, one would negotiate a sum for the initial option period (which may or may not be applicable against the purchase price) and for the second, third and other optional periods. There is a standard one year for an initial option period and extensions of six months or a year for the second and subsequent optional periods. There are no standards as to price, which depends upon what the market is willing to pay.

Second, one would specify what is required to exercise the option, in other words, notice and payment of the purchase price only; or commencement of principal photography plus payment of the purchase price; or a third-party commitment (from a director or star, perhaps) plus payment of the purchase price or other conditions.

If the work is "written for hire" instead of a speculative screenplay, other questions arise. Does the purchaser have the right to hire an additional writer (other than the writer of the screenplay) to perform writing services during the option period? Is any portion of the option money applicable against money payable to the initial writer for his services? If the initial writer performs services during the option period, is any portion of that money paid to him applicable against the purchase price?

The full purchase price is then negotiated, which may include a deferred amount, payable upon an event that may occur in the future, such as the start of principal photography of the picture.

Next, one must determine if the writer is entitled to participate in revenues generated by the film. The purchase price and profit participation must be negotiated with great care at the outset of the transaction, when the buyer's appetite is at its zenith.

Participation usually ranges from 5% to 15% of something. That "something" may be net profits, gross, gross after breakeven (or earlier), or some other formula. Each of these positions within the revenue stream must be carefully negotiated with the help of an accountant and legal counsel. (See "Elements of Feature Financing" by Norman H. Garey, p. 139.) Otherwise, a great deal of leverage is lost, and in the negotiation of subsequent covering documentation there is a give-and-take that usually takes from the writer.

One key issue is whether the participation is levied on 100% of revenue—gross, net or otherwise—or on that portion of the revenue that is retained by the producer. In the former there is a "wholeness" to the participation; in the latter there is a "halfness" because the producer rarely receives more than 50% of 100% of the net profits (or some commensurate amount of gross). Therefore 10% of the producer's net may be 5% of 100% of the net of the picture. The magnitude of profit participation is usually tied to the screenplay credit determined by the Writers Guild. A shared credit generally reduces the writer's participation by half.

Other niceties regarding the participation include a definition no less favorable than that of the producer, or other person with leverage; auditing rights and other concerns.

In a transaction for a spec screenplay there are additional questions about reserved rights (covered below); payments from possible theatrical remakes or sequels; payments from possible television remakes or sequels; additional compensation for release of a television film in theatres in the United States and/or overseas; payments for a miniseries or television series or spin-offs. For example, if an initial television series (such as *M*A*S*H*) is produced, is successful, and characters are spun off into a second series (like *Trapper John, M.D.*), what is the compensation?

It might be helpful at this point to explain the concept of *reserved rights* or *separation of rights* as it affects the writer client. The Writers Guild, through negotiation with management, has established, among other minimums, that certain rights are the property of the writer or writers who receive story credit in motion pictures and television. The precise definition of these rights and the conditions that surround them are available in the Writers Guild Basic Agreement. The rights *separated* out and *reserved* for such a writer include dramatic stage, radio, live TV, merchandising and publishing. (For more about merchandising, see article by Stanford Blum, p. 407. For details on exploiting book-publishing rights, see the article by Roberta Kent and Joel Gotler, p. 113.)

When making development arrangements for a theatrical feature, a two-hour movie for television or a television mini-series, publication rights are treated as a valuable point. Naturally the agent wants to make a publication deal as early as possible in order to have the necessary lead time to get the book published well in advance of film production.

Since the financing studio contributes millions to the development and production of the film, they insist upon a sizable chunk when a publication deal is made. The agent resolves this in many instances by giving the studio 20% of net publication revenue after agency commissions have been deducted. In a typical transaction, our client recently retained 80% of net publication revenue and the studio retained 20%. We were fortunate that the studio allowed our commission off the top rather than simply off the client's share. (Sometimes commissions can go as high as 25% in foreign publication deals. Since the base agent has to share with the agent abroad—to motivate that overseas agent—half of our usual 10% commission is not enough. Occasionally the overseas agent receives as much as 12½% while we receive a like amount, for a total of 25% of commissions.)

Publication revenue includes the publisher's advances as well as royalties earned after recoupment of advances. An advance is simply a royalty paid in advance. If a client has received a $300,000 advance from a publisher for writing a book after a studio deal has been made, the studio gets 20% and the client gets 80%; the payments generally will be $100,000 upon signing the contract, $100,000 when the manuscript is delivered and $100,000 upon publication. From the $300,000 the agent takes his 10% commission, and the balance is divided between the author and the studio in the 80/20 relationship.

For the writer who has not yet determined whether he or she wishes to write exclusively for motion pictures or for publication, it's useful to be aware of the economics of both industries and some of the attendant selling problems. In the case of a first novelist, less money is available for a book than for a screenplay written for hire and based upon an original idea. If the novel is enormously successful, the economics even out. Also, writing a book produces certain collateral economic problems that relate to the producer-buyer's attitude toward the risk involved in acquiring the book rights and then investing even more risk capital in the development of a screenplay. When a development deal is made exclusively for a screenplay, there is only one investment and one assessable risk. As an example, if a book seems desirable for film purposes, it must be either optioned or acquired outright, depending on conditions imposed by the author. The producer knows that in addition to the money spent for the option or outright purchase, he still has to arrange a substantial additional investment for development of the screenplay. The same story might have been developed for somewhat less money as a screenplay. Obviously it's not unwise to write a book; it simply adds a dimension of time and risk for the entire process, which might diminish marketability of the property.

Returning to our hypothetical negotiation, one additional issue to be resolved is whether the writer has any rights of *turnaround*. This would be triggered if the producer or studio fails to produce a film after a certain period of time, whereupon all rights revert to the writer, allowing for resale of the property, contingent upon repayment of invested costs by the subsequent purchaser.

When writers were called upon to write material years ago, frequently they weren't paid if a producer didn't like the result. Further, there was no machinery to guarantee the writer credit on the screen for what he had written. The Guild has now established that no writer may speculate his writing for an employer or potential employer; he must be paid for what he writes. Also, every writer must receive credit for what he has written. Because it's a highly subjective business, one writer's work may be combined with that of another in the course of a rewrite or an adaptation of a story. In such a case, the Writers Guild, not the studio or employer, is the final arbiter as to who receives credit.

If arbitration seems necessary, the Guild will submit proposed credits to selected member writers, who will read the material and recommend who gets what credit. In this arbitration procedure, the writer is given an opportunity to examine the list of potential arbitrators and eliminate a reasonable number of those who might have some bias.

No discussion of the literary agent can be complete without references to packaging and auctioning.

*Packaging* is the process of assembling several creative elements—

director or producer or star, or any combination—with a screenplay offered to a financing source in order to enhance the transaction. To achieve this, the larger agencies, such as Triad Artists, Creative Artists, ICM or William Morris, are at an advantage because they have a wider talent pool from which to choose. Because it places a certain pressure on potential buyers, packaging is sometimes maligned. After all, a producer-buyer of a screenplay alone would then proceed to "package" it, in effect, by assembling creative elements from a pool wider than that of a single agency. But from the writer-client's perspective the process can be constructive. For example, we recently sold a screenplay that, in my estimation, would not have sold had it not been coupled with a star of some magnitude. The film went on to be successful. But there are other instances where agents have put clients together in packages that have ultimately failed at the box office. The executives who bought these packages had the ability to "pass" at the time of the offering; failure of the films led to criticism of the agency packaging process. It all comes down to imaginative coupling, subjective judgment and the courage to pass or proceed.

Before discussing auctions, I want readers to realize that far more screenplays are sold without auctions. These are screenplay transactions in which the agent believes there is only one correct buyer, especially suited to the material—perhaps a producer or a director with a specific track record who is willing to pay a fair and competitive price. These may be viewed as commonsense deals or preemptive buys.

Before an auction, the first decision is whether the property is worthy, since auctioning can have a reverse impact. As quickly as the market can heat up for a property, it can cool down. Assuming a screenplay is auctionable, strategy is then drawn up, including a list of targeted buyers; elements that might be added, as in packaging; and how marketplace response will be communicated from buyer to buyer. The idea of the auction process is to create an atmosphere of insecurity among the buyers. If there is more than one bidder, it reassures the others that their judgment is sound.

If the agent is confident, perhaps a deadline for bids is stated. A set of rules may call for the highest bid to win; or the author will determine among all bids so that the highest bid will not necessarily win; or if there are ties, the author will consult with the offerors and determine among them. A price doesn't have to be stated; the market sets its own price in most cases. If a plateau is bid, there may be a "topping procedure," whereby each potential purchaser is advised of the figure and has an opportunity to bid more, and an escalation occurs. At some point the buyers will sift themselves out. The auction process has been around for years, but is being used more recently because the literary marketplace has become very aggressive. Million-dollar deals for screenplays are occurring more often.

The agency business today is expansive, and there has been an expansion in the market calling for services of writers, directors and actors, since more films are being made. In features the essential market is composed of the major studios. In television, buyers for two-hour movies are the three domestic broadcast networks—ABC, CBS, NBC—the Fox Network, and cable channels TNT (Turner Network Television), the USA Network (owned by Paramount and MCA), HBO (Home Box Office), Lifetime, the Family Channel and the Disney Channel.

There is real growth potential in the international market, especially in selling talent and literary material, and we have dedicated a department at Triad to the development and exploitation of that market on a prospective basis. For example, coproduction in overseas television is on the rise because of pressures upon producers in the domestic TV area from higher costs, flat network license fees and a reduction of the value of markets after network exhibition.

The creation of Triad Artists in 1984 was a response to trends observed in the early 1980s from the vantage point of our literary agency, Adams, Ray & Rosenberg. In projecting our position in the industry forward, it was clear that market forces such as the growth of the cable business, the increased value of ancillary rights, the expansion of global entertainment and growing foreign investment in the Hollywood community called for us to aggressively expand our franchise and grow through acquisition or merger to properly position ourselves for the future. I decided that a merger with two other independent agencies would complement our expertise. Triad Artists was formed from this group. One agency was active in the talent field, DHKPR (David, Hunter, Kimble, Parseghian and Rifkin), and the other was involved in booking musical acts in venues throughout the world, Regency Artists. The personalities fit, and Triad has emerged into a full-service agency of some 250 people, bicoastal, representing clients from below-the-line (editors, cinematographers, production designers) to motion picture and television packaging, motion picture and television talent, literary rights, music and a host of other services. Each agent in each department is delegated responsibility for covering segments of the marketplace. Then there are agents responsible for each client who coordinate the career and relationships of that client with other agents within the agency, with emphasis on personally updating the client, and social growth with the client. There is a constant exchange of market details among our talent agents, literary agents and packaging agents, which feeds back into staff meetings. Minutes of these meetings are distributed to all departments. Through this intense internal communication and strategic expertise, Triad provides leverage, recognition and placement for our clients within expanding entertainment markets throughout the world.

# THE STORY EDITOR

*by* ***ELEANOR BREESE,*** *who has served as both executive story editor and story analyst for companies throughout the entertainment industry, including the Walt Disney Company, Columbia Pictures, Lorimar, CBS, ABC, Brut Productions, Wizan Productions, David Wolper and Warner Bros. A writer of novels, screenplays and short stories, she is a member of the Writers Guild of America, West, the Story Analysts Guild and Women in Film, and has presented seminars about screenplays.*

***New or unproduced writers are encouraged to write complete screenplays . . . even if this must be done on speculation . . .***

The title *story editor* is gradually becoming a misnomer in large production companies because the person holding the title doesn't really edit or develop stories, screenplays or other literary material; that is the function of the production executive and the producer working on a specific project. But the story editor does perform an essential service involving all literary material circulating within a company by acting as a sort of creative traffic director for all submissions and coverage, consulting with producers, writers, directors and production executives and overseeing a staff of readers. The story editor serves all of these creative executives by handling, in the language of the business, "submissions of material." After all, literary material is the coin of the realm.

A basic requirement of the job is the recognition and appreciation of good dramatic writing and a thorough familiarity with writers working in movies and theatre, including credits. Most creative executives in the movie business carry with them a mental storehouse of writers' and directors' credits, based on hours of moviegoing and television viewing. This familiarity with talent and credits becomes part of their lexicon. It's not unusual for a story editor, production executive or agent to come home with a pile of scripts to read at night or over a weekend. To place the story editor in perspective, let me "widen to a long shot."

There is a normal daily flow of submissions to a film company from writers and agents. Submissions come in a variety of forms. The one technically closest to production is the screenplay, averaging about 120 pages of dialogue and description. Over the years the industry has developed an accepted screenplay form, which is a departure from the dialogue and description form used by dramatists for stage plays. (See excerpts from screenplays of *Butch Cassidy and the Sundance Kid*, p. 87, and from *The China Syndrome*, p. 257.) The novice screenwriter is advised to follow this form in order to be competitive. A screenplay is rewritten in a sequence of drafts, so one may bear the reference *first draft, second draft, revised second draft* or no such reference at all on the title page. A *revision* is generally a reworking of a prior draft that is not as much work as a full draft; a *polish* is even less work, usually the making of minor corrections in and additions to a prior draft. These terms are often used in a business sense to define the point at which a writer is to be paid. In a *step deal* with a writer, for example, he or she may be paid for a first draft, one set of revisions and a polish, or any variation thereof, and the agreement may stipulate payments to be made upon delivery of each step. If a screenplay is being developed in-house, such title-page references are instructive; if one is submitted from the outside bearing the legend *first draft,* it cannot be verified as such. An agent or producer might submit a fifth rewrite as a first draft to create the impression of freshness.

Projects may be submitted variously in the form of *treatment, story outline,* or *format,* which are like short stories of varying lengths. What distinguishes them is that they are not in screenplay form. They convey the writer's story in prose style, perhaps with some examples of dialogue, concentrating on story points, sequence of events and characters. This form of presentation is often used by established writers as a shortcut method in selling their stories to a financier-distributor, who would then finance the writing of the screenplay by the same writer. New or unproduced writers are encouraged to write complete screenplays, however, even if this must be done on speculation, that is, without advance compensation. This way the new writer communicates writing ability as well as story sense. If a new writer were to submit a treatment only, there would be no evidence of screenwriting ability, especially in the handling of dialogue. This also makes sense from a business standpoint, for the new writer's bargaining power is enhanced with the buyer if a screenplay, rather than a treatment, is to be acquired.

Books are submitted from agents or producers in various stages of prepublication. The latest stage would obviously be the hardback or paperback edition, prior to release to bookstores. Working backward, an earlier step would be a bound copy of the *uncorrected page proofs,* available several months before publication and possibly including errors, used to gather early reviews or for selling subsidiary rights, such as motion picture rights. A typewritten *manuscript* is the earliest form in which a full book would be available. There is rivalry among some story editors, especially in New York where most publishers are based, to have access to desirable manuscripts at the earliest possible time, so that a producer or financier may have an advantage over the competition.

Other types of submissions are as varied as any source material for movies can be. *Urban Cowboy* and *Saturday Night Fever* were based on magazine articles; *Dick Tracy, Batman* and *Teenage Mutant Ninja Turtles* were from comic books; *Ode to Billy Joe* was from a song.

The story editor receives literary material from either outside the company (from agents and other creative personnel) or inside the company (from production executives or creators under contract) and oversees coverage of the material by a reader. *Coverage* refers to a reader's synopsis that is the objective retelling of a story in condensed form. Rendering a synopsis, which may fill from 2 to 12 pages, including a reader's comment, is the main function of a story department.

Many writers deplore the system of readers writing a synopsis of their work that is placed in a story file, available for possible destructive reference forever, but this is the only way yet devised to cope with the mass of material submitted to any single company for consideration. It is simply a matter of saving an executive's time, a shortcut method to sift through the material, and an aid rather than a decision maker. A synopsis is useful to an executive in deciding whether to read a project and is seldom a substitute for firsthand reading of something under serious consideration. Writers suspect that an unfair evaluation may be made by a reader who happens to be in a bad mood or who may be predisposed against a certain kind of story.

Often projects are covered at competing companies. It might seem that this duplication of coverage could be avoided by using a pool of readers serving all companies. But the impulse to secrecy and the intense competition in the acquisition of literary material make this impracticable. Naturally employers expect readers to be discreet about material under consideration.

Just as experts in art or cooking develop taste in their fields by exposure and experience, a good reader develops judgment by reading hundreds of screenplays, viewing movies regularly and discussing aspects of evaluating and reporting projects with the story editor and other readers. As he reads, there is a movie going on in the reader's head; scenes are visualized and dialogue is heard. The reader hears and sees a work as a possible theatrical film (different, attractive, commanding, big), as a television movie (an effective personal story that would be intriguing on the small screen) or as a pilot for an episodic series.

Following years when readers were paid only a few dollars for reading, synopsizing and evaluating material, a Story Analysts Guild was established in 1954 within the IATSE, the motion picture multiunion, with a graduated pay scale determined periodically by contract negotiation between management and the union. If the workload in a story department exceeds the amount the staff readers can cover quickly, the editor may hire other readers from the industry experience roster of the Guild for a minimum of one week's work.

The daily routine of a story editor can best be described by reviewing the mail that arrives, which can be broken down as follows:

1. Material from company executives. This includes submissions to them from agents representing a writer or a whole production package, from producers who hope to engage the company in a project and from acquaintances and friends of executives.

2. Scripts or books from individual producers working within the company framework. Sometimes a fast synopsis is required, and overnight service is given on such priority requests. Usually no more than a week passes before coverage is returned to the executive or producer along with the material.

3. Books from the company's New York office, which are submissions from East Coast agents as potential films. Some books are covered by readers in New York, and the synopses are sent to the West Coast. (Writers live all over the country, and while life experience elsewhere can be enriching for a writer, I believe writing for the motion picture market usually requires the writer's presence within a reasonable distance of the marketplace. A writer cannot judge current film interests by released films playing in theatres or on the television screen, for these represent buying decisions made eight months to two years earlier.)

4. Scripts or books submitted directly to the story editor by agents or friends. These may be accompanied by personal letters with information about a new writer or something of the history of the project. Agents who are selective in sending material to a likely buyer are appreciated, while the few agents known for their scattergun approach of submitting projects indiscriminately receive low priority for coverage. Projects from the editor's friends may receive personal attention, but recommendation to management is made on the same basis as other material.

5. Projects that come in "off the street" or unsolicited; this is, not represented by agents. These must be accompanied by a release signed by the writer. If an unknown writer who does not have an agent wants his work to be read by a production company, it is advisable to request and sign a legal release to protect the company from a lawsuit in case they pass and later produce a film on a similar theme. Many companies will not read unsolicited projects. Material submitted by a licensed agent is protected by agreement between agents and film companies, so a release is unnecessary.

With all submissions, the appearance of the material is a factor; with the work of a writer who is not established, appearance is sometimes the deciding factor between initial attention and indifference. A screenplay submitted to a potential buyer should be clean and fresh, not dirty or rumpled; readably typed in the established screenplay form, from 100 to 130 pages in length; and as free as possible from distracting typos and errors of spelling. Samples of screenplay form may be found in textbooks or mass-market paperback versions of certain movies.

After sorting the submitted material according to priorities and by the speed of coverage requested, the editor decides which reader reads what project. The never-settled debate among writers over whether it is better to have a prominent agent and risk getting lost in the shuffle of volume or a less prestigious agent who will provide more personal representation intrudes on the issue of priorities of material to be covered. The story editor is inclined to give precedence to the "important" agent, even though the other agent may be more careful in selecting the target buyer and generally works more closely with the writer client. So the debate continues.

The reader's coverage is returned with the material to the producer or executive to whom it was first submitted. (An internal system of identification attached to the property prevents mix-ups.) After reading the synopsis, the producer may want to read the material thoroughly or merely skim it. Or the synopsis may simply aid him in returning the project to the agent or author with an appropriate letter. Copies of all coverage are distributed internally to executives and producers, unless someone stops such a distribution to protect a confidential submission. It's possible that another producer in the company may read a certain synopsis, spark to it and ask to read the material itself.

Sometimes an executive or a producer may ask for a *selling synopsis* for presentation to a director, star or network. A reader writing a selling synopsis would not make creative changes in the story, but would emphasize, without comment, its exploitable aspects. The language of a synopsis should not reflect editorial comment by the reader, as in "a silly chase ensues," but should objectively convey the style and substance of the material. The reader's separate comment following the synopsis can be more personal but should be an evaluation of the material specifically for the market.

When material submitted to the story editor is turned down, it is returned to the agent or writer, usually with a note. A story editor's helpful comments in a rejection letter will sometimes result in an appreciative response from the writer. A noncommittal rejection leaves a writer baffled, wondering if there is anything to be done to improve the salability of his work.

There are other services the story department performs. An executive or a producer may ask for a copy of a certain book from a bookstore or publisher. There may be a query about the availability of film rights in a literary property, in which case the name of the book's agent can be obtained from the publisher's subsidiary rights department and the answer learned from the agent. Further, a producer may ask the story editor to do research on a subject being considered or ask for a list of suitable writers for a specific project. There are also daily calls from executives requesting old coverage from the story files.

For early scouting of potential movie projects, publications containing capsule synopses of forthcoming books are read in the story department, such as *Publishers Weekly* magazine and *Kirkus Reviews*, a private reviewing service. The industry trade papers, weekly *Variety* from New York, and *Daily Variety* and *The Hollywood Reporter* from Los Angeles, are read to keep up with the activity of people in the business and to follow film grosses; the Nielsen ratings are studied as an index of television-audience response. The story editor's rounds include being aware of plays produced locally, lunching with agents, writers and other story editors and attending special screenings and parties. The annual budget for the story department is projected through meetings between the story editor and company budget officers, and it is the responsibility of the story editor to see that the operation of the department stays within these budget guidelines.

It used to be that all story editors were heavily involved in the development of projects, meeting often with writers and executives, helping to evolve a screenplay through draft after draft. This is still done by a story editor in a small production company, where only one or two projects may be in work at one time. But today in the major studios or a large independent, the individual producer chooses his writers and works with them to develop a final script. In some companies, the term *story manager* is now used for the management of traffic of material in and out of the story department, while a *development executive* looks for scripts and is involved in their journey toward production. The designation *story editor* more aptly describes the story editor on a television series who works with the producer of the series in determining the format, choosing material from episodes submitted by agents and rewriting purchased material for the series with or without the collaboration of the original author. The movie business could do with a better term to distinguish between the TV series story editor, who really does edit scripts, and the studio story editor, but so far none has evolved.

# EXPLOITING BOOK-PUBLISHING RIGHTS

*by* **ROBERTA KENT,** *who has been a literary agent since 1972, first as a publishing agent in New York, then as a literary agent in Los Angeles. Currently, she is with STE Representation, Ltd., a talent and literary agency with offices in both Los Angeles and New York and affiliations in London, Paris, and Rome. She represents novelists, screenwriters, television writers, directors, and producers.*

*and* **JOEL GOTLER,** *who began his career at the William Morris Agency in New York in 1971, then moved to Hollywood to work with the legendary agent H. N. Swanson. In 1987 he established his own company, Los Angeles Literary Associates, a full-service agency for novelists that also represents dramatic material to the major studios and networks on behalf of various New York agents and publishers, before becoming a partner in Metropolitan Talent Agency in Beverly Hills.*

***An important lesson is that the prices usually quoted as sales to paperback houses for movie tie-in rights are guaranteed advances against royalties to be earned by sales. . . .***

The value of book-publishing rights in connection with motion pictures can be traced from one of three sources: manuscript, original screenplay or existing novel.

With an author's manuscript the agent will either auction or nurture. An auction is a calculated risk, applied to a presumably hot book, in order to avoid alienating the deep-pocketed movie buyers expected to bid. To auction movie rights (one of the subsidiary rights) of a novel, manuscripts are sent to competing parties (production companies or financier-distributors), a deadline is set and bids come in. This rarely used process can raise prices into the millions. Michael Crichton's novel *Jurassic Park* was sold in 1990 for $2-million plus points to Universal in a deal that included the author's commitment to write the screenplay and anticipated Steven Spielberg to direct for his Amblin Entertainment. The landmark deal in this area was made in 1979 when United Artists bought motion picture rights for the novel *Thy Neighbor's Wife* by Gay Talese for $2.5-million plus points. The book went on to become a major best-seller, but no movie was ever made.

The strategy of nurturing a movie sale is far more usual. When a writer-client completes a manuscript, the agent must decide when to offer movie rights to prospective buyers. It can be offered to a publisher and a movie source simultaneously; or, more typically, after the publishing deal is made; or closer to publication, in galley form, to take advantage of favorable reviews. Timing and strategies vary depending upon instinct, intuition and an assessment of buyers and the marketplace.

If an original screenplay is commissioned by a producer or written on spec by a writer and then purchased by a producer, there may be an effort to make a book deal before a financier-distributor gets involved. In this case the screenplay will be submitted to publishers (hardcover or softcover) for evaluation. If the publishers feel that the property will not succeed on its own as a book, they may want to review it when a movie deal is set (involving commitment of financing, a director and stars). This is most common. But if the producer must wait, the issue of lead time becomes a factor.

It's appropriate to review the mechanics of publishing at this point because as a movie project gets farther along, time pressure becomes more acute. A novel can take anywhere from three to eighteen months to write. The manuscript is turned in to the publisher, who starts the editing and then the printing process. It takes anywhere from six to nine months to get that hardcover book into print and another month to distribute it to bookstores. A paperback generally comes out nine months to a year following the publication of the hardcover edition. Usually timing calls for a hardcover novel to be in bookstores at the time the picture is shooting, while the paperback tie-in will be published just about the time the motion picture is released. A paperback original requires about as much lead time for publication as does a hardcover, to allow for design, typesetting and printing.

To return to the hypothetical example, assume the producer is having no success in preselling hardcover or softcover publication rights to the original screenplay. When he makes a deal with a financier-distributor, however, it's a new ball game. Now the mass-market paperback houses (Bantam, Dell, Fawcett, Avon, Jove, New American Library, etc.) become interested because the screenplay will be a movie and therefore has new value. The financier-distributor usually takes over from here and makes a novelization deal with a paperback house that involves the screenwriter and producer. Essential to the movie tie-in deal is the acquisition of the movie logo (called *artwork*) and stills, which must be obtained from the distributor.

Before the elements of such a deal between financier-distributor and paperback publisher are described, the issue of who writes the movie tie-in for an original screenplay must be covered. Under the Writers Guild of America Basic Agreement, the Guild can tentatively name who will receive "story by" or "written by" credit prior to formal designation of screen credit, and that writer benefits from separated rights (including publication rights) and thereby has first chance to make the novelization deal. This would require his dealing with the producer for permission to use the artwork and stills from the movie as well as with an interested publisher. If the Guild-named writer does not make a novelization deal within a stipulated period of time, it falls to the producer to do so with another writer and then pay the Guild-named writer a percentage of the adjusted gross receipts from such a novelization deal. For example, a paperback publisher could pay a $100,000 advance to a studio for the novelization of an original screenplay that the screenwriter declines to novelize. (The Guild requires the studio to pay $3,500 separately to the Guild-named screenwriter in that event.) Then the studio may pay a novelization specialist a flat fee of, say, $15,000 to do the job. The Guild permits a deduction of $7,000 from the studio's share covering a use fee for the artwork, which the studio pays to itself before reaching "adjusted gross receipts" of $78,000 (subtracting novelizer's fee and the fee for the artwork from $100,000 publisher's nonrefundable advance). Under the Guild rules, the Guild-named writer will receive 35% of the adjusted gross receipts, minus the $3,500 advance, or $23,800 in this example, a sum higher than the novelizer receives.

The elements to be negotiated in a novelization deal with a softcover house include the *advance,* which averages $50,000–$60,000 but can be as high as six figures or as low as $5,000. Usually everyone involved in the project receives a slice of the advance: producer, financing entity, screenwriter and writer of the novelization. Next there is a *royalty,* which is generally 6% of the cover price for the first 150,000 copies sold, 8% thereafter. The royalty is also divided between the entities, but only a novelizer with strong bargaining power will share in this amount. The *territory* usually covers the United States and Canada only, but sometimes worldwide rights are sold, and the advance reflects that. Ideally *timing* is planned so that one month before the picture's release the book is on the stands and the public can become familiar with it. Other negotiated points include completion of the manuscript and coordination of the artwork. Obviously as soon as there is a logo design and production stills, the distributor should send them to the publisher in an effort to reduce the necessary lead time. If a publisher has paid a substantial sum to a financing entity for novelization rights, the contract will usually state that the financing entity will supply artwork at no extra charge.

There had been movie tie-ins for years, but when David Seltzer's 1973 book version of *The Omen* (written after his original screenplay) sold over 3-million copies, yielding royalties in six figures, the book-publishing world and movie-makers recognized the increasing value of coordinating book sales with movie sales to their mutual benefit.

After the huge success of *The Omen* novelization (a made-up word that offends many novel writers), there followed years of escalating sales to paperback houses that wanted to publish book versions of screenplays projected to be successful films. Movie tie-ins that once brought in $3,500 as a total advance from a publisher began attracting prices of $100,000 and more. The market became as cutthroat as the paperback competition for a hardback best-seller. Agents held auction sales, and every few months new price records were set. If *The Omen* heralded the market boom in 1973, *FIST* was the turning point in 1978. The $400,000 sales for Joe Eszterhas's powerful novelization of his original screenplay was associated with an unsuccessful film in the marketplace. Just as the movie company was shocked when it did not do well on the film, the book company was appalled when book sales were low. This taught the publishing industry that the only reliable value of a movie tie-in was as an adjunct to the picture. The result has been that guaranteed advances have leveled off to an average of $50,000–$60,000 for novelization rights.

An important lesson is that the prices usually quoted as sales to paperback houses for movie tie-in rights are guaranteed advances against royalties to be earned by sales. High advances anticipate high sales. The paperback house pays this amount, which is nonrefundable, to the holder of the novelization rights. But whether the advance is $7,500 or $75,000, the money earned on a book (the royalty) will be the same if it sells 2-million copies. The only difference is the spread of the initial advances spent. On a strong sale of 2-million copies the advance, whether large or small, will be earned back, and the extra earnings will be converted to royalties. Today publishers have decided to rely more on the strength of the movie tie-in to attract sales (and therefore royalties) on its own, with decreased advances paid, rather than hoping that royalties from high sales will equal a strong advance.

One other factor that bottomed out the movie tie-in market was that publishers were buying novelization rights to television movies, but people were not buying these books. The only television program that can usually succeed in selling tie-in books is the long-form mini-series.

The original screenplay is the second type of source material that can be exploited in book publishing. A third type is the existing novel, which can be issued in paperback as a tie-in, timed with the release of the motion picture.

A publisher generally makes comparatively little money publishing and selling a hardcover book. The primary source of income for a hardcover house is the sale of subsidiary rights, such as book-club and paperback rights. Revenues from these sources are usually divided by giving 50% to the author and 50% to the hardcover publisher. Movie rights are generally held by the author. Paperback rights are sold by competitive auction, which accounts for the huge sums often involved. Each year several sales are made in the million-dollar area. However, a book must earn back that advance in sales before a dime of royalty is collected. There was an estimate that for Bantam Books to earn back the $3.2-million advance to Judith Krantz on *Princess Daisy,* 9-million copies had to be sold. Usually a best-seller in hardcover is one that has sold between 25,000 and 50,000 copies; a moderate-sized first printing in paperback is 80,000; if a paperback has sold a million copies, it is selling phenomenally.

On a prestige book, subsidiary rights deals are generally made when the book is in galley form, any time from two months to two weeks before publication. The paperback house, movie producer and studio will work very closely on the timing of the book. If the film will be a long time in the making, the paperback house will publish one edition and then reissue it at the time of the movie's release. The difference is that the reissue would contain the artwork and stills that coincide with the marketing of the film.

There is a symbiosis that can take place among the sales of the various subsidiary rights. A book that is a book-club selection with large paperback sales will certainly be offered as an attractive and expensive movie sale. Conversely, if a book has been sold as a movie, the amount paid for the book by a book club or paperback house increases; the pressure goes both ways. The publisher pressures the agent to sell the movie rights to enhance the subsidiary value of the book, and the movie buyer pressures the agent or publisher to sell the rights to enhance the value of his screen property.

One issue that arises involves advertising costs. If a hardback publisher has benefited from a large paperback sale made on the heels of a major movie sale, the producer can ask the hardback publisher to share those benefits with the movie by committing sufficient money to successfully promote the book, "creating a best-seller." The hardcover publisher can argue that the big movie sale was made on the strength of the literary material alone and can suggest that the producer contribute to the book's advertising budget to boost sales potential. This issue is unresolved.

Another wrinkle is that the studios have gone into the book business. Disney owns Touchstone Books; MCA Universal owns Putnam and Berkley; Paramount owns Simon & Schuster; Time Warner owns Warner Books. But the studios and their book-publishing subsidiaries don't cross-pollinate as much as one would expect.

To review, if a movie's source material is an original screenplay, it can be sold as a hardcover or softcover book before the movie becomes a reality. If the source material is an existing novel, a subsequent paperback edition can become a movie tie-in timed with the release of the picture. If a project originates as a magazine article, either the author of the article will expand it into a book or a book will be derived from the eventual screenplay. *Saturday Night Fever,* which originated as a *New York* magazine article, became a novelization based on the screenplay. *King of the Gypsies* is an interesting case because it began as a *New York* magazine series by Peter Maas, who made a book and movie deal simultaneously. Bantam bought the softcover rights and sold hardcover rights to Viking in a reverse of the usual procedure. Coinciding with the placement of the book rights, the motion picture rights were sold to Dino DeLaurentiis. Other movies with origins as articles were *Testament* and *Funny About Love.*

Yet another way a studio can exploit book-publishing rights is to finance the writing of a novel. This way the studio gets a screen property for a relatively small investment, is involved in its earliest development, can help in obtaining publishing deals and participates in the revenue from all deals made. This strategy was in vogue in the late 1970s and early 1980s, but fell out of favor when certain deals resulted in disappointing novel sales, usually canceling the film version.

Book-publishing rights can be further exploited in the form of photo novels, "The Making of . . ." books, large-sized art books and calendars. *Star Wars, Close Encounters of the Third Kind, E.T.*, and *Dick Tracy* offer examples of the possibilities.

The future of publishing will find more paperback originals being developed, either from original screenplays or from novelists. The publishing business is going to become more dependent on nonprint avenues of promotion, such as radio and television. Our society is becoming less a printed-word society and more an image society, and publishers will be more dependent on those images to sell the word. The business is as brisk as ever and is always adapting to change. Books on tape carved out a new niche in the market. Eventually customers may be buying disks that plug into the personal computer and a hologram will read aloud.

III
THE MONEY

# MOVIES, MONEY AND MADNESS

*by* ***PETER J. DEKOM,*** *who is a partner in the Los Angeles law firm of Bloom, Dekom and Hergott, specializing in entertainment-related legal and business matters for a diverse client base of actors, directors, producers and executives. A graduate of Yale University and the UCLA School of Law (first in his class), Mr. Dekom has served as an adjunct professor in the graduate school of UCLA, teaching legal and business principles to film, business and law students. His numerous articles have appeared in such diverse publications as the UCLA and Fordham Law Reviews, in a series of books issued by the Practicing Law Institute and in* American Premiere *magazine. A member of the California bar, Mr. Dekom has lectured extensively all over the world on entertainment issues. He is on the board of directors of several companies, including Imagine Films and MCEG, and is president of the board of directors of the American Cinematheque.*

***It is one thing for an executive to ignore the numbers having considered them carefully, but it is quite another not to understand their meaning and to decide independent of mathematical analysis. . . .***

As film libraries and studios continue to change hands, bidding up the asset value of films made in the past, it seems impossible to lose money in the movie business. Yet there are a few facts, connected to a lower profitability for a number of major studios in spite of the reported huge grosses, that bear focusing upon.

As of this writing, the average major studio feature budget is approaching the $26-million mark. The average cost of opening a film wide in domestic theatrical release hovers around $12-million. When one adds interest and overhead (yes, folks, these are real hard costs), the average film has to generate well over $40-million just to break even. That figure must come from theatrical film rental, not box-office dollars (rentals are about half of box-office), and the net revenues from home video, not the retail selling price of cassettes (wholesale on cassette averages around 65% of retail).

With price escalations going through the roof, studios are all saying they want more product, more films to distribute, to fill up their monstrous distribution mechanisms, to fatten their libraries, to build their asset base, to improve their cash flow. One may very well ask, "What's going on here?"

1. *The profit margins in the motion picture business are coming down.*

The internal rate of return (return on investment), like interest on a checking account, is a measure of profitability for a specific business. In some high-volume companies, the profit margins are very slim. For example, in the retail grocery business, profits range from 1% to 3%, and money is made by sheer volume. In other businesses, such as the soft-drink business, internal rates of return can exceed 40% (as in the case of the Coca-Cola Company). Generally investors who are looking for high-risk opportunities (typical of venture-capital situations) expect rates of return in the 25%–30%-or-more category. Everyone knows that the motion picture business is risky; shirts that people have lost in this industry would easily fill the entire Sears retail chain. So, typically, when people are asked how profitable the movie business is, those with a modicum of business understanding will suggest that the rates of return must be well in excess of 30%, given the risks involved and the intense media hype about Hollywood cash flow. But this is not the case.

In the early days of the movie business the rates of return were vastly higher than 30%, sometimes hitting as high as 100%, given that the studios controlled production, distribution and exhibition and held contract players to remarkably "low" salaries relative to the revenues that were coming in.

Talent profit participations were virtually unheard of (the Marx Brothers did some pioneering in obtaining a significant piece of their films), and studio moguls grew rich on incredibly high rates of return. As profit participants began to appear on the horizon, and as residuals began to chip away at the rock pile since the 1960s, the internal rates of return at studios began to drop sharply, but were still in the 30%–50% range for many years.

Today, looking at the motion picture groups of the various studios (including development, production, distribution and overhead of all kinds, but excluding their television wings and any ancillary nonmovie operations), the motion picture business generates internal rates of return of between 0% and 20% or more, with the average (and mean) somewhere in the 8%–15% area. For people generating 8%–15% on certificates of deposit, mutual funds and other relatively low-risk investments, the fact that a motion picture company—taking the huge risk of making and releasing movies—is generating rates of return on that order of magnitude must come as quick a shock.

While it is terrific to look at a single successful motion picture, studios also have to take the losers into consideration. Absent an occasional aberration, no consistently active major studio (taking its overhead into account) today experiences higher than 30% internal rates of return on its motion picture group as a whole. And given current trends in talent cash costs and back-end participations, it is likely these numbers will decrease, especially as studios increase production output, causing them to incur higher production costs because of the higher volume but not seeing the offsetting revenues until sometime in the future. (If we were to focus solely on the cash invested in the individual film slate, as opposed to running the studio as well, we would see that this specific return on investment would of course be higher than these corporate rates that are more reflective of reality.)

As a corollary to the drop in the internal rate of return, the time period needed for most studios to recover the cost of their investment, including interest, has become longer. In the early days of the business 100% of the investment was recovered in just a few months after theatrical release.

Today studios rely on the windows of home video, pay-television and syndication for recoupment and, hopefully, profitability. This could take years. In fact it is these vast libraries of preexisting film and TV product that are generating the cash flow to keep many studios afloat or at least in a linear and predictable cash-flow position. Thus studios are very anxious to build up the value of these libraries, and their distribution mechanisms are capable of handling many more movies in the ancillary markets than they are currently processing. However, there are some serious questions as to whether the major studio theatrical marketplace can handle the increased volume product that this foretells. As new markets open up overseas, as new methods of home delivery expand, as new technologies add new potentials to existing libraries, the desire to make product to fill that perpetual hopper increases exponentially.

Given this, there is one key question with a number of potential responses. With the rates of return dropping and the risks increasing, what are the likely economic solutions to this problem?

A. *The general-store model.* The grocery chain has a very low internal rate of return, but makes up for this through volume. Hence, if a studio were making fewer dollars for every film it produced (based on capital invested), it might simply make more films. This simplistic analysis seems to have captured the imagination of many studio executives and, when coupled with the need to build the ancillary values as described above, creates a veritable feeding frenzy. Volume does have the benefit of amortizing the relatively high cost of maintaining distribution and production operations over more films, but it also forces management to split its time among more projects, allowing a lot more product into the system with a lot less control. Another aspect of this approach dictates that by creating more product, a studio takes up more shelf space and squeezes out its competition, a theory that seems to have been disproven on more than one occasion.

B. *Reduce costs, increase revenues.* This form of solution goes along with the "buy-low and sell-high" philosophy of doing business. It is great, until you try to figure out how to apply it to motion pictures. In network television this philosophy gave rise to reality-based programming, the revenues from which have been either quite good or at least equal to those of dramatic programming. Is there a parallel development in the movie business? Since the market does not appear to be inclined to an entirely different kind of motion picture, the focus here would be on cost reduction of the existing format. But this is extremely difficult, considering the competitive salaries of desirable writers, directors and performers, in a limited talent pool.

C. *Sell the beast.* While rates of return have been relatively small, the price-earnings ratios in the motion picture industry have been remarkably good. Companies with huge asset bases (large libraries, real estate holdings, ancillary businesses) cloud the issue somewhat, but on a pure motion picture industry basis, it might well be better to sell the company (assuming solvency and some solidity) and take advantage of the high price-earnings ratio rather than hold on to a losing beast. In fact in recent years, studios have appreciated at an annual rate averaging 17%, and it looks like that trend is here to stay (on a long-term basis with normal market fluctuations), simply because there are very few studios and a lot of people interested in Hollywood, whether for strategic reasons (Matsushita and Sony) or simply the glitz and glamour of a high-profile industry.

D. *Diversify.* This area is fraught with risks, particularly in the age of interest uncertainty. However, for extremely well-capitalized companies, the opportunity to take advantage of parallel businesses can be quite lucrative. Both Universal and Disney have demonstrated this philosophy extremely well with their studio-tour/theme-park philosophies. There, characters created in motion picture divisions are exploited in another extremely heavy positive cash-flow business. While the need for initial capitalization in theme parks/tours is very high, well-placed strategic investments (as opposed to head-to-head competition, which seems to be *de rigueur* these days) can generate untold future dollars. Likewise, these types of synergies are present in record companies, television and radio stations, and the very profitable basic pay-television networks.

E. *Take greater risks.* This seems to be directly contrary to the theme running above, but in fact every time somebody avoids a future risk by taking a present payment, they have in effect shifted that risk (and some *profits*) to someone else. This happens, for example, when someone presells video or overseas rights. In these instances the person who is taking the risk (the video or overseas-rights buyer) will now have to take a larger piece of the pie, reducing the seller's upside accordingly, in order to justify the risk taken. Since we have already noted that the profit margins in the movie business are slender at best, further diluting the profitability of a motion picture must, necessarily, yield lower revenue streams to the filmmaker.

F. *Focus on marketable concepts without big stars.* Where budgets can be contained and the concept is marketable, this is a risky, but not unreasonable approach. But a "bad" big movie with stars may well make money, while an unsuccessful concept movie without stars will definitely lose money. This approach takes guts, but usually the record-breaking films offer the audience something new or something they haven't seen in a while. Sometimes you just have to swim upstream.

In the case of major studios, avoiding risks (by taking serious downside protection) is simply not a business plan. If a film is so risky that one wishes to shift that burden to someone else, then the studio in question should simply not make that movie unless it is a joint venture with cheap capital. If the management has insufficient confidence in its own abilities to choose and distribute motion pictures, perhaps they should find solace in another industry. The business of business is risk.

2. *Costs, costs, costs!*

By looking at the landscape, one must conclude that budgets cannot continue to spiral upward at the same rate as in the past. If anything, there should be a period of contraction and consolidation in film costs, if for no other reason than to accommodate the larger volume of product being produced.

For those who have wondered where the difference between the old and new internal rates of return has gone, just look to individuals: studio executives, directors, writers and, most of all, highly paid megastars. Hollywood is and always has been an industry of millionaires and aspiring millionaires, the latter perhaps severely underpaid and the former among the most highly compensated individuals on earth, exceeding the highest levels of corporate America. The list of $10-million-plus actors (applied against dollar-one gross) would probably appall a newcomer, but salaries have escalated way beyond that point, especially for sequels and presold materials.

Where did this price escalation come from? Why does the average major studio action-adventure piece cost well above $30-million? Certainly, below-the-line costs have increased, but the real cost factors lie in the highly compensated talent pools (the above-the-line personnel), mainly actors and directors. Seven-figure scripts are no longer aberrations, and many directors make at least $1-million per film. (Directors' salaries north of $5-million are not so uncommon these days.)

The recent feeding frenzy in the overseas marketplace has certainly added fuel to the cost-escalation fires. Companies that specialize in financing their motion pictures by looking at presales overseas well understood the value of a big name in terms of hard dollars. It did not take long, however, for the agents and representatives of this special talent to recognize the premium being earned by these foreign-driven production companies, mostly independents distributing through major studios. The awareness of these premiums, and the willingness of foreign-driven independents to pay increasingly high sums for this talent, led to a bidding war that generally moved acting and directing salaries into a new, stratospheric level. Even marginal actors, whose recognition was just beginning overseas, found wages rising faster than the blood pressure of a number of Hollywood major-studio executives.

As Hollywood studios escalated their budgets, thereby increasing their risk in hard-value terms, they wanted some kind of "insurance" that they could "open" their films. Historically big stars and a few limited mega-directors were able to draw the audience in for that crucial opening weekend, and from that point on, the film's quality (fueled by word-of-mouth) would support the remaining run of the movie. Hence, studios purchased "opening insurance" by hiring these mega-stars for their product, even though their prices had recently been bid up by the foreign-driven production companies. It got more expensive all around.

The higher costs for talent are shifting the internal rates of return away from the studios and into the purses of these creative individuals and, gratifyingly, their representatives. People often ask agents and lawyers how they can continue to extract increasingly high salaries for creative personnel. As one studio executive put it, "Don't you understand what you're doing to this industry?" But the role of the negotiation process is necessarily an adversarial one. The agent or lawyer who decides not to take a higher salary for his or her client because it may be "bad for the industry" is likely to find himself or herself in a different business. As long as there is a willing buyer, representatives will continue to guide clients as best they can and secure the highest prices obtainable. If the studios object to these escalating costs, they should exercise restraint and not require these creative individuals to be involved in their movies.

Pressures are even mounting from net-profit participants, who also want a piece of the pie in spite of extremely lopsided contractual definitions. One such participant, Art Buchwald, has even managed to convince a trial court that such net-profit definitions are unconscionable and represent contracts of "adhesion." Clauses that charged expenses on an accrual basis but included income on a cash basis, included gross participations as negative-cost items that accrued both overhead and interest, and provisions relating to the charging of interest and overhead against each other were rejected by the court, which decision, as of this writing, has yet to be affirmed or overturned on appeal.

There are solutions to the price-escalation wars, even though some have long-term fuses. They are painful and risky, but something needs to be done.

A. *Increase the talent pool.* Looking at the supply-and-demand curve, it doesn't take an economic genius to recognize that if studios insist on a limited pool of acceptable directors and actors, and if they are increasing production of movies, then the prices for that limited talent pool must escalate. Thus it is in the studios' interest to increase the supply of actors, writers, producers and directors whom they deem acceptable.

The techniques for this are myriad. The first is a Disney technique, which recognizes that once a person is a star, even if his or her fortunes seem to have changed, he or she is still generally recognized and valued by the public. By uniting an attractive concept, well marketed, with a recognizable name (although perhaps a face from the recent past), Disney has been able to generate considerable grosses. By obtaining options and multiyear agreements with these same actors, Disney has also ensured itself of a talent pool for the future at generally reasonable rates. These performers have now rejoined the acceptable talent pool.

Second, studios need slowly to move directors from the television divisions, in a managed manner, into features. The path is simple: from directing episodic series to long-forms to lower-budget features to higher-budget features.

Third, pairing a skilled and charismatic actor who has low recognizability with a mega-star (probably in more than one movie) will eventually bring that skilled actor's abilities to the attention of the public. Although they are coming to see the mega-star, they cannot help but notice the secondary player, who may eventually become valued as a "star" as well. If the studios' executives have good creative instincts, they can recognize this type of talent and experiment accordingly. Further, they could secure this actor's "loyalty" either as a contract player (as in the old days) or through a series of successive options on more films. Even if the contracts and/or options are renegotiated to higher prices, they will never mirror the prices that could be paid in an open bidding war. There is one fly in this ointment, though, since big stars usually want "names" to play opposite. It may take moving a potential star into third position a few times to make him or her sufficiently acceptable as a second lead to the top-liner on a subsequent film.

B. *Develop screenplays internally.* Studios that wait for packages to materialize at their door are likely to suffer from one of three possible scenarios:

(i) They are vulnerable to an outrageous bidding war in which the package and/or script winds up costing the studio a vast multiple of what it would have cost to develop alone in the first place;

(ii) they receive prepackaged mediocrity that may or may not have been passed on by other studios; or

(iii) they learn there are almost no real packages anyway (that is, there are vastly fewer fully cast, ready-to-film projects than the public perceives).

Effective internal development is the key to success of virtually any studio (or production company, for that matter), although this is not the sole route for the acquisition of film product. Strong relationships with proven creative talent are another important facet to studio success. Internal development requires creative executives who, instead of doing favors for their friends, marry hard concepts (many of which are created by the executives themselves) with good workmanlike writers and, potentially, the directors who will be asked to shepherd the productions into existence. At the concept stage, discussions can even be had with marketing people to learn whether the potential movie can be sold. This is not to relieve the creative executives from the responsibility of fighting for quality, even when marketing may disagree. It is simply to encourage a greater level of budgetary and marketing responsibility in the executives who develop such properties.

By developing a script internally, a studio clearly avoids the bidding war that would probably accompany an independent screenplay that was strong enough to be the basis of a green-lighted movie. Also, if a studio happens to own a script that a star insists on doing, there is even more bargaining power in the negotiation of compensation for that performer. Studios that develop an internal stable of competent producers who truly manage development and, ultimately, production also increase the studio's ability to control costs of its product.

C. *Consider "rent-a-system" distribution deals.* For those studios short on capital or needing additional product, they can focus on a short slate of high-quality movies by opening up their distribution systems to respected creative filmmakers—and their outside financing groups—who are prepared either to cofinance the production/distribution risks with the studio or to put up all the risk dollars involved, including prints and advertising, in a separate slate of studio-type films. Both Largo Entertainment and Morgan Creek are examples of companies that take such financial risks and look to the studios principally for distribution only. These "rent-a-system" deals, where the outside financing entity literally "rents" the studio's distribution network (usually only domestically), do not provide high upside dollars for the studio. After all, the studio often negotiates a downscaled distribution fee in such cases (averaging 20% as opposed to a 33% average fee applied to studio-financed product) and maintains no net-profit position in these outside-financed films. But this type of deal does offer a studio an easy way to amortize the overhead costs of maintaining production and distribution operations and provides dollar-one gross participations with no cash risks. Any studio relying principally on this form of product would ultimately fail, however, since the cost of maintaining operations cannot be covered solely in this way.

D. *One mega-star may be enough.* Many studios will load a motion picture with several mega-stars plus a very expensive director. Rather, since this is basically "opening day" insurance, having one such actor should be sufficient to get people in the door. Two or more such actors, receiving high cash advances against their dollar-one gross participations, would generally make the motion picture top-heavy, too risky and economically unjustifiable absent extraordinary circumstances. (In some cases a top star asks for great co-leads as a form of failure insurance for the insecure mega-star.)

E. *Creative financing partnerships.* Extremely high-budget motion pictures might be an area for joint ventures between studios and outside equity and/or overseas distribution ventures. If breakeven is being pushed that far back, and if the risks are truly that high, then the studio either should not make the movie at all (probably the better choice), or, if it must, it may want to share the risk with an outside entity. There is enough money "out there" available for this type of coproduction, but unfortunately studios cannot arbitrarily choose the deals in which they wish to include outside money and those they do not. Hence these coproductions frequently involve producers and/or separately funded production companies that have developed these extremely costly films themselves and are more likely to participate with the studio in their production and financing, but the upside to the studio is accordingly reduced.

F. *Know the math.* Creative executives must understand the mathematics behind film production and distribution. People who have the power to make deals and green-light development and/or production must understand the economics involved in their decisions. This encompasses legal and business-affairs personnel too. Unfortunately too few executives, especially in the creative areas, really understand these economic points. Obviously top executives at successful studios can probably recite mathematical models in their sleep (if they are getting any).

A bit of training might go a long way. Sophisticated computer modeling is available today at all major studios. The projections, giving varying revenue assumptions, are extremely accurate and are getting better all the time. This probably does not give too much solace to those writers and directors who really like "off-the-wall" material, but it does equip a studio executive to make a meaningful decision. It is one thing for an executive to ignore the numbers having considered them carefully, but it is quite another not to understand their meaning and to decide independent of mathematical analysis. Certainly there is a time and place to disagree with the number crunchers, but at least one ought to see their opinions before writing them off. How many times have studio executives green-lighted movies with truly uncertain futures where breakeven required more than $100-million? This should never happen.

3. *The negative side of increased product demand.*

While each of the major studios is well positioned to market ancillary rights, regardless of volume, it is a very big question as to whether any studio is capable of properly marketing and domestically distributing annually the 24 to 40 releases theatrically each is claiming it would like to release. This requires the marketing organizations of each company to create an entirely new advertising, promotional and publicity campaign for a theatrical release occurring, on average, every week and a half to two weeks. Decisions as to which films should be supported and at what level become intertwined in a cacophony of excess volume.

In the area of large media advertising buys, for instance, the major studios are forced to escalate their releasing costs to pay for advertising campaigns and saturation media buys in order to distinguish their movies from the flood of other product from the other studios. This is contrary to the economics of scale that usually accompany such large media buys. Competition for magazine covers, critics' reviews, and so forth, escalates, and the costs of supporting those mechanisms go up accordingly. Publicity and promotion, the much-undervalued stepchildren of the marketing wing, are also flooded with a product flow beyond their ability to handle.

How much of this very complicated marketing strategy can actually be coordinated by a single marketing organization within the studio? And if multiple marketing divisions are established, how does one keep them from competing with each other and reducing the impact of each other's advertising in the general marketplace? The use of independent-contractor ad agencies will surely increase, but in this arena the jury is still out. Can the studios actually handle the theatrical marketing, at least domestically, of this increased volume of product?

4. *The trends toward diversification and consolidation.*

There are two basic methods by which companies increase in size:

A. *Growing the core business* through success and, perhaps, adding new businesses that are developed totally internally; and

B. *Acquiring preexisting product or businesses* through purchases and/or mergers.

Internal growth in studios appears to be stimulated by bringing together marketing and distribution operations, creating greater interrelating among the development and production executives and their distribution counterparts. As talent moves freely across television and motion picture borders, there is increasing reason to consolidate movie and TV divisions under the helm of a senior executive with an eye for moving television performers into the theatrical arena, and taking movie product and developing it in the TV arena as well.

Growing new businesses is extremely time-consuming (it may take years to catch up to the competition) and may require some kind of a jump start with at least a minor acquisition. Different players approach the problem with different solutions: Sony acquired a record company (CBS), while Disney is growing one. Growing businesses requires high amounts of capitalization (without offsetting assets), with significant losses anticipated in the early years. This long development period, and the patience to withstand the pressures for instantaneous success, is part of the process. Disney and MCA grew studio tours and/or theme parks; Fox grew Fox Broadcasting; Turner grew TNT; and many new companies are on the drawing boards of other studios. If successful, the payoff is tremendous.

The latter, external method allows companies to carefully examine their targets before acquisition and see (a) how the new company would fit into the existing structure; (b) what savings could be effected by consolidating overlapping functions; and (c) what portions of the new company can be sold off to help pay for the purchase.

The risks in such acquisitions are of course manifold: An adverse change in the marketplace (as in the collapse of many junk-bond deals) could raise interest rates and change inherent values of properties that were going to be spun off for purposes of paying for the acquisition; governmental intervention to prevent anticompetitive mergers (less of a problem these days); insufficient due diligence (in examining the target company before acquisition), leading to some unpleasant "surprises"; the opening of golden parachutes costing more than originally bargained for (again, a fault of inadequate due diligence); an unexpected, but successful, shareholders derivative action; finding oneself in the middle of a bidding war having placed the target company "in play" and thus open to other suitors; and the list goes on.

For companies following this route there are a few admonitions. First, the concept of being big for bigness' sake is not enough. Management cannot use the generic argument that to compete in the global marketplace of large conglomerates, one must become one. The "world player" argument common today, absent a specific business plan with staged growth, is simply an excuse to spend money and increase personal power. This type of thinking leads to excessive debt and a monolithic bureaucracy that is virtually impossible to govern.

Mergers and acquisitions in the movie business must be driven by a precise strategy as to what will transpire following the consolidation of the entities in question. Executives must determine what the new business plan will be, where they expect the key synergies, and how best to consolidate, eliminating overlap and waste between the two companies. For example, if a company with a strong theme park acquires another company with a lot of well-known comic characters, those characters should be in the theme parks as quickly as can be done under strict quality-control measures.

It is an unfortunate reality that, in many cases of merger and acquisition, jobs are lost. However, the increase in efficiencies should be such that the new entity is even stronger than the previous two and will serve to better protect the jobs that remain.

In the late 1980s it was very much a major studio trend to purchase movie theatres. Each studio believed that once the feeding frenzy commenced, they would be at an extreme disadvantage if all the other major studios owned theatres and they were left out in the cold. This motivation, principally fear (almost always a bad reason, by itself, to act), generated sales of theatres to studios and their affiliates at twelve to fourteen times net positive cash flow, one of the most alarming outlays of cash in recent memory. Paramount (later partnered with Warners) bought Mann Theatres; TriStar bought Loews Theatres, and MCA acquired half of Cineplex Odeon. Today the value of theatres has plummeted, with prices much more in the five to eight times net positive cash-flow range. In short, the major studios could have generated vastly higher profits by simply buying certificates of deposit than engaging in the theatre-buying feeding frenzy. The fear factor was simply never justified.

5. *Trends and the ancillaries.*

No discussion of movies and money can be complete without a look into the future. In the 1990s we are seeing many new technologies (see the article by Martin Polon, p. 451) with a lot of interesting initials. Will DBS (direct-broadcast satellite) threaten pay-cable? Will HDTV (high-definition television) render obsolete the videotapes and laserdiscs we have accumulated? Will laserdiscs themselves or another smaller device replace videotape as the medium of choice for rentals and purchase? Will basic pay-television become an increasingly necessary substitute for syndication of motion picture product? What will happen to the video marketplace if pay-per-view increases in volume? Will the telephone companies be permitted to compete head-to-head with cable operators, and vice versa? The questions go on forever, and any good long-term strategist has to be worried about the answers.

The cable industry has long argued that its franchises are vastly undervalued. Notwithstanding the very real fear of re-regulation (when they cry poverty), they argue to the investment community that their products generate huge cash flows from basically captive audiences. They quiver from fear at the thought of new delivery systems, particularly any controlled by the telephone company, which currently has access to virtually every home in the United States. Yet how many cable subscribers are disappointed in the quality of the service (poor reception and many breakdowns) and the outrageously increasing costs of the system as a whole? In Los Angeles monthly cable bills can easily exceed $70, an extremely high price for the typical family. Thus the thought of having a hand-size satellite dish receiving over 100 channels from a high-powered satellite transponder for a price significantly less than what is being paid for current cable service is very appealing. The lack of cable competition due to local neighborhood franchises has certainly made the pay services expensive, but what would happen if they faced true competition from a readily accessible direct broadcast satellite? This is a tough market to second-guess.

With the price of laserdisc players dropping to the low $400 range each, perhaps in the coming years the laserdisc, with its superior picture quality, will become the medium of choice for people buying videos until even smaller, microchip and/or small-disc technology takes over. Discs are easier to store, do not deteriorate nearly as rapidly as tape, produce the highest quality digital audio sound (the same as compact disc) and generate image quality that can be as high as 40%–60% better than that currently available on videotape. The only drawback is that one cannot record on a laserdisc without very expensive equipment; hence, it is a playback medium.

The battle between pay-per-view and home video has not really become nasty yet, because pay-per-view systems still only reach about 14-million subscribers, many on antiquated ordering systems. But as pay-per-view systems become more sophisticated and can deliver viewing on demand, complete with the ability to interrupt, one would expect to see a significant if not total erosion of the home-video rental business. Revenues on a per-picture basis (wide-release major studio product) is in the mid-range of six figures, with escalations into seven figures for high-performing product.

As the domestic marketplace becomes glutted with TV movies, made-for-cable movies and theatrical films, the domestic after-market for syndication is reaching a saturation point, with per-picture revenues (exclusive of the hot performers) plummeting. As Fox Broadcasting, HBO, TNT, Viacom, the Family Channel, the USA Network and others manufacture more movies for first-run showing in their respective media, likewise, the after-market gets quite crowded, and the supply-and-demand curve will be and is causing a drop in prices.

Enter basic-pay (stations like USA, Lifetime, TNT, etc.) and the superstations (WTBS, WWOR, WGN). In many cases studios are now skipping the initial window of syndication and placing their product directly onto these basic-pay services for significant seven-figure license fees. To companies like Universal and Paramount, which jointly own the USA Network, the ability to play something on their own basic cable service is reassuring insurance in this era of decreasing free television syndication values. Basic cable has come to the rescue of a number of TV series that were unable to find syndication homes and is now doing the same for new packages of currently produced feature-length motion pictures. The trend is likely to continue and probably will help ensure those who are still in the syndication market that at least there is some outside competition to keep the syndication-level prices from eroding totally.

The battle in pay-television, on the premium services, escalated when MCA Universal recently skipped the pay window on a number of successful features and resurrected high-end network (CBS) deals instead. Will this trend continue? Do studios feel intimidated that HBO, one of the two major pay services, is owned by a competitor, Time Warner? Stay tuned.

6. *Capital Is king.*

The 1990s opened with a very strong emphasis on those companies that are adequately capitalized with a reasonable debt service. Undercapitalized companies are unlikely to be able to face the challenges of the nineties and survive into the next century. To this end, we may in fact see a reversal of the trend of huge conglomerate organizations, diversified in many directions. Rupert Murdoch has begun to sell off certain assets in the News Corporation empire, and Martin Davis certainly led Gulf + Western from a fully diversified holding company into an extremely well-focused company, Paramount Communications, in a relatively short span of years.

Growth is likely to be focused in core and related businesses, and diversification is unlikely to continue into unrelated areas. Acquisitions will be strategic and synergistic rather than reflecting growth for growth's sake. Will companies that are thinly capitalized with heavy debt service be able to limp on or will they be mercifully taken off the market for friendly mergers or acquisitions involving significant recapitalization? The issues abound.

In the independent world there is a resurgence of a firm and healthy marketplace for smaller companies. But as these companies expand, the temptation to move their product line into direct competition with major studios, capitalized many times more than any independent, is a temptation that has led, in the past, to the demise of a number of companies. Independents must continue to conserve their capital, keep their budgets low and by all means avoid the money-eating ravages of continual and sequential wide releases for their film product. Arguably an independent distributor should release a film wide (400–2,000 or more theatres) only under two conditions: (a) if the movie is a sequel to an already successful film; or (b) if there is a built-in marketplace for the product in question. In all other cases independent product should be nurtured into the market on a platformed basis. Direct head-to-head competition with major studios at major budget levels should be avoided, because it does not take many failures to drag an independent company down.

Is the movie business healthy? On the whole we may be in for some material shake-ups in the near future, but it has and will continue to survive and produce jobs and cash flow.

The trend toward an increasing number of independently financed companies funneling their product (at least domestically theatrically) through the major studios is likely to continue. Where such an independent company is founded upon established talent with an average per-picture performance track record higher than that of the average major studios, the rates of return will be sufficiently high to justify such equity investment. This is further enhanced by the reality that an independent does not have to manufacture product to feed a distribution monster, but can concentrate on making movies in which it completely and totally believes. Careful cost management can also make such companies more profitable.

However, I see many potential companies out for funding that are offering investors noneconomic deals and/or insufficiently proven talent to justify an equity investment. In the high-pressure world of raising private capital, many investment banks and individuals have gotten on the bandwagon in the hopes of getting a good finders fee for setting up deals that, in all likelihood, will find their way to the motion picture graveyard where the corpses of Atlantic Releasing, New World Pictures and the old Cannon, just to name a few, rest under newly turned soil. And just as has occurred in the past, upon the failure of a few of these independently financed companies, Wall Street and independent equity investors will say that this form of investment, as a whole, could not have worked in the first place, and the well will run dry once again. It will be a little-noted fact that those independently financed companies that were well formulated with good business plans and strong management (like New Line, Imagine and Castle Rock) will continue to manufacture top-quality films and be quite profitable. Then the investment community will find a new vehicle to apply to motion pictures. And so it goes, as the movie business makes its way into a new century possessing the same enigmatic qualities that make it attractive in the first place.

# ELEMENTS OF FEATURE FINANCING

*by* ***NORMAN H. GAREY,*** *who was a partner in the law firm of Garey, Mason & Sloane, based in the west side of Los Angeles. He represented actors, directors, writers, producers and production companies in the motion picture and television industries. Mr. Garey graduated from Stanford University and Stanford Law School, lectured at Stanford, at the USC Entertainment Law Institute and at the UCLA Entertainment Law Symposium, and was a member of the Los Angeles Copyright Society and the California bar. He died shortly before publication of this book.*

*An important lesson is to define at what point one's gross or net-profit position is achieved. . . . Another lesson is not to use phrases like* net, *or* gross, *but rather to define net* of what *or gross* after what. . . .

The matter of feature motion picture financing usually begins with the philosophy of the entrepreneurial producer. There are many established producers, for example, who are not particularly interested in buying or optioning a big, best-selling book or a finished original screenplay by a "star" screenwriter because these are very costly items, and many producers don't wish to spend a lot of money out of their own pockets. They would prefer to be original and either create ideas themselves and hire writers to develop them or induce financial sources to hire these writers. Or a producer may want to buy or option a screenplay that comes in from a relatively unknown source, perhaps a young writer who is a graduate student at UCLA or USC. It may be that the writer who is a bit better established, and who has an agent, in the normal professional course of events makes a submission of his material through the agent to the producer, who reads it and decides to buy it.

First, the producer's lawyer will commission the requisite title search to make sure that the material (and hopefully its title as well) can safely be used and that there aren't any conflicting claims, at least as disclosed by the copyright records in Washington. Next, the producer will decide how much to spend in option money and what kind of price to establish with the writer's agent. If it's a finished screenplay that the writer has written on speculation, he deserves some reward for it, which is generally reflected in the purchase price due on option exercise. The price would be constructed not only in terms of the cash payment due upon exercise of option and/or commencement of photography, but also (in many cases) in terms of contingent compensation as well.

As an example, suppose that the writer is established and has at least one produced screen credit and that the producer is negotiating for an original screenplay. If the producer is putting up $25,000 for a year's option with the right of a renewal for a second year for another

$25,000 (usually the first payment applies against the purchase price, and perhaps the second does as well), the purchase price may easily be $250,000 or more. That second option period for the second $25,000 may be necessary because it takes that long to mount the production, to secure the elements necessary to put the financing in place, and to know that there will in fact be a movie that then warrants exercising the option and paying the $250,000.

On the "back end," this deal may provide that 5% of 100% of the picture's net profits will be payable to the writer. But both the cash payment due on option exercise and the 5% of 100% of net could be made to depend on that writer's having received sole screenplay credit. If another writer has to be brought in, then the first writer arguably hasn't completely delivered the value expected. The Writers Guild, in making its credit determination, may find that there was material in the shooting script that was created by the second writer, who naturally has also been paid. Some flexibility must be allowed in constructing the first writer's deal for the potential necessity of bringing in a second writer and for the economic consequences of that eventuality. Thus both the cash and the contingent compensation of the first writer may be reducible in the event that he shares screenplay credit.

Suppose, as another example, that the writer is even more well established than the one just described. Perhaps he wants a lot more than $25,000 option money up front. Maybe he wants $50,000 to $75,000 up front in option money for a year, or even wants his work purchased outright rather than optioned at all. The producer doesn't want to spend that much out of his own pocket and doesn't want to go to a studio for it because that will mean giving too much away at the back end of his own deal. He may then establish a partnership with the writer, whereby the producer's profit points and the writer's profit points (and perhaps even their cash compensation) are pooled and divided in some way. Under such an arrangement, the writer takes a risk along with the producer, but he also stands to be rewarded along with the producer if they're successful. This doesn't mean that the pooling and division necessarily result in a 50/50 split. It could be 66/33, 75/25, 60/40 or some other percentage. But the pooling and division partnership has become a rather frequently used device to accomplish a deal between a producer and a writer or the owner of a piece of material when the producer doesn't want to lay out a lot of money and the writer doesn't want to accept short money up front without a commensurate back-end reward for doing so.

A more typical deal is the for-hire development deal, in which a producer creates the idea and will either hire a writer out of his own money or go to a studio for development money to hire a writer. In this deal the writer will frequently ask for a large cash fee as well as for participation in the profits of the picture and the proceeds of the subsidiary or "separated" rights. These negotiations are frequently difficult ones. The writer may conceive that he is the originator of the material since he is creating certain of the characters and their interrelationships, even though usually the story line and the plot premise at least have already been created by the producer. There is often some up-front ego conflict in these cases about who's contributing what. But these matters can be resolved by allowing the writer for hire to participate to some degree, beyond what the Writers Guild Minimum Basic Agreement requires, in the proceeds of the subsidiary or separated rights (such as theatrical and television sequel, print-publishing, and merchandising rights), or by rewarding the writer with a net-profit participation, which the Guild agreement does not require at all. If a significant cash fee is being paid up front, then the writer generally doesn't deserve all that; he's performing an employee's function and taking no risk at all. If, on the other hand, he is taking less than his normal front money and is contributing a great deal, then perhaps he does deserve to participate in these back-end rights and/or in the movie's net profits.

Assuming that the screenplay is of very good quality, the producer will then want to go to the marketplace and seek production financing based on this material. He might approach either a financier-distributor (a major studio) or one of the major independent production financiers. There are five or six independent production companies that finance pictures but do not distribute their own product, and therein lies the distinction between them and the so-called majors. If the material is absolutely sensational, the producer may be fortunate enough to be able to sell the "minipackage" consisting of script, rights and producer's services. Knowing what the costs are, the producer should be in a position to construct a price, both in terms of guaranteed cash compensation (part of which may have to be earned out by producer services rendered over a future period) and contingent compensation, whether measured in terms of gross receipts, gross receipts from some future breakeven point, or net profits.

On the other hand, it may be that the material, while very good, is only very good when allied with certain other creative elements. In that case, it may be necessary to do some prepackaging. This may involve negotiating for a certain director, either because his name will help attract financing and/or cast or because the material may need further development. In the latter case it's wiser from a creative standpoint to involve the director who is actually going to make the movie in further developing the screenplay. It may also be desirable to attach some actors and thus complete a major package for delivery to the production financier; a complete package will improve the producer's deal vastly.

The philosophy that one must recognize and observe in making these decisions is that of the risk-reward ratio: The farther a producer has moved the project—that is, the greater the investment he has made in terms of time, effort and money and the more finished the product he is delivering—the more he's going to get for it, both in cash and in back-end contingent interest. If he's in a position to bring in what is nearly a completed movie that just hasn't been shot yet, he's going to be able to make the best deal with the financier. However, there are some risks to this approach, both strategic and financial. That's what makes the game interesting. Perhaps the elements one involves may turn out to be liabilities. The director may be terrific for the material but turn out to be unacceptable to the financier. Or, although everybody may have thought that the actor's last picture did very well and therefore that he should provide a big advantage, a picture that he made a while ago may come out in the meantime and turn out to be a disaster, whereupon no financier wants him anymore. The producer may have to drop one or both of these gentlemen from the package, politely and without creating any legal or economic liability in the process.

Sometimes it's necessary, in order to involve a director or an actor, to put up *holding money* or option money, just as one would do to secure rights to a literary property. An actor's time and a director's time are valuable, and they will not always pledge their availability to a certain project without knowing either that there is financing that guarantees their future compensation or that there's some money payable now for which they're willing to give the producer a temporary hold on their services. Naturally the independent producer's lawyer must negotiate separately for the services of each of the creative people in the package before going into the final negotiation with the production financier.

Thus if the producer is in a position to bring a finished script to a studio, and the studio has not taken the development risk (because the producer has done so), he can command a far better deal from the studio. If he's willing to take it a step farther and put down some money to tie up a director and/or actor, then he's entitled to even more from the studio. Perhaps that entrepreneurial producer can also independently raise production financing from private sources, either those who, as in the tax-shelter days, were seeking a write-off but also liked the movie business for its glamour, or from genuine equity financiers. If he can take the project all the way to completed film and then merely seek distribution for it, the producer will make an even better deal.

There are other potential sources of production financing to consider. Assume the producer is beginning to assemble those creative and money elements necessary to physically produce the film. Those sources of money may be *territorial;* that is, he may sell off the distribution rights for all media in certain territories of the world in order to raise money out of those territories with which to produce the movie. These guarantees received from overseas territories can be taken to banks and discounted for cash, which is then used to finance the movie. The producer may wish to sell off or license in advance the right to receive income from certain ancillary rights sources, such as sound-track album recording and publishing rights to the music used in the movie. It may be that the merchandising rights are extremely valuable; if it's a high-technology picture, there's a lot of gadgetry that can become good toys. The merchandising rights may also be presold for guarantees that can be turned into cash. Book publishing is another possible source. A publishing house may be willing to put up money in advance of the material being secured or written. This investment will come at the earliest possible time if the producer is going to presell the publishing rights.

One technique that has been used to finance many low-budget films is raising money through *limited partnerships* formed by individual investors. Each investor is a "limited partner" (in that his liability for losses is limited to the amount he invests, and he incurs no personal liability beyond the amount) whose contribution, when aggregated with those of his partners, totals some proportion of the total production cost of the film. As a unit, these investors share a percentage of the "money's" share of the film's net profits (usually at least 50%) equal to the percentage that their total investment bears to the total production cost of the picture. The remaining net-profit shares are retained by other "money" (who get the balance of that 50%) and by the producers and other creative elements, who get the other 50% (or less) of the net profits. Broadway shows are traditionally financed in this manner. The individual investor can be attracted to films by means of the limited partnership, which finds doctors, accountants, farmers, realtors and a whole cross-section of society drawn to the glamour of movies.

If the producer is financing the picture privately, without going to a studio, it will be absolutely necessary that he secure a completion guarantee from one of the traditional sources (unless his own financial resources are so substantial that he can and will place at risk his own general assets). There are a handful of recognized, professional completion guarantors. (See article by Norman Rudman, p. 216.) A *completion guarantor* is a kind of cost insurer, or insurance company. He agrees (for a premium usually computed as a percentage of the picture's production cost) that out of his resources or financial contacts he will guarantee the money necessary to complete the picture if the money that has been raised from other sources turns out to be inadequate, for example if the picture runs well over the original budget. The producer must satisfy the completion guarantor that he has a good track record for cost responsibility going in and that the director also does, because the director is in a real sense the general contractor in the field. He's responsible for building the building, and if the costs are going to run over, it'll probably be because of him. The completion guarantor will want to put his own supervisor or production manager and location auditor on the picture to watch what's going on, in order to protect his risk. He may insist on the right to take over the picture, to take it out of the producer's and the director's control, if it's going substantially over budget. There may be penalty clauses in the completion guarantee stating that if the completion guarantor is required to come up with money, that is, if his guarantee is actually called upon, he has the right to invade the contingent compensation (or in some cases even the cash compensation) of the producer or director or both. The controls that the completion guarantor will insist upon may be at least as onerous as those that a major studio would impose upon a producer if the studio were putting up the money. The source that's ultimately answerable financially is the one considered entitled to impose those controls. The completion guarantee then has to be approved by the basic financing source(s) because they want to know that there's a real guarantee in place; they don't want to be confronted with the unwelcome choice of either accepting a three-fourths completed film after having put up all of the money or having to put up money beyond their original commitment in order to complete the picture.

Another way for a producer to finance a picture is through a *negative pickup*. It may be that he hasn't been able to raise all of the production money from nonstudio sources, but only a part of it. If there's a completion guarantor willing to back the ultimate cost of the project, the producer can then go to a studio (or other production financier) where the management believes in the picture enough to take a limited financial risk on it. They will agree to guarantee payment of part or all of the negative cost upon delivery of the negative. The studio is not taking a production risk because if the negative is not delivered, the studio has no obligation. On the other hand, if the negative is delivered but the picture is not very good, the studio is still obligated to pay the "pickup" price. That negative pickup guarantee by the studio can be taken to a bank and turned into cash.

One other way for a producer to proceed is to go directly to a bank and secure a production loan on the strength of all the other guarantees that he has assembled. The bank loan itself can be "taken out" (discharged or paid off) by a negative-pickup guarantee from a studio. That's another little wrinkle in the financing package. Basically the bank's loan is interim financing, and the negative pickup is permanent financing or the "takeout loan," as it would be called in real estate parlance. The financing devices used in the motion picture industry are strongly similar to those used in the real estate development and construction industry. Making a movie is analogous to building a building on a lot. The literary property is the lot, the piece of real property. The movie is analogous to the building. The screenwriter is the architect, the director is the contractor and the producer is theoretically the owner. The major studio (if one is involved) can variously represent a lending institution, equity partner and/or leasing agent for the finished building (in its capacity as distributor).

If the producer has completed the film entirely from nonstudio sources of financing and is now seeking distribution, he would screen it for a major studio distributor. If it was well received, the producer could possibly command a deal for the film in which the studio would advance all or more than all of the negative cost, thereby perhaps generating an immediate profit as a result of simply making the deal. He might further obtain a guarantee from the studio of a certain number of dollars toward print and advertising expenses. This guarantee is very important because the commitment of exploitation money frequently indicates the seriousness of the distributor's intent. Even the best film is not going to be successful if it's advertised poorly or inadequately promoted.

Next the producer will probably be able to command some kind of gross-receipts participation deal. This is structured so that once the studio has recouped its expenditures (its outlay for prints and advertising and the advance made to the producer to cover the negative cost outlay, enabling the producer in turn to pay off the various loans taken out along the way), the proceeds derived thereafter will be divided between the producer and the distributor on a gross basis, without any further deductions. The division could be 50/50 after recoupment by the studio, or it could be an alternating 60/40 deal. This might call for a 60% split in the studio's favor until the studio has recouped all of its outlay (prints and ads, its advance to the producer, etc.) with the percentage then shifting in the producer's favor up to a further point based on gross receipts before flattening out at 50/50 or some other split. There are varying combinations, depending on bargaining power.

If the producer does not require any advance whatsoever from the studio because money from foreign territorial sales and other sources has financed the picture as well as all print and advertising costs, he can hire the studio as pure distributor on a reduced distribution fee basis. There is conceded to be a variable profit factor in the standard 30% U.S. domestic distribution fee, so that it is possible, when the studio has taken no financial risk, for the producer to hire the studio as salesman only ("leasing agent" in the real estate analogy) and expect the studio to distribute the picture at a less-than-normal profit, for example at 22½% domestically instead of 30%, and at 32½% instead of 40% for any previously unsold foreign territories.

Producers raise production financing themselves to avoid having to go to the domestic distributor earlier than necessary because the farther down the road one can go alone, the better the deal will be with that major studio distributor at the end of the line. The best position for a producer is to be able to walk into a distributor with a completed film under his arm and say, "Okay, we'd like to have you as a distributor. What kind of guarantee of prints and advertising are you willing to make, and what kind of reduced distribution fee are you willing to quote?" The more studio money the producer accepts and the earlier he accepts it, the greater the risk he asks the studio to take and the more the studio will expect to be rewarded for it. Again, the risk-reward ratio operates. If the financier-distributor is asked to take the entire production risk, the best the producer can expect—assuming he's assembled all the elements and has paid the entire development cost—is usually a 50/50 net-profit deal. For that much risk the financier-distributor is generally considered to be entitled to at least 50% of the net profits, and perhaps more, and to insist on the standard distribution fee of 30% for the U.S. and Canada.

The producer's profit percentage will be reducible by the profit participations he has had to give up along the way to raise the money or to obtain the creative elements. The writer will probably receive 5%–10%, the director 10%–15%. Immediately the producer has given away perhaps 25% of 100% of the net profits, half his share, and that's with virtually no cast. If there is major casting, he may be giving away gross participations that, at best, come off the top and significantly reduce what constitutes net profits in the first place and, at worst, may also come to some extent from his end of the net profits. With studio financing, the producer may end up with somewhere between 10% and (if he is relatively fortunate or has been able to negotiate a net-profit "floor" as a protection against further reductions) 20% of the net profits, with some gross participations off the top that reduce the net-profit pie before it can be cut up.

To understand this in practical terms, one should track the money flow briefly. A ticket dollar comes in at the box office to the exhibitor during the first week of the picture's run in a major city. The distributor has been able to negotiate typical terms from the exhibitor, a 90/10 split in favor of the distributor. First, we must deduct from that dollar the exhibitor's house nut (or expenses) of around 10%, so we're down to 90¢. If the 90/10 deal calls for the distributor to receive 90% of the remainder, that's about 80¢ out of the dollar that has come in at the box office; 80¢ of that dollar constitutes the distributor's rentals. In the U.S. and Canada the distributor is entitled to a 30% distribution fee, which is 24¢, so now we're down to 56¢. The 30% distribution fee is to pay the distributor's own nut, or internal expenses, and is conceded to contain a profit factor the amount of which depends on how many pictures the distributor has in exploitation in a given year and thus how easily the fixed costs are covered that year. Advertising and publicity expenses are very high, especially television advertising; they generally run at 20%–25% of rentals, and these expenses come next. If our rentals were 80¢, deduct 25% of that for advertising, or 20¢, bringing our 56¢ down to 36¢. We haven't even considered other distribution expenses, such as the cost of striking prints, taxes, dubbing, the MPAA seal, shipping and transportation. With all these the 36¢ will be reduced to less than 30¢. Now the negative cost of the picture must be recovered. If it's supposed that the negative cost was roughly one-third of the total dollar that came in, we've got nothing left. Thus, if we're lucky, we've covered the negative cost of the picture, but we probably haven't. We're probably going to have to wait significantly far into the next dollars that come in to break even and to begin computing net profits, if any.

If we are barely into net profits, the studio is entitled to 50% for putting up the production money. They're the investing partner in this sense, not the salesman, in the sense that they've received their distribution fee earlier. If there were 2¢ left out of that original dollar (let's be generous to ourselves), the studio gets 1¢. But wait—if there was a gross participation out to a major star, then that actor's participation would come off the original 80¢ that we took in at the box office. If he received a 10%-of-gross participation, there would have been another 8¢ gone before we even got this far. So instead of having 2¢ left, we'd be minus 6¢; we'd be unrecouped.

If you take that dollar and multiply it by many millions, it illustrates the difference between that actor's being a gross participant and that same actor's being a net-profit participant. If he was a 10% net-profit participant, that actor would wind up with nothing or with his minor share of 1¢. But because he is a popular actor and has a good enough agent, he commands a 10% gross participation from the first dollar and comes out of that sample dollar with 8¢ (which is far more than anyone else except the distributor) rather than (maybe) a fraction of 1¢.

There are a small number of directors in the world who command first-dollar gross participations of a significant kind, and hardly any writers who can. However, there are many actors and directors, and even a few writers, who are given gross participations from breakeven, that is, once breakeven is achieved. So once we got down to that 1¢ and beyond, which is theoretically breakeven, then and only then (on the next dollar) would this hypothetical actor or director or writer participant's gross participation be payable.

An important lesson is to define at what point one's gross or net-profit position is achieved. If it's gross from first dollar, that's very different from gross after breakeven or some other break point. In our hypothetical example, the gross-from-first-dollar participant would have gotten 8¢. The gross-from-breakeven participant would have gotten nothing or nearly nothing because we barely reached breakeven out of that last cent of the dollar. Another lesson is not to use phrases like *net* or *gross,* but rather to define net *of what* or gross *after what.*

The foregoing has used the producer with a track record as its focus. The new or impoverished producer is going to get stung when he goes to a studio because he has no leverage. In this case there may be a small cash fee and an even smaller profit participation, certainly not 50% or anything very close to it.

# VENTURE-CAPITAL STRATEGY AND FILMDALLAS

*by* ***SAM L. GROGG,*** *who holds a Ph.D. in popular culture from Bowling Green University and has executive-produced several motion pictures including* The Trip to Bountiful, Patti Rocks, Da *and* Spike of Bensonhurst. *Mr. Grogg was the managing general partner of FilmDallas and currently is the president and chief executive officer of Apogee Productions, one of the premier visual-effects producers in the world, having won the Academy Award for its work on the legendary* Star Wars. *In 1989 Mr. Grogg and Apogee's chairman, John Dykstra, formed Magic Pictures, a theatrical motion picture development and production company.*

*FilmDallas sought to offer a better deal to its investors, one that compensated them for the risk they were taking and the time value of their money. . . .*

The 1980s saw the beginning of a significant geographical shift in the traditional centers of the motion picture industry. Hollywood was not displaced as the industry's mecca, but new production centers began emerging outside of southern California. The industry trade magazines dubbed the phenomenon "runaway production" and ran story after story on the economic threat that such budding regionalism represented. But the trend persists and remains in the 1990s as more of the motion picture industry recognizes the contributions—economic and creative—of new production centers in Texas, North Carolina, Georgia, Florida and Vancouver. Along with the regionalism comes an increasing interest on the part of the financial interests in these new production centers to find ways to become involved in the activity and its economic potential.

The newly developing production centers acted as beacons during the 1980s to draw entrepreneurs from across the country to consider involvement in the upswing of the booming film industry—cable had matured, new superchannels were leading new markets for movie sales, videocassette rentals were growing at phenomenal rates and the international market was on the rebound. This was new information to attract new investors, and soon the motion picture industry became a viable, albeit minor, facet of the standard investment portfolio. For the entrepreneur the challenge was to create an investment strategy that provided these new investors with a sense of comfort and understanding that would allow them to participate.

It was a fairly simple goal—create a venture-capital vehicle, modeled on the mutual-fund concept, to invest in motion picture development, production and distribution. By taking positions in a variety of investments, the fund would generate returns that could be utilized in a "roll-over" fashion in order to participate in subsequent opportunities and to generate a meaningful annual rate of return for its investors. That idea formed the basis of FilmDallas Investment Fund I. The application of the structure led to significant investments in several motion pictures, among them the Academy Award–nominated *Kiss of the Spider Woman,* for which William Hurt won the Oscar for Best Actor, and *The Trip to Bountiful,* for which Geraldine Page won the Best Actress Academy Award. Eventually the fund, its affiliated entities and successors would participate in the financing of development, production, acquisition and distribution of fifteen motion pictures over a three-and-one-half-year stretch during the mid to late 1980s.

## THE STRUCTURE

FilmDallas Investment Fund was structured as a private limited partnership organized under Regulation D of the Securities and Exchange Commission code. The limited-partnership structure was chosen in order to limit financial liability on the part of investors to the amount of money they would invest. While limited partnerships in movie investments had, in earlier days, offered some economic advantages through tax credits, such advantages had long since disappeared, and the FilmDallas investment offered no such advantages in its limited-partnership structure.

The partnership eventually consisted of three general partners and thirty-five limited partners. All of the partners qualified as accredited investors meeting the SEC "accredited investor requirements" under Regulation D. The general partners offered the investments and commissions were paid to brokers or other sellers. Likewise, the general partners were responsible for the management of the partnership while (as with all limited partnerships) the limited partners were not involved in any aspect of the partnership's management in order to limit their liability to the extent of what they had invested.

Limited-partnership units were sold in the amount of $50,000 per unit. The minimum amount of units offered was 50 (or $2.5-million) up to a maximum of 100 units (or $5-million). The initial offering closed at its minimum, but FilmDallas-related entities subsequently raised over $20-million in development, production, acquisition and distribution funds.

The original limited partnership offered investors the choice of either paying cash or financing their purchase of units. For those who chose financing, the partnership "brokered" an arrangement between qualifying investors and a bank whereby the investors could utilize the proceeds of a three-year promissory note at a reasonable interest rate in order to purchase their units. Investors could elect that proceeds from their investment be distributed directly to the bank in order to satisfy their personal obligation.

The three-pronged strategy of the limited-partnership structure, the minimum-maximum funding level and the alternative regarding purchase by means of a promissory note was consciously designed to keep the investment simple, attainable and attractive to investors. The limited-partnership structure was and is one of the most familiar investment vehicles and is used in many other industries, such as real estate and energy. The minimum-and-maximum level allowed for a reasonable target to attain capitalization (and therefore closing) of the partnership. The promissory-note alternative allowed the investors to make their purchase without dipping into their liquid capital.

More central to the overall strategy regarding creating the venture-capital fund was the realization that most efforts to raise financing for motion picture production and related activities turned on the ability of the investor to analyze the potential likelihood of success of a single motion picture. Hundreds of limited partnerships and other investment opportunities had been formulated whereby an investor was asked to make a decision regarding whether or not a particular combination of screenplay, talent and budget level might form the basis of an enterprise that will realize a profit. It was forcing this "decision" on the part of film investors that was at the heart of most failed efforts to raise private investment capital for movies.

As we refined the initial plan for FilmDallas, we monitored the experience of many other film investments that had circulated among the investment community to which we were about to appeal—sophisticated Texas business people. There was an appetite in that community to invest in high-risk/high-reward activities. The Texas investment community had been built on backing wildcat oil-well drillers and entrepreneurial speculators. In many respects investment in motion pictures easily qualified for such an investment mind-set. However, for one reason or another (maybe it was just because land and energy had been around longer than motion picture images flickering on screen), this aggressive investment community had not backed any meaningful percentage of investments in the motion picture industry up until the advent of FilmDallas. The reason was simple: The potential investors did not want to make the decision regarding which movie might be successful. Our strategy attacked this collective aversion by offering a plan that took the decision regarding a particular investment in either a motion picture or some aspect of the development, production or distribution of a motion picture out of the investors' hands. Our "mutual fund" concept presented a structure whereby an investor's capital would be administered, and these decisions would be made, by an informed manager. The manager would invest that capital according to a set of specific criteria that were designed to minimize risk and diversify opportunities for profit. Although the investors did not have to make the decision regarding what movie to back, they did have the opportunity to examine in detail the plan by which their investment capital would be managed.

## THE INVESTMENT CRITERIA

The general strategy inherent in a mutual fund—managed capital designed to be allocated among a number of potential opportunities in order to realize an attractive rate of return—was refined into a set of general criteria that ruled the actions of FilmDallas's management. In quick summary the criteria were as follows:

1. The fund would not invest more than $500,000 in any one project or endeavor. The plan was to diversify the fund across a number of projects, spreading the risk in order to create several opportunities for success and to offset those investments that would fail.
2. The total amount of overall financial commitment per investment could not be more than $2-million, with the FilmDallas contribution representing no less than 25% of any total financial commitment, nor more than 50%. A $2-million overall budget for a motion picture or some other aspect of the development, production or distribution process kept the overall amount of capital to be returned at a very modest level. Also these limits regarding a minimum investment of 25% and a maximum of 50% assured that the FilmDallas investment would maintain a significant enough position in the overall invested project in order to influence it and, at the same time, be limited so that the FilmDallas contribution would not bear all of the risk of any one project. Therefore a major tenet of the fund's overall management philosophy was to identify investments and positions in those investments where the FilmDallas contribution could maintain an appropriate amount of power yet share risk with others.
3. At least one-half of all funds available in the investment fund were required to be invested in activities that could take advantage of production resources around Dallas, Texas. A regional film production center, with the Dallas Communications Complex of Irvine, Texas, and its studios at Las Colinas as a centerpiece, was emerging at the time. *The Trip to Bountiful* and *The Dirt Bike Kid* were shot at Las Colinas. This requirement allowed half of the fund's investments to be monitored close to home rather than in far-flung locations, a detraction of most private investments in the film industry (more often than not, the filmmakers who raise money for motion picture production want to be spending that money as far away from their investors as possible). The FilmDallas plan sought to keep investments close at hand and to also take advantage of relationships its management had developed with the regional film community.

There were several other "internal" criteria that the fund used to analyze investments: requirements of distribution guarantees or pre-sales to cover large portions of the potential invested picture; approval of all key creative elements of any project in which the FilmDallas investment took a large percentage; and review and evaluation of all significant contracts that might influence the individual investment, among other factors.

Financial judgment was left as uncompromised and as unprejudiced as possible when investment opportunities were analyzed. Initially the Fund hoped to remain passive once individual investments were negotiated and consummated. However, as time went on, the Fund became more and more active in the management of a particular investment throughout the development, production and distribution process. And in the end the Fund did evolve into a more traditional producer's role, requiring that all of the standard elements and ingredients of the motion picture production process be weighed as decisions were made, in order to protect and enhance the future of the particular project/investment.

## THE INVESTORS' "DEAL"

Investors in FilmDallas received a very favorable participation in the Fund's returns. The allocation of proceeds was divided as follows:

| | Investors' Percentage of Proceeds | Managers' Percentage of Proceeds |
|---|---|---|
| From $0 until recoupment of original investment | 99% | 1% |
| From 100% to 300% of original investment | 80% | 20% |
| From 300% to 500% of investment | 60% | 40% |
| Thereafter | 50% | 50% |

A typical split between "Managers" (Producer) and Investors in a film deal was 100% to the investors until recoupment of original investment and a 50/50 split thereafter. FilmDallas sought to offer a better deal to its investors, one that compensated them for the risk they were taking and the time value of their money. And, more importantly, weighting of the deal in favor of the investors demonstrated the confidence of the fund managers that their plan would produce results benefiting all parties favorably. (Too often the promoters of investments take positions in the revenue stream that raise concerns regarding their willingness to stand aside until their investors have had an opportunity to be well compensated.)

## THE INVESTMENTS

The planning of the Fund's management required that diversification of the capital be a primary goal. Also, the "roll-over" objective called for analyzing the timing of potential investments versus their projected revenue return in the hope that such return would replace funds invested and therefore extend the life of the investment pool. A combination of short-, near- and long-term investments were sought in order to provide returns within six to nine months, twelve to eighteen months and over twenty-four months, respectively.

An understanding of the projected cash flow resulting from the exploitation of theatrical motion pictures was essential to this planning process. Investing in theatrical-distribution costs (instead of production) provides the prospect of a relatively short-term return, since monies are received out of gross receipts rather than from a return of production negative cost, which is, by nature, a long-term investment. A typical motion picture distributor recoups money in the following order:

- The Distributor's Fee (from various formats)
- Then Distribution Expenses (typically advertising and print-duplication costs)
- Then Cost of Production (advances and/or negative cost)
- Then Profits

Investment in distribution, therefore, garners a return positioned ahead in the recoupment stream of a production investment, which comes at a later time, depending on the quantity of revenues from the various markets.

The initial deals of the fund sought positions at various levels of the recoupment stream in order to diversify the timing of potential revenues:

*Choose Me,* directed by Alan Rudolph and starring Genevieve Bujold, Keith Carradine and Rae Dawn Chong, was FilmDallas's first investment. $500,000 was invested in two positions connected to the financing and distribution of this motion picture. $300,000 was invested in the cost of distribution, which in this case was structured to provide a percentage return of gross receipts even before distribution fees. $200,000 was invested as equity in the picture, along with an overall participation in all net revenues resulting from its exploitation throughout the world. The first part of the investment began returning revenues almost immediately, while the overall net-profits participation returned monies over years. This investment made in 1985 had returned approximately 60% of the original amount invested as of 1988, the year full accounting was concluded in the Fund.

*Kiss of the Spider Woman*, directed by Hector Babenco and starring William Hurt, Raul Julia and Sonia Braga, was the Fund's second major investment. FilmDallas invested $500,000 in the acquisition for North American distribution rights in all media. The investment provided a sliding percentage participation in "adjusted" gross receipts, defined as those receipts to the distributor after deducting distribution expenses only. No distribution fee was deducted for purposes of the calculation. This investment, made in early 1986, returned, by 1988, approximately 200% of the original investment.

FilmDallas's financing and production of *The Trip to Bountiful*, directed by Peter Masterson and starring Geraldine Page, Rebecca De Mornay and John Heard, followed on the heels of *Kiss of the Spider Woman*. FilmDallas invested $500,000 in its development and production. The Fund also produced the picture and served as a general partner of a separate limited partnership, which financed the entire production. The Fund received its pro-rata share of all proceeds resulting from the exploitation of the film throughout the world. The investment was made in early 1985, and by 1988 had returned approximately 175% of its original value.

*The Dirt Bike Kid*, directed by Hoit Caston and starring Peter Billingsly, was also one of the Fund's initial investments. The Fund contributed $500,000 to the production, representing approximately 45% of the film's total budget. The Fund also participated in its pro-rata share of net revenues resulting from the exploitation of the picture throughout the world in all media. This investment was made in late 1984, and by 1988 had returned approximately 70% of its original value.

Overall these initial investments represented a variety of financial structures, analyses, expectations and results. Each returned and continues to return on different relative schedules and in different amounts. And each represents a clear choice in terms of product type and market focus.

After all the planning of the investment vehicle was completed and the Fund established, the most difficult challenge facing the Fund's managers became the question that all the would-be movie investors do not want to answer—which movie to back? The selection of the above four investments was, obviously, critical to FilmDallas's business plan. In the end the managers relied on the same approach to the "creative" choices as to the creation of the fund, its structure, its investment criteria and its management. That approach was, again, simple: Try to be "smart" in making what are often irrationally influenced decisions. Nothing is more illogical, it often seems, than the success or failure of a motion picture. Good movies fail and bad movies become hits on a fairly regular basis. So the attitude of FilmDallas was to emphasize credibility of the product.

Now, what does that mean—"credible product"? Well, it meant avoiding investments in projects with titles like (I kid you not) *Motor Home From Hell* or *The Oklahoma Weedwhacker Massacre*. The high-concept exploitation film was not the focus of our plan, even though our investment criteria emphasizing low-budget fare could easily have led us in that direction. FilmDallas invested in the "specialized" independent motion picture designed to fill a clear need in the marketplace for sophisticated and quality entertainment. While we all knew that our chances of financial success were as speculative as moviegoing taste would allow, we made sure that the films in which we invested would not be easily forgotten. Six Academy Award nominations and two Oscars in that first flight of investments gave FilmDallas the credibility to expand its activities greatly and to demonstrate that a modestly conceived enterprise can have a significant influence on and make a contribution to film history.

# ANALYZING MOVIE COMPANIES

*by* **HAROLD L. VOGEL,** *a first vice president at Merrill Lynch & Co. in New York. Mr. Vogel holds an M.B.A. in finance from the Columbia University Graduate School of Business, an M.A. in economics from New York University and is a Chartered Financial Analyst (C.F.A.). For almost two decades he has been recognized by* Institutional Investor *magazine as one of the top analysts in the field and was rated first in seven recent years. He serves on the New York State Governor's Advisory Board for Motion Pictures and Television and is the author of the text* Entertainment Industry Economics *(second edition, Cambridge University Press, 1990).*

***. . . intelligent and honest people can have significant differences in opinion as to how best to account for their incomes and costs. . .***

If you really want to see lots of movies, you probably shouldn't become an investment analyst. I should know. I've been analyzing the business of making movies for over twenty years, and I can rarely find enough time to see all the films that I'd like to see. Moreover, even after I've seen a film, I've discovered from experience that my personal reactions and opinions are quite irrelevant when it comes to predicting how the shares of the company that produced or distributed the film might be treated in the stock market.

My interests notwithstanding, there are, in fact, many other variables that can affect the profitability performance of an entertainment company and its common stock. For example, Hollywood's typical hype and publicity for an important release may sometimes cause share prices to run up well before a picture opens. Often, then, it may be advisable to sell the shares even before the first weekend's grosses are reported. Yet, at other times, a film that almost no one has heard of becomes, initially through word-of-mouth advertising, a smash hit that continues to generate significant income for a considerable time after its first screening.

Much of this underscores the uniqueness of investing in movies as opposed to investing in industries that manufacture hard, finished goods, such as cars or computers. Entertainment industries instead produce things you can't hold in your hands, or taste, or smell: they produce things that you *experience* and that you emotionally carry away with you long after a movie's last reel has unspooled or a recording's last song has stopped playing. As the history of the movies and of other forms of entertainment has shown, people will pay a lot for such experiences.

In 1990, for instance, people in the United States paid a little over $5-billion for movie tickets, $10-billion for home-video rentals, $7-billion for music recordings and $17-billion for cable-television services. And over the last forty years the rate of growth of demand for these products and services has exceeded the real growth rate of the overall economy by at least two or three percentage points. But just as important, in other countries—whenever and wherever large, young, middle-class populations emerge—the same pattern of above-average growth has inevitably evolved. Currently at least 3% or so of total personal-consumption expenditures in the United States are for entertainment products and services (compared with around 6% for clothing and 12% for transportation).

Of course, in looking at all of this, it is easy for an entertainment-industry analyst to at first be overwhelmed: standard cookie-cutter accounting and forecasting methods as taught in the business schools don't seem to work very well when applied to this industry. In the movies, just to take an obvious example, each completed film is produced with unique elements and cost structures that will never be identical to those of any other film. Indeed, from its first date of release, each film is in and of itself a marketing experiment, conducted in real time and without recourse to a second chance. Brand names of stars and directors, unlike those for laundry soaps or sodas or cigarettes, have relatively limited shelf lives. And even a distributor's brand name doesn't matter anymore; with the possible exception of Disney, no one goes specifically to see a film distributed by Fox instead of, say, Paramount.

But if the top-down forecasting methods that are so useful, perhaps, in forecasting demand for housing or horseradishes don't seem to work very well in entertainment, what then?

The answer—that there is no single, simple answer—is not very comforting to those who seek decimal-point precision for their projections. Not the fanciest spreadsheet analysis in the world can predict how an entertainment company will perform relative to the economy. And that's because entertainment-industry investments have instead proven over time to be far more sensitive to the deftness of a company's management in the formulation and execution of long-term growth strategies than necessarily to changes in broad economic variables (including, for example, measures of real disposable income, household wealth and interest rates).

In other words, if a company has demonstrated an ability to *profitably* attract and retain creative talent, the analyst might then reasonably assume that future performance can be extrapolated from the successes of the past. More to the point, proficiency in the care and feeding of many large (if not sometimes irrationally inflated) creative egos is in and of itself a decisive management skill.

Still, that alone does not a good investment make. What is also needed is a little luck and pluck, and the accumulation of a significant film library and/or music catalog that is, more precisely, a pool of rights. For the investment decision-making process, it is the expected stream of cash receipts (cash flows) that are to be generated from exploitations of this pool of rights that then becomes most important. Once such cash-flow estimates have been discounted back to net present value through the application of interest-rate mathematics, and then further adjusted for a company's debt load and risk profile, the analyst may finally have a basis on which to make a recommendation.

All of this would, of course, be relatively straightforward and simple were it not for the uncertainties in forecasting the size and regularity of such future cash receipts, or income ultimates, as they've come to be called by accountants. Any number of things including fads and fashion styles and shifts of consumer sentiment can affect the size and probability of receipt of such income ultimates. And so, the evaluation of entertainment-library values remains yet more an art than a science.

In fact, nowhere does this notion of financial artistry become more evident, it would seem, than in accounting for film-company profits—a subject on which cynicism abounds. Here, though, it might be remembered that the basic purpose of the accounting rules is to formulate a method through which revenues and expenses can be properly matched so that readers of financial statements may obtain a reasonably accurate picture of a company's current financial condition.

Unfortunately, however, this is often more easily said than done because intelligent and honest people can have significant differences in opinion as to how best to account for their incomes and costs. For example, if a studio licenses a television program or a feature film to a television network and receives the promised license fees in cash installments, should such revenues to the studio also be recognized in corresponding installments, or should they be immediately recognized all at once?

And what about the costs of making the programs or features? Should the production expenses that may have been incurred a long time ago (and when the buying power of a dollar or when the currency exchange rate was different) be somehow charged against the current period license fees, or should these expenses have been fully charged against revenues that had been previously received? How long, anyhow, should a movie studio wait before it finally recognizes that the production boss's hot picture of the year is really a dud that will *never* recover its costs?

The problem here for the outside investor, and also for the financial analyst, is that the so-called income-forecast method of accounting used by the industry (and codified in Financial Accounting Standards Board [FASB] Statement No. 53) relies to a great extent on a *management's* estimates (possibly far into the future) of how much cash income a film or television series might generate. And there are few, if any, means through which a management's estimates can be directly challenged or validated by interested third parties.

But that's not all there is to worry about. The method, under FASB No. 53, also allows for considerable management discretion in the way marketing and other costs should be expensed: there is room for controversy, and for approaches that could be deemed liberal or conservative with respect to applications of the rules.

Up to a point, analysts can get around this problem—and adjust their opinions accordingly—by paying careful attention to the balance-sheet footnotes and by determining the accounting-treatment policies that a management intends to follow. Some companies might, for example, decide to charge off all national advertising expenses during the initial domestic theatrical run. Others—arguing with some justification that the domestic home-video release schedule also benefits from such advertising—may decide to charge off only a part of such costs against the initial theatrical run. And the expensing of foreign marketing costs may be a different story entirely.

Accounting treatments for individual profit participants are, of course, also related to those seen on the corporate level. But disputes concerning individual profit-participant shares always seem to be the loudest. From the outside analyst's point of view, the most essential element in these situations is that the expected value of a profit participation will normally be a function of the relative negotiating clout of the parties *before* a project is begun.

Yet, no matter what the accounting variances, it is the availability and predictability of cash flow that is the analyst's most fundamental concern. And during the 1980s—a period in which the cost of making and marketing major feature films rose at twice the overall rate of inflation (at compound average annual rates of 11.0% for negative costs and 10.4% for prints and advertising)—the industry's sources and uses of funds changed dramatically. Early in the decade home video was but a tiny proportion of a revenue stream that was mostly derived from exhibition in theatres. By the end of the decade home video accounted for over 40% of total receipts. Moreover, by the end of the decade proportionately more of the industry's available cash flow was being absorbed by marketing and production costs and interest expenses than had been absorbed ten years prior. As shareholders of many small new independent companies formed during this period sadly discovered, it is often the common-stock investor who bears the largest risk of budget-cost overruns and cash-flow shortfalls.

In fact such cost pressures have raised the barriers of entry into the business and have caused the industry to seek new sources of financing—most notably, in recent years, from Japan. The pressures have also given birth to the widespread notion that bigger is better and that you have to go global to survive.

Though it is too early to determine whether these notions are indeed correct, the plot, as they might say in Hollywood, has noticeably thickened. It may just be that entertainment companies function best on a scale that is relatively small and familiar as compared with the relatively impersonal scale that is needed for large industrial enterprises to flourish. Also, it is not at all clear that the cultural mind-set that is required to run an efficient factory floor, or that is required to design a new electronic device, has anything much in common with that of, say, running a studio or writing a screenplay.

In fact it would seem that the promises of major synergies from the merging of consumer-electronics hardware manufacturers such as Sony or Matsushita with software companies such as Columbia/TriStar and MCA, respectively, have yet to be realized: near-term costs may have thus far outweighed the benefits. It would also seem that the primary thrust of these electronics-manufacturing/movie-company mergers would be to use software to sell more new hardware, rather than to use hardware to sell more new software (although the purposes are not mutually exclusive). It is quite revealing to note, for example, that giant MCA, with estimated 1990 revenues of $5-billion, only accounts for less than 15% of Matsushita's total revenues.

None of this, however, is to deny the importance of new technology to the production and distribution of entertainment. Already, on the distribution side, we've seen how the home-video market has come from nowhere in the late 1970s to now account for the bulk of the industry's revenues. And developments in computer technology have already had a profound impact on every phase of production and distribution—from special effects to editing, and from the enhancement of television signals to the latest of multimedia materials. (For details on future technologies, see article by Martin Polon, page 451.)

Of course, whenever a new technology for production or distribution is about to be introduced, or is in the early stage of growth, there is a natural tendency for investors to become excited about the future of the film business. Films and other program materials are supposed to be like old wine that can be endlessly repackaged and resold in new bottles. Yet although such investor enthusiasm is often warranted over the long run, it might be remembered that many new products and services are, by nature, more substitutional than incremental. Network television viewing in prime time has, for instance, been significantly reduced as cable networks and independent local stations have proliferated. And it would be fair to say that in the absence of home video, attendance at movie theatres in the 1980s would probably have been much higher.

Will the growth of pay-per-view cable services adversely affect home-video retailing? Or will the emergence of commercial television in Europe affect the demand for other entertainment products and services? How profitable will the new channels of distribution based on new technologies be? Those are the kinds of questions that remain open, and that help keep analysts employed.

Years ago, when the business was a lot simpler, it sometimes seemed that analysts and investors could decide on whether to buy or sell movie stocks just by reading a film's weekly grosses in the trades. But because most studios have become parts of much larger corporate entities, and because entertainment has become a truly global, capital-intensive, technology-driven business, the process of investment analysis has now become much more complex, and much more demanding. Even so, however, with all the never-ending action, on-screen and off, it's still fun.

IV
THE MANAGEMENT

# A CHAIRMAN'S PERSPECTIVE

*by* ***MIKE MEDAVOY,*** *chairman of TriStar Pictures, based in Culver City, California. A 1963 graduate of UCLA, he served as a motion picture agent for ten years, representing talent who became some of the leading artists in the industry, before joining United Artists Corp. in 1974 as senior vice president in charge of West Coast production. Four years later Mr. Medavoy became a partner and cofounder of Orion Pictures Corp., where he was one of the driving forces behind many critical and box-office successes, including the Academy Award winners* Platoon, Amadeus *and* Dances With Wolves. *Throughout his career, Mr. Medavoy has been active in civic, cultural and educational organizations. He was appointed chairman of TriStar in 1990.*

***You learn quickly to be cynical. Yet, hope springs eternal and films get made, and the good ones change people's lives for the better. . . .***

TriStar Pictures was founded in 1982 by three partners, CBS, Home Box Office and Columbia Pictures Industries. Notice that each of the partners was involved in different media; they were therefore not directly competitive and could license the rights to TriStar Pictures in their respective media: CBS for network television, HBO for pay-TV, and with Columbia handling theatrical distribution. Over the years, however, the triumvirate changed. CBS opted to disengage from the company in 1985, HBO reduced its position over time until it was a minority owner, and Columbia became the dominant shareholder. TriStar and Columbia merged in 1987 to form Columbia Pictures Entertainment, then controlled by the Coca-Cola Company. The Sony Corporation purchased CPE in 1989 from Coca-Cola and the public. Today TriStar Pictures and Columbia Pictures divisions operate as self-contained financier-distributors with the same parent, renamed Sony Pictures Entertainment in 1991.

As chairman of TriStar, all areas report to me: creative, financial, legal, business affairs, sales and marketing (advertising and publicity). Domestically, TriStar distribution and marketing is separate from Columbia, using different personnel for the first-run bookings, which garner most of the theatrical revenue. (Subsequent runs of both TriStar and Columbia product are handled by Triumph, a division of SPE.) Overseas, the distribution entities are combined. On-site executives for theatrical distribution report to Ted Shegrue, president of Columbia/TriStar International Film Distributors. Ted and other overseas executives handling all other formats report to Arnie Messer, president of Columbia/TriStar International Releasing Corp. Both of these gentlemen and their respective organizations report to me on all TriStar-related product.

Our corporate structure has Sony Corporation of Japan at the top, above Sony USA, which oversees Sony Pictures Entertainment, run by Peter Guber. SPE is the parent company of TriStar Pictures, Columbia

Pictures, Columbia Television and Loews Theatres (see article by A. Alan Friedberg, p. 341). As a board member of SPE, I attend quarterly board meetings in Culver City along with representatives of Sony.

When one is appointed chairman of a motion picture studio, it is vital to quickly become familiar with the existing personnel and with the full production slate, ranging from completed pictures back to projects in development. There are immediate decisions to make, such as moving a screenplay to a subsequent draft or deciding on a picture package being offered by an agency. Each of these decisions has a financial impact. The job requires, in my opinion, the blending of two main talents: One is artistic and the other is business.

When I came on board at TriStar, excellent people were already in place, so I made no changes in the key personnel. Since I've always had a management philosophy of building from within, I chose someone already on staff as the new head of production. The first order of business was to read through all the screenplays the company had invested in, to decide which were valuable enough to go forward and which to put into turnaround. Sometimes one is validating prior management decisions, sometimes not. There were also some completed pictures, which I could not help; there were others that I could.

For example, I was happy to find *Hook,* a screenplay that had been developed at the studio, and proceeded to interest Dustin Hoffman. Knowing that Dustin and Robin Williams wanted to work together, I then went to Robin and finally approached Steven Spielberg to direct. This is the kind of direct pursuit of creative talent that a studio executive must employ in order to make the best possible movie from what starts as 120 pages of imagination.

The *Hook* deal is unique, and may further the way for studios to share more readily in the earliest profits (or losses) with major talent who might otherwise prohibitively inflate budgets with their going rate of cash compensation. It is no secret that movie budgets are skyrocketing, limiting the return on investment. The *Hook* model is one way to employ the highest-priced talents and control costs while sharing more generously in the success of a picture with such talent. It's a classic business trade-off.

Here's the concept: With a picture cost of $X-million, the two leading actors and director together receive Y% of the distributor's gross from first dollar up to $X-million of film rental, reduced to a smaller percentage of the gross from $X-million to $Z-million of film rental. From $Z-million on up, with a hiatus, their share goes back up to Y%. Notice that the dollar plateaus represent *gross from first dollar* of film rental, that is, of the first monies TriStar receives, even before deducting a distribution fee, making the studio true partners with the talent. In fact the distribution fee, usually a 30% deduction from film rental, is postponed to later on in the revenue stream. This also means that our fee and recoupment come from a proportionately smaller share of the traditional revenue pie. For example, out of the first $X-million that flow to TriStar as the studio's share of box-office on *Hook,* Y cents of each dollar goes to Messrs. Spielberg, Williams and Hoffman. From the balance, the studio recoups advertising and print costs and deducts its distribution fee. Naturally this is the kind of deal struck only if one is confident of a very large box-office gross. Is it a substantial risk? No less risky than paying the principals their going rate of millions of dollars in cash fees up front. The result is more of a partnership.

I have operated with the same management philosophy since joining United Artists in 1974, carried through to Orion Pictures, and now at TriStar: to find the best filmmakers and stories around and to let the creators make their films. Great leeway is given to filmmakers who have already demonstrated their worthiness through their reputation and track record; this is true as long as the financial end of the deal is maintained. Nevertheless we do exercise a level of control; after all, we are custodians of the money. Using 1990 as a sample year, the average studio-picture budget was virtually $26-million. Add to that $12-million per picture for marketing, and you can see the enormous risks each picture decision carries. We often have strong suggestions for our filmmakers, and we never abdicate our right to approve screenplay, director, producer, principal cast and the budget of every film we are involved with.

As one of the people in Hollywood who can say yes, it's important to have strong relationships with both talent and agents. I was an agent for ten years and was able to launch the careers of many filmmakers. But every top manager has his or her own way of working. For example, I have always immersed myself in the work of the most active writers, directors, producers and principal performers, through seeing their movies, reading their screenplays and meeting them personally. I also keep abreast of new talent in the same way. That's the creative side. From the business standpoint, having negotiated or been responsible for the negotiation of countless talent and picture agreements, there are references to cash and expertise tucked away in memory that are always called upon. Formulas are connected to faces connected to credits. The creative and business disciplines blend via a sort of mental checklist, which evolves over the years and is the source of intuitive reactions when putting together a motion picture deal. I have always been a hands-on executive, so that when a snag develops in a negotiation, I like to have a personal sense of the parties, in order to anticipate an outcome and will often get directly involved, to iron out issues (or to confirm intransigence).

This kind of job requires a certain toughness, clothed in a real affinity for talent, along with respect for the investment and for the audience. The business is highly seductive, and executives are under constant pressure to maintain relationships. Yet the high cost of making and marketing pictures and the level of control by agents and lawyers has made the road more perilous. For example, sometimes talent and their representatives are insulted if they are not offered their last deal, even though that picture was a failure. As a matter of fact they usually want more. Also, perks have become a new big problem in Hollywood. (The private jet is a "must" for stars today, almost more important than the deal—it's absurd.) Cost cutting is on every studio executive's mind. The challenge is not whether to do it, but how best to do it. From the studio standpoint, there has to be a concerted effort to hold the line; it is irresponsible to do otherwise, especially with the press spotlight on costs and returns and with the extraordinary competition within the industry.

Since making movies involves relationships of creative people under stress, sometimes there are problems. Perhaps a picture is in its second week of shooting and behind schedule. Usually the production imposes corrective measures and rights itself, because most picturemakers are very cost-conscious and one doesn't want the reputation of being irresponsible. Sometimes a director's intent does not correspond to the best interests of the picture. In most cases, this can be ironed out through persuasion. Other times an approved budget must be altered to coincide with unforeseen production problems on an otherwise terrific shoot.

Because "the buck stops here," my job can call for confrontation if someone is not performing properly. I've been responsible for about 250 pictures and I've had to fire only one director. This is not an easy task, and I resist it, exhausting every avenue until I know it simply must be done. A word of advice: Always tell someone the truth yourself, however unpleasant, rather than hide behind someone, such as an intermediary. This is one way to maintain relationships with people; they will always know where you stand. Often that is not reciprocal. In the movie business, talent can hide behind agents and lawyers, and sometimes this can cloud relationships. There's nothing like a one-on-one conversation between principals to clear the air. Each picture is unique, and it takes experience and instinct to shepherd the people and money to best advantage.

Over the years there are certain lessons learned. For example, don't "shoot a start date," that is, don't rush a picture into production to accommodate a release schedule. Choose talent you trust and who have good track records. Then it becomes like sports, with streaks and slumps. Don't refine a deal to the point of absurdity. There must be practical give-and-take on both sides, but sometimes a negotiation is prolonged for reasons involving impressing a client or demonstrating clout, and this can jeopardize the project. Nobody goes to the theatre to look at a deal. Other times representatives are intransigent in holding to an inappropriately high price. I'd rather put all of the money on the screen rather than waste it.

There are lessons to be learned on a personal level as well. Although it is a ferociously competitive business, those who run it are highly intelligent and caring. Over time one learns about who is trustworthy and who is not; who will have a conveniently faulty memory; who is likely suddenly to want to change a deal point at the last minute. That kind of game playing can be found in any business. However, there may be more narcissism attached to this business, since the product is entertainment and we are dealing with artists. You learn quickly to be cynical. Yet, hope springs eternal and films get made, and the good ones change people's lives for the better.

Interestingly the intangibles about the movie business don't change: it's a huge gamble (and the money at stake keeps growing); there is a global audience out there to please, and there is no magic formula to accomplish this; by its nature entertainment is always in flux (stars of today are not necessarily those of tomorrow); today's decisions are really projections, since the product will reach the screen a year or two from now. Some companies do better one year and worse the next. Decision making is constant, and it all comes down to choices. As a measure of how mysterious these can be, most Academy Award Best Pictures I was involved with were turned down by others, often more than once: *Rocky, One Flew Over the Cuckoo's Nest, Amadeus, Platoon, Dances With Wolves.*

The movie business is a seven-day-a-week vocation, twenty-four hours a day. Although there is no typical workday, there is structure to the job, and scheduled meetings help coordinate the infinite details. Each area of specialty has problems to be solved and questions to be answered, and it comes down to how best to respond to each of these issues in the most productive, fastest way. Time is money. There is nonstop, constant decision making, ranging from whether to buy a certain property, to when will be the best time to release a picture, to how many advertising dollars will be spent to support it. There are levels, of course, to all of these, and it's critical to be up-to-speed on all of the latest-breaking elements.

This is accomplished through a heavy schedule of meetings, from 9 A.M. to 7 P.M. each workday. And the day does not end at the office; most evenings are spent working as well. There are three production-staff meetings each week, Monday at 9:00, Wednesday at 11:30 and a Friday lunch meeting. Two of these meetings involve every key executive in the company—sales, business affairs, marketing, production, music—some fifty people going over the status and planning for each picture. In detail the meetings cover a production report, a postproduction report, a business-affairs report (as to what deals are being made) and a development report (in three stages, based on how close a project is to being green-lighted). Because it's a small business and information travels fast, at staff meeting we also learn what is going on in other companies and what new trends are being observed.

The Monday and Friday sessions also anticipate and report on weekend reading. Reading is the lifeblood of any company, since it triggers the key yes-or-no decision on literary material. The first priority is projects we have an investment in: Has the project been improved? Do we proceed to a new draft or abandon it? The next priority is submitted projects. Our story department and executives receive around 450 scripts and books a month, from all kinds of sources, including agents, lawyers, and filmmakers. Never underestimate the amount of reading required of a production executive; it is overwhelming. And although one often consults coverage to aid at a lower level of decision making, at the highest level there is no substitute for personal reading. The buck stops here.

On a weekend reading list there are usually six or seven screenplays. At the same time we are always awaiting word on offers out to certain people to become part of our projects. The reading decisions being made now are for pictures that may shoot from four months to a year from now, to be released perhaps a year to two years from now. This lag time must be factored into the decision making.

Sometimes an agent will force a swift decision by conducting an auction on a property. I've never believed in getting into a bidding war; people will want to do business with TriStar or they won't. Instead, philosophically, we have borrowed the old studio concept of the thirties and forties: to put an array of talent under contract to the studio who want to be part of TriStar history. The result is some long-term deals with filmmakers such as Robert DeNiro, Roland Joffe, Danny DeVito, Barry Levinson, the Hudlin brothers, Jon Davison, Cher and others. We also have Carolco's product going through our pipeline. There are three instances wherein TriStar shares a talent deal with Columbia: for Dustin Hoffman, James Brooks and the Zucker brothers.

The idea behind long-term deals is to create an environment that allows people to work together or separately, building a program of pictures around them in a nurturing environment over the long term. Generally our benefit is the exclusive rights to pictures they generate through their companies for a given period, and their benefit is the security of a home, instead of having to go out and sell each project individually. Naturally I am very pleased that they are attracted to the company, and perhaps to my track record as an executive who has been involved in six Academy Award–winning Best Pictures.

The concept of long-term planning that Sony adheres to is not one that had been comfortably applied to the movie business. But it can be, and that is our approach: to try to maintain a continuity of management, a sense of futures, a confidence and security for the talent who have long-term deals with us. How? By offering quick answers on projects, providing career guidance, honesty and looking out for their best interests. Something like a quick answer, be it yes or no, should not be underestimated. Nothing is more frustrating for a filmmaker than the "Hollywood No," which is silence. It is debilitating and insensitive.

What is extraordinary about motion picture management is that, although a lot of it seems to come down to instinct, timing and luck, experience over many years is required to develop the proper intuition. For example, the decisions to couple certain talent with specific material can appear to be instinctive but are based on having made such "marriages" countless times before and observed the outcome. Similarly, good timing in the releasing of a movie can appear to be fortuitous, but actually stems from a specific expertise, having tracked releases and grosses over the long term. As to luck, that is certainly an intangible that comes into play, but really as a bonus, once the best choices are made, based on experience.

As noted earlier, in refining decisions, business-financial and creative disciplines must be applied. This duality of business-and-creative impacts on every judgment. Since so much of this is subjective and intuitive, how can it be taught? Through experience (repeated for emphasis), and through encouraging capable executives who have demonstrated an instinct for it. I came to United Artists with expertise as an agent and strong relationships with talent, but little background in production. There I learned from people such as Arthur Krim and Eric Pleskow. As a result, it later became important for me to help younger executives gain confidence in their careers, through encouragement and plain hard work.

This dovetails with how I relate to production executives. There is a discipline that must be followed by these staff members who receive submitted material. Their job is to pass on or pursue a given project with intelligence and speed. If a project goes forward to the development stage, I encourage a production executive to think through the entire artistic and cost structure of a project before presenting it. A story may be in the form of an idea at this point, but part of the presentation should include cost estimates of budget, possible creative elements (writer, director, cast), marketing and time of release. This is an example of the coupling of financial and creative disciplines noted earlier. In the production phase, two executives are assigned to each picture. They must intimately know how their pictures are progressing, by being in touch with the set and viewing dailies, and reporting on progress at the meetings. A production executive visits the set at ng film. These changes will probably cut costs in some areas, rices in others, and change the configuration of the "picture edefining how much revenue will be derived from which for-'ew challenges abroad will find overseas markets expanding, vith possible tariff, tax and political problems. Competition will the European Community strengthens, containing 100-million otential customers than the United States. And once the Eastern untries settle their economic problems, they will emerge as new s.

business is globalized, with four of the majors in international What is already emerging from Hollywood is the international whose filmmakers, stars, financing and subject matter come l over the world. Examples include best pictures such as *Char-Fire, Gandhi,* and *Dances With Wolves*. But while the creative, ng and transmission aspects of the business will be influenced nt wherever it emerges, the business of making films will con- be based mainly in Hollywood because of the sheer concen- of filmmakers and distributors.

o talent, there will be fluctuations in negotiations, depending ow profitable the business is and how much bargaining power command. Over the years I observed the following equation: re players there are, the more competition there is for what is pool of proven talent, resulting in higher prices. The reverse, haracterizes the business today, should also be true: The fewer itors there are, the tougher it is to maintain prices. Logic would a leveling off in major talent fees. But logic doesn't always the movie business, and this hasn't proven to be the case. here are fewer players, and talent fees continue to rise.

ollywood is to survive as an industry that purveys popular it must adjust to such economic issues as increased costs. But, e person said, "if money is the mother of all businesses, then on is one of its children." We, as leaders of this industry, must nt in preventing this; after all, we are creating history.

least once, a postproduction executive visits at
is a problem, there is more than one visit.
involved if there are problems.

A creative executive develops instincts by
and reading as many screenplays as possible.
go way beyond this, with personal reading co
literature. Since films are a popular culture tha
society, this kind of background is always help
executive develops an instinct that identifies wi
the very rarefied and difficult form of the scre
rectors and performers who possess specific styl
the screen. Added to this is an appreciation of
ing relationships and business responsibilities of
this is applied to the selection of people to join t
of a movie within specific creative and cost pa
that takes years to absorb.

To review, the selection of a project cent
approvals: screenplay, director, producer, princi
picture will be green-lighted once all five are a
sequent decisions flow from these: Physical-pro
planning their work (see article by Paul Maslansk
people begin conceptualizing as to marketing co
article by Robert G. Friedman, p. 291); sales pe
jected release date (see article by D. Barry Rear
given moment we are revising our release schedu
pictures, mapped out one year in advance. The in
Christmas playing times are like tentpoles. The r
built around them and is subject to change, depe
progress.

Regarding distribution, I daily meet with Bil
sales. On Tuesday mornings we join with the m
and go over the details of release scheduling and
tures. Every picture has its own unique quality, o
determines what the marketing is going to be. Tr
dios, handles everything from big pictures, such
ones, such as *Bingo,* each of which requires specia
ing a picture is a very unique art. Just as in prod
an entire process to be carefully thought out, from
from artwork concepts to media campaigns, inclu
essary to reach the target audience.

As to the future, new markets and new techno
to have a great effect. (See article by Martin Polon
is contemplating high-definition television, and pay
a much larger influence. Prints will probably be re
video delivered to theatres by satellite, with visual

# MANAGEMENT: NEW RULES OF THE GAME

*by* ***RICHARD LEDERER,*** *a marketing consultant in the film industry, who has served as vice president of worldwide marketing at Orion Pictures and as a consultant to Francis Coppola's Zoetrope Studios in the areas of film marketing, advertising and publicity. A twenty-five-year veteran of motion pictures, Mr. Lederer was vice president of worldwide advertising and publicity for Warner Bros. from 1960 to 1975, during which period he served for a year as a production executive at the Warner Bros. studio. He produced* The Hollywood Knights *for Columbia Pictures, coproduced with John Boorman* Exorcist II: The Heretic *for Warners and has lectured widely in the field of motion pictures.*

*Evaluation of what the audience will accept is difficult at best and impossible at worst. . . . The only business that comes rather close to it . . . is fashion. . . .*

The industry is changing, but only to the extent that it always has throughout its history. It has never been static; it has always reacted in one way or another to new conditions. It has never stood still as a communication form or—to a lesser degree—as an art form.

Yet it is too easy to assume that some violent upheaval has taken place, that a new art form, a new audience and a whole new set of business rules are at hand. We must in fact observe that pictures made today—some of which are the most successful films in the history of the industry—are traditional in terms of their dramatic content. What has changed is the filmmakers' technique. New technology, such as high-speed film, very fast lenses, radio mikes and portable cameras and lights, have given filmmakers the mobility to shoot a picture more realistically and with more visual excitement than ever before. The audience raised on television, and on television commercials in particular, is comfortable with shortcuts in visual storytelling, allowing more information to be imparted in less time. Thus movies move a little faster today and are more "cinematic" than they used to be, but the stories are essentially the same.

It should not detract from the expression of social concern or the aesthetic possibilities of film that major companies who have an economic interest in the business must continue to regard movies primarily as an escapist entertainment form. The management of these publicly owned companies must show a responsibility to shareholders and consequently to profit-and-loss statements. These are the realities a major studio must observe, and it therefore follows that the studios will be making pretty much the same kind of films they have always made. But this hardly implies that a slow evolution is not constantly in progress. Taste is more advanced and technique is more sophisticated than in the past. The nation has matured with regard to what it will accept and what it will tolerate in the arts and literature. Audiences have accepted, for example, a far more candid and explicit screen exploration of sexual relations. Essentially, however, the movies are simply dealing with human problems in a more realistic fashion, and the degree of realism and detail should not suggest a trend away from the basic escapist-entertainment appeals.

We must not confuse, however, *what* is made with *how* it is made. Having said, in effect, that there is a great deal of stability in the fundamental nature of movies, I must now turn to the factors that set motion pictures apart from other businesses. One of the greatest of these is the uncertainty of the marketplace. Many other industries have accurate indications of their market when they set out each year. The home appliance industry can judge its potential sales and make sound business decisions regarding refrigerator styles, models and the number of units to manufacture. Unfortunately the movie business has not enjoyed that degree of predictability since before the Consent Decree, when a major company owned its own theatres and consequently knew where its marketplace was. It knew how many films it could make a year and that those films would be exhibited in regular fashion. Today, with some studios owning controlling interests in certain domestic and foreign movie theatre chains, perhaps some order will be returning to this uncertain marketplace.

A second major factor that distinguishes motion pictures from all other businesses arises from the enormous impact of individual creative talent upon production cost and upon market success. Evaluation of what the audience will accept is difficult at best and impossible at worst. This is a major aspect of the "old game," and new rules are not really changing it. The only business that comes rather close to it, I should think, is fashion, where trying to judge what styles will sell next year—how to tailor an inventory, how much cloth to buy, how much to cut and so forth—is a bit of a guessing game. But movies are the super, number-one guessing game. Making movies, as one old-timer put it, is "not an industry, but a disease."

Let's consider the role of talent in this game. Ultimately movies are a product, and the product comes in a package that can be more important than what it contains. The package can be more or less attractive, depending on the names that are associated with it. A producer who has a fairly good action script, for example, can make the film for $20-million with a good actor and $25-million with a top actor. That extra cost is something he must think about in terms of actual return. Is it worth the extra millions? Will the film do that much more in business as a result of overinvestment in the top actor? There are, after all, only a few actors who seem capable of delivering a larger audience than the ordinary actor, and a good deal of that happens overseas. Escapist entertainment is still the major attraction around the world, and although the "star system" is in the past in America, many a picture is made only because a certain actor will commit to do it.

To some degree, the contemporary audience dictates the type of film Hollywood will produce. It is sad but true that movies have always been an imitative—not an innovative—industry. Miscalculation abounds. In the 1970s, everyone read the new demographics and learned that we have a young audience. Their immediate reaction was to plan and make films that were geared to the tastes and interests of young people. This in turn resulted in all kinds of unsuccessful films. It has always been that way, and I don't suspect it will ever change. (See audience statistics, p. 388.)

An extraordinary contribution made by the younger generation of moviegoers is the capacity for repeat business. Children enjoy this, and it's become part of their experience, like going back to visit a friend. Repeat business is the key to huge successes such as *Batman, Teenage Mutant Ninja Turtles, Beauty and the Beast* and, of course, movies with a numeral after the title. Proceeds from these pictures have redefined the higher limits of profit potential, and new pictures are constantly challenging these heights.

In the 1990s we have the phenomenon of the graying of the American movie audience, tracing back to such successes as *Cocoon* and *Parenthood.* The baby boomers who are now in their forties and fifties have grown up receiving most of their information from a screen. Weaned on television, they found a symbiotic relationship between TV and the movies, and going to the movies continues as part of their lifestyle. Parents bring babies to movies, and some theatres may introduce child care. Families often arrive at a multiplex together, divide to see different movies, and meet later for the drive home. Elderly couples who had enjoyed *Driving Miss Daisy* are returning to the moviegoing habit.

America's love affair with the mall places the movie theatre in a most compatible environment. Equipped with the finest projection and sound, multiplex screens offer a variety of selections. Patrons park in the safety of the mall garage, see a movie, then go next door to a café serving cappuccino. If they arrive to find their first-choice movie is sold out, they are likely, since they're in a moviegoing mood, impulsively to see a second choice. In the summer of 1990 both *The Hunt for Red October* and *Ninja Turtles* went on to gross over $100-million at United States box offices, and their audiences were entirely separate. This is very healthy for the business; there's enough to go around.

It is a fact that people in the industry, in their frantic efforts to analyze audience desires, deny the history of movie entertainment as well as their own instincts by jumping on the topical bandwagon. Management's alternative, if there is any single approach to planning successful pictures, is to try to make interesting films without regard, necessarily, to whether they really are geared to a certain type of audience. A major studio committed to doing ten to fifteen films a year should be trying to make marvelous stories—films that are interesting, different, special.

The evidence is there. Successful films have always been well-directed, well-written, well-made films about something that a majority of the people can relate to or empathize with. But this does not mean management should avoid taking risks on what is new, fresh and has a specialized appeal. In 1970, when I was at Warner Bros., *Woodstock* was a gamble that worked for us. Nearly twenty years later both *Miss Daisy* and *Ninja Turtles* were turned down by rival studios.

Miscalculations are possible, of course, under the best of circumstances, but I think they are even more likely to occur because of industry developments such as the controls upon management that conglomerate takeovers introduce. In such cases there are predictable changes, and my guess is that they are usually for the worst. We must remember that until the 1950s the industry was still in the hands of the so-called pioneers. Good, bad, or indifferent, right or wrong, they were a unique breed in American business life. Their backgrounds were dissimilar. None of them came out of film schools. Many were immigrants, barely teenagers when they arrived in this country. With no academic training, they went into various businesses and happened to be around when movies were born. All had an innate sense of showmanship, an instinct about this country and a prescience about the entertainment that movies would become.

These were very special people. When their kind passed on, new management or ownership replaced them, and some significant things happened. In nearly every case where the new management was a conglomerate, the company became overly business-oriented. This new breed, in the best American tradition, was made up of well-trained business-management graduates who were used to systematized and highly structured business organizations. They knew everything on the business side of how to run a company. After one look at a movie company, they found it to be amorphous and seemingly running amok. They were aghast. Their strict sense of business training was offended, and their impulse was to systematize and structure, to make the company, in their terms, "make sense." This often led to near disaster.

Out of four examples of conglomerate takeovers—of United Artists, Columbia, Paramount and Warner Bros.—only two were truly lasting. Transamerica's acquisition of UA in 1967 resulted in a clash of management style causing the top UA executive team to break away and create Orion Pictures; Transamerica sold UA to MGM in 1981. Coca-Cola purchased Columbia Pictures in 1982 with much fanfare, only to reduce its position to half a few years later and sell out entirely when Sony bought Columbia in 1989. Under Charles Bluhdorn, the Gulf + Western purchase of Paramount Pictures was successful, with the movie company taking on such corporate importance that G + W changed its name to Paramount Communications in 1989. Warner Bros. was acquired by Seven Arts in 1967, and there followed mostly unsuccessful films. It was only when Kinney Services, under Steve Ross, bought the company from Seven Arts in 1969 and restaffed it that Warners was revitalized. Veteran Kinney executives preferred business with no inventory, and Kinney had been thriving with an office-building-cleaning business, funeral homes and parking lots. But Steve Ross saw the potential in the expanding world of entertainment, mixed movies with music and television to great success, changed the parent company name to Warner Communications, and merged it with Time Inc. in 1989 to form Time Warner Inc.

It should be pointed out that whatever problems "new management" had and still has in trying to fathom the mysteries of production and in developing systems that are functional, it has had extraordinary success in revolutionizing the art of film distribution. Correctly identifying the movie market as a mass market (in most cases), it moved aggressively into the area of network-television advertising with heavy print and radio support. Management capitalized on this gigantic ad expenditure by having up to 2,500 prints of the movie playing simultaneously around the country. This was not as simple as it sounds. Exhibitors who traditionally demanded and received exclusivity in major markets had to be persuaded to allow their competitors to share in the release pattern. Huge advertising budgets, which often exceeded the negative cost of the picture itself, had to be approved. The same applied to the heavy cost of release prints. This is just one example of the change in marketing technique brought about by modern management, and the results are encouraging. Successful films today are achieving film rental returns never dreamed of only a decade ago.

The old-timers were born gamblers; the new people are forced into being gamblers, and they're uncomfortable with it. They're businessmen, and no good businessman likes to gamble; rather, they like to insure their bets whenever possible. But "insurance" such as high-priced stars or high-priced directors can be disastrous, as in *Heaven's Gate, Ishtar* or *Leonard Part 6*.

There is an interesting cycle in management that has repeated itself recently in the movie business. It occurs when a new generation of production executives takes over a studio. They are usually somewhat unfamiliar with certain phases of the business, particularly marketing and distribution. They approach their head of sales or marketing and ask how many movies should be made this year. The answer is, "If you made six successful pictures, that's all the playing time I can handle; if you make twenty failures, I haven't got enough." The insecurity that runs through this new generation of studio executive stems not from deciding what pictures to make, but how many. Because of the powerful impulse to make movies, they reach for material a more conservative management might turn down, due either to story content or to the high numbers involved. But they are hungry to make movies, so they make risky, innovative decisions. In the three years it takes from making the decisions to tracking the results at the box office, this generation of studio executives usually has some phenomenal hits on its hands; the gambles have paid off. But they are also watching some of their profits dissipate because hefty participations were built into certain artists' deals on these pictures. This executive team is now rich and successful; they're no longer the gamblers they were three years earlier. They may even begin to think they know what they're doing, which is a terrible mistake in this business. They start making tougher deals, minimizing outside participations in order to protect their profits. They're not hungry anymore.

At this point another group of new production executives takes over another studio, and they are as hungry as the first group was three years ago. An agent approaches the first group with a strong package, asking high fees and participations for his or her clients in the package. The first company refuses to make such a rich deal, so the agent goes across the street to the second studio, makes a deal with the eager executives and the cycle goes on. Interestingly studios that have been forced to gamble on material out of desperation or need have generally been successful. Intuition and luck play enormous roles in the business.

Another new challenge in the management game is to harness the great potential of overseas revenue. As an example, the box-office of *Back to the Future II* was stronger outside the United States than within, with a global theatrical gross of approximately $200-million, divided $130-million overseas and $70-million domestic. Further, the excitement generated by an American movie-launching spreads quickly around the world via satellites and the global reach of CNN (Cable News Network). To capitalize on this publicity, studio management has moved up the worldwide release patterns of movies. Until recently pictures opened overseas six months to a year or more after their domestic debuts. Today the timing of international release dates generally follows the American launching much more closely.

Unfortunately movies are not a business in the strict sense of the word. Studios demand unique talent and unique understanding if they are to be run effectively. The successful major motion picture studio of the future will be one that manages the following: (1) a reorganization of its outdated physical structure to bring inflated above-the-line salaries and production overheads down to reasonable scales; (2) the development within its creative manpower of a "sure nose" for potential motion picture material that will be popular worldwide; (3) the ability and know-how to attract the proper talent in the industry to these various projects; (4) the diplomatic skill needed to cope understandingly with the creative temperaments and excesses of these gifted producers, directors, writers and stars while at the same time imposing upon them realistic and responsible fiscal controls. Utopia? Maybe. Admittedly a nearly impossible set of conditions—but there are clear signs that management is slowly meeting this challenge.

# THE FILM COMPANY AS FINANCIER-DISTRIBUTOR

*by **DAVID V. PICKER,** an independent producer based in New York who has served as president of three financier-distributors: United Artists (1969–72), Paramount's motion picture division (1976–78) and Columbia Pictures (1986–87). A graduate of Dartmouth College, Mr. Picker joined UA in 1956, rose through the ranks and was responsible for such pictures as* Tom Jones, *the James Bond series, the Woody Allen movies and* Midnight Cowboy, *among others. At Paramount his pictures included* Saturday Night Fever, Grease, Days of Heaven *and* Ordinary People. *After a stint as president of feature films at Lorimar* (Being There, An Officer and a Gentleman), *Mr. Picker produced movies independently before joining David Puttnam at Columbia as president and chief operating officer* (Hope and Glory, School Daze, Punchline). *As a producer his credits include* Lenny, Smile, Bloodline, Oliver's Story, The Jerk, Dead Men Don't Wear Plaid, The Man With Two Brains, Beat Street, The Goodbye People, Stella *and* Livin' Large.

***A company cannot remain stable making $40–$80-million movies every year; the risk is simply too great. . . .***

Since 1950 the major studios established in earlier years have had to seek new roles and functions for themselves within a changing industry. United Artists created what may in many ways be regarded as a blueprint for film companies that sought to combine an already-existing enterprise with the skills of successful financing and distributing of motion pictures.

In order to understand how this new identity was established, we need to begin when Arthur Krim and Robert Benjamin bought control of United Artists from Charles Chaplin and Mary Pickford in 1951. The company had been losing money. But by late 1952 United Artists was in the black, and Krim and Benjamin were in a position to consider financing their own productions. Their plan was to finance pictures by dealing directly with the creative forces who make them. Their initial concept involved the extension of creative autonomy and a percentage of profits to the filmmaker. The company's interest was to secure all distribution rights and a share of the profit in the film. This *modus operandi* was clearly contradictory to the policy of the major studios that owned all their films and—by keeping editing powers to themselves—did not relinquish creative control to individual filmmakers.

This idea could not be given any real test, of course, until United Artists built a strong financial base. But as a result of a succession of good pictures, the concept that Krim and Benjamin had initiated became a way of life for the company. Such early successes as *High Noon, The African Queen* and *Moulin Rouge* resulted from deals with filmmakers. By the mid-1950s they had also initiated production programs with various film companies, one of which was Hecht-Hill-Lancaster, and this resulted in such remarkable films as the Academy Award–winning *Marty, Trapeze* and *Sweet Smell of Success.* Later they established a relationship with the Mirisch Company that lasted some sixty pictures.

Throughout its history the management of United Artists gave creative filmmakers the right—within various approved frameworks of budget, script, cast and director—to make films as they wanted to make them. In exchange for that right, which was revolutionary for Hollywood, UA was able to attract many of the top filmmakers in the world.

Other companies eventually caught on because there was nothing essentially unique in what UA offered, with the exception of its own management techniques. The Krim-Benjamin philosophy of extending creative freedom to the filmmaker in 1952 led to the general industry approach of today.

It may be informative to consider in some detail the way this philosophy has been translated into operational realities for other financier-distributors. We can usefully examine what happens to a dollar that comes in at the theatre box-office window, following its course backward from there in order to see how a company makes its money.

Let us assume that half of the average box-office dollar is retained by the exhibitor and half is turned over to the distributor of the film. That distributor does not share in the receipts from concessions in the theatre, which are exclusively the exhibitor's (see "The Theatre Chain: American Multi-Cinema" by Stanley H. Durwood and Gregory S. Rutkowski, p. 352). The 50¢ that comes to the distributor represents 100% of *film rental*. When a picture has "done $40-million," that does not mean its box-office gross is $40-million, but that the film rental earned by the distributor is $40-million. The figure in fact represents, depending on various deals in various theatres throughout the world, the sum that comes to the financier-distributor.

A percentage of this money is charged for distributing the film: 30% of the gross for the United States and Canada, more overseas. Out of each dollar of film rental paid to the distributor domestically, 30¢ is retained as its distribution fee. In addition, distribution costs are deducted, including prints, advertising and interest as well as other expenses. From the "film rental dollar," then, 30¢ is taken as distribution cost and another 20¢ is taken to cover prints, advertising, interest, taxes and distribution expenses. (These figures are approximate.) What is left is known as the *net producer's share*, which is used to help pay off the loan secured to fund the negative cost of the picture. If a picture costs $25-million to finance and distribute, then the 50¢ returns that constitute the net producer's share must add up to $25-million before the picture approaches breakeven. But that rarely happens from theatrical alone. Studios need the huge revenue stream from home video, coupled with strong receipts from overseas, to theoretically reach breakeven.

At this point all profits are split between the financier-distributor and producer on a basis that can vary from 50/50 to 80/20, usually favoring the financier, and depending upon bargaining power. Because of the great delay in reaching profits (if any), many more "gross deals" are being made, wherein gross participants get money "off the top," without any concern for what the actual profits are. Various formulas may find a director or star receiving a percentage of the gross from first dollar, a percentage of the gross after an agreed-upon, fixed breakeven formula is reached (but before profits are divided), or a percentage of the gross after a multiple of the negative cost is reached. In these cases "gross" is the distributor's film rental, not the theatre gross. For example, a certain actor might be given $2-million cash against 10% of the gross. If the film rental on the picture totals $40-million, that actor will have received $2-million in advance against the total $4-million he will earn on the film without regard to distribution costs, prints, advertising or any other costs. A *percentage of the gross* deal, it might well be said, is a favorable deal if a picture does well.

To summarize: If the net producer's share from all sources equals the amount of overall production risk, the company is in good shape. If it falls short, and losses can be recouped from the various distribution fees without threatening a basic financial position strong enough to carry on the organization, then they are also in good shape. Where they can get into trouble is with a motion picture that costs a great deal of money and grosses nothing at all. This is why it is so dangerous for any company to sink an enormous amount of money into one picture. If the picture fails, they not only have lost the distribution fees that keep their establishment going but also have no distribution-fee profit to pay off the loan against the picture. And of course they lose the net producer's share as well.

In 1967 Transamerica Corporation acquired UA in an effort to diversify into the leisure-time field, just as Gulf+Western had acquired Paramount a few years before. The move gave United Artists an umbrella of enormous importance, since it was now part of a corporation with over a billion dollars in assets. With the advent of non-movie corporate management, the gap between executive and filmmaker and between corporation and filmmaker became increasingly wide. In the case of the Krim and Benjamin team, certain policies set down by the parent company were so unsettling that, in 1978, they left United Artists and established Orion Pictures. Transamerica got out of the movie business in 1981, just as the Coca-Cola Company, which had purchased Columbia Pictures in 1982, left the business by selling to the Sony Corporation in 1989.

In analyzing how movie companies are run today, conventional wisdom calls for long-term management to establish an operating philosophy and to hold to it. This has certainly been successful at Disney among the majors, and New Line Cinema among the independent distributors. Disney, for the most part, holds firm within certain cost parameters, and struck gold with *Pretty Woman*. New Line distributes a wide assortment of pictures in a lower budget range and broke through with the *Ninja Turtles*. Other companies are certainly looking for locomotives, but have no clear philosophy as expressed through their movies. Until such discipline and philosophy on a long-term basis are established, the market is chaotic, because there are always players who are prepared to outspend the others to prove they are in the business or to own what they think is the sure hit.

The result is that costs escalate beyond control. A company cannot remain stable making $40–$80-million movies every year; the risk is simply too great. And regardless of negative cost, marketing expenses are astronomical. (See the article by Robert G. Friedman, p. 291.) This inhibits pictures of a special or unusual nature, because such risky pictures cost millions to open and play to a national audience. But if a company defines its philosophy, is smart, lucky, disciplined and is prepared to say no to those tempting, high-cost packages, the risk of making movies can be reduced.

Cash flow is a common problem for companies in today's movie business. Disney enhances cash flow through their theme parks; Columbia/TriStar and MCA Universal have been enhanced through acquisitions by Sony and Matsushita, respectively. But how long these global acquisitions will remain in place in the face of a mercurial business is anybody's guess.

In reviewing how the business has changed over the years, agents are at the forefront. They have become stars, and are the key power players, as opposed to the clients they represent. The deal has become more important than the product, the dealmaker more important than the filmmaker. And the movies have become somewhat formularized as a result. An example was the spate of produced sequels (now in decline), since sequels represent a form of insurance rather than creative ingenuity. Management's justification was that the audience was saying, "I like what I know." But making high-priced sequels meant that those dollars were not available for something the audience didn't know, and the risk to pursue that was too great.

The training ground for the business has changed as well. Where do new filmmakers come from? Mostly from film schools, rather than television or the studios, as in the past. Studios look at students' work while they're still wet behind the ears, and pay enormous sums of money. A young writer coming out of film school can be paid $800,000 for a screenplay and is expected to deliver as if he's been making movies all his life. There are very few talents who can survive that kind of pressure.

The way the majors do business hasn't changed in thirty years. Any producer who makes a development deal does so with eyes wide open, has more or less muscle depending on the elements brought to the deal and knows that the financier-distributor will own the lion's share of the project. This hasn't changed at all. Anyone who wants a more favorable position must bring substantial financing to the table and make an output deal with a given studio.

Although this is still a relationship business, what has changed is the nature of those relationships. Today they are based on deals, money and power, as opposed to a filmmaker deciding to make a long-term association with a studio to find artistic and commercial expression in the best atmosphere. The Mirisch brothers made sixty-six movies at United Artists, Stanley Kramer made ten. Today it is difficult to find someone who will make two pictures in a row for the same company. Do you want a relationship today? Pay more and you have it.

It's a seller's market, and because of this there are no longer any deal terms that are sacrosanct. If a seller has a project that is in demand, companies such as Columbia, Paramount and Fox can step up to competitive deals. Issues such as overhead, viewing of dailies, final cut, and the net-profits definition can favor the seller if his project is desirable. Bargaining power decreases if the project is something the financier-distributor would be willing to develop on a less-competitive basis. In this case the seller will face tougher terms and definitions.

A producer faces the same basic dilemma that confronts any financier or distributor. What an audience is going to want to see, how they will respond to a given motion picture, and all the variables that take place while a picture is in production—these factors are beyond analysis. One difficulty in this regard is the lag between the time the decision is made to finance a script or go into production and the time the picture is released. In addition, there has been a polarization in taste. In practice this means that the successful picture is more successful than ever, but the unsuccessful picture is more unsuccessful. There used to be a base audience; you knew you could count on certain numbers for almost every picture. That audience simply does not exist anymore. How can we know that the $25-million that is going to be spent is being invested in a product the mass audience will want to see a year hence?

In closing I might say that we are in a time that is more precarious, but at the same time more exciting, than ever before. Audiences are extremely unpredictable, and their decisions often have nothing to do with a film's merit. When millions are being risked in an effort to choose those few pictures that audiences will decide to see, it can become a pretty scary business. The trouble with our business is that nobody trusts anybody in it. The distributor doesn't trust the exhibitor. The exhibitor doesn't trust the distributor. The producer doesn't trust the creator. The creator is sure the distributor is putting in invalid charges against his picture. The financier is positive that the creator has spent forty-three unnecessary days in shooting the picture. Despite all this, somehow or other we wind up with films that people sometimes go to see.

As to the future of the pictures I've green-lighted over the years, the legacy will always be there on the screen. But it is poignant that the library of United Artists pictures from the late fifties to the early seventies, for instance, is no longer related to anyone who had any involvement in making those movies; it's in the hands of money people who paid for the rights to that library.

The management atmosphere at UA in that period was warm and familial, which sparked a unique allegiance between filmmakers and executives. The only company that comes close today is Castle Rock. The five principals are friends, and really enjoy each other. As long as they can control their own destiny, they have a better chance of turning out product they can take pride in than a larger corporate entity. Today's movie management style is more in line with that of corporate America, which does not engender that kind of warm feeling. Players in the business must find a way to make that work for themselves. This business was run for many years by people who were tough and idiosyncratic, but who grew up in a business that they loved and understood. Today it's really not a motion picture business; it's a product business that happens to be film, which then has multiple financial uses. The focus is not on the filmmaker and the product (except in publicity terms), it's on the results, the money that can be made with it, what deals are required to get it, and how to achieve the best possible deal. That doesn't necessarily bode well for the content of the material on the screen.

# V

# THE DEAL

# THE ENTERTAINMENT LAWYER

*by **NORMAN H. GAREY**. For a biographical note on Mr. Garey,* *see page 139.*

***A lawyer in the entertainment field must . . . get along with people under difficult circumstances, and sometimes . . . know the client perhaps better than he or she knows himself or herself. . . .***

An entertainment lawyer's practice and his relationships with clients may differ significantly from those of other practitioners. The emphasis is on servicing an individual and thus on the relationship between an individual lawyer and an individual client. These relationships are often lifelong or at least career-long and go deeper than traditional lawyer-client relationships in other areas of legal practice.

The clients an entertainment practitioner has are typically creative, volatile and quixotic. A lawyer in the entertainment field must have the ability to adjust and adapt to disparate personalities, to get along with people under difficult circumstances, and sometimes to know the client perhaps better than he or she knows himself or herself. Many clients realize this and often ask for or expect advice that is not, strictly speaking, legal advice.

In the feature motion picture business there are several individuals besides the lawyer who typically influence the client's life and career. A performer will usually have a business manager–accountant responsible for financial counseling and planning, an agent, perhaps a public relations counselor and sometimes a personal manager as well. It's interesting how these interpersonal relationships overlap and how they affect one another in the decision-making process that goes on for each issue in the client's life; it often seems that an actor's life is run by committee. A producer will usually have only a lawyer and a business manager or accountant; a director or writer will have a lawyer, business manager and in most cases an agent.

If the client is a performer, a writer for hire or a director for hire—somebody who simply renders services for compensation—it is the primary function of the *agent* to seek, find and negotiate the basic terms of employment for that individual; it is not (and should not be) a lawyer's responsibility. The lawyer should be working closely with the agent, and ideally they should have a good personal relationship with each other. The lawyer (and the business manager) should be kept apprised by the agent of the client's career prospects, including what possibilities are being explored and what the status of each of these explorations or negotiations is; in other words it should be very much a team effort.

However, if the client is a producer, an "entrepreneurial" or "promoter" producer, the lawyer's involvement should generally be primary. Frequently the agent doesn't have the formal background in business organization, accounting principles, economic theory or literary rights in the legal sense that a lawyer has. And all of these areas of preparation are necessary for effective representation of a producer or of an entrepreneurial writer or director who also acts in a packaging or producing capacity.

If a client simply renders services for hire, it's not necessary that the lawyer hear from the agent until the agent has struck the basic deal for the employment of the client, whether as writer, director or both. When that is done, the lawyer becomes involved in the negotiation to assist in refining the deal, particularly in developing the formulae that relate to the net-profit or other contingent compensation position and definition. A lawyer generally shouldn't have to get involved in the up-front negotiation, which includes such major terms as cash fee, services to be rendered, time periods involved in the rendition of those services and the basic rights to be granted (if there are rights to be granted). All of the selling and positioning that it takes to get a client a job in the first place, to create the "want to buy" on the part of the buyer, are an agent's function. Upon the lawyer's entry into the negotiation he should learn the history of the negotiation from the agent and then generally speak to the business affairs person with whom the agent has been negotiating—the buyer. The business affairs vice president is usually the primary negotiating representative of the buyer, if the buyer is a studio financier-distributor or a major independent production company (see article by Stephen M. Kravit, p. 204). If the buyer is an individual producer, the attorney in question may deal with the buyer-producer's own outside lawyer. In any event the buyer's representative and the lawyer will then refine the deal further, and the documentation will commence. The lawyer generally ought not to get involved in the negotiation until he has seen something in writing, whether it has come from the buyer's representative in the form of a deal memo or from the agent for his client in the form of a deal letter that informs him rather clearly where things stand at that point. It's part of the agent's task to memorialize in writing the basic terms that he has negotiated.

If the creator-writer client wants to buy the rights, for example, to a magazine article on which to base a story in a script, it is his lawyer who will typically supervise the buying activities, such as acquiring the rights and checking and clearing them. But when it comes to selling the completed screenplay of the pure creator, that more properly is an agent's role. If the agent is experienced, knows the buyers and will listen to a lawyer when it comes to constructing price, he will then perform the primary selling function.

In the case of a screenwriter selling his writing services to adapt a novel for a producer-buyer or studio-buyer, the documentation would be generally in the form of an *employment agreement,* in which the writer-client is acting as a writer for hire; or a *loan-out agreement,* in which case he has his own internal corporation formed for various corporate and tax purposes, which loans his services to the buyer. However, if that same writer plans to render writing services in the adaptation of the magazine article he's already bought, the documentation would characteristically take a bifurcated form. There would be an employment agreement prepared with respect to the rendition of his future services, and there would be a rights option and acquisition agreement prepared with respect to the property rights being conveyed; in this case the rights to the existing magazine article.

For any pure creator, whether writer, actor or director, there are employment agreements that focus on the nature of the services to be performed. These agreements include a definition of the responsibility assumed; the period of time during which the client is expected to render his services; whether and, if so, to what extent he's to be exclusive during that period of time (i.e., whether he can engage in work on other projects); and the money involved. A myriad of questions can arise within those general categories. If he's not to be exclusive, what is his availability? Does he have to be on first call, giving first priority to this project? Are there options in favor of the employer-buyer for any further services? How much is to be paid for the services? How, and over what period of time, is the guaranteed compensation to be paid? Regarding contingent compensation, is there a *deferment,* meaning a fixed amount of money paid on a contingent basis if a certain time event, profit event or other event comes about? How does the deferment relate in order of priority to any other deferments? Is there a contingent participation (gross receipts or net profit)? If there is a gross-receipts participation, when does it accrue? If it's a net-profit participation, how are net profits defined and determined? Is there a reducibility factor; that is, is the client's participation reducible by participations granted to others?

Then there's the important area of controls. This isn't quite as important in an actor's or writer's deal, because writers and actors generally are considered under the control of the producer or director for whom they work. The director generally is considered to be under the control of the producer, but not always. The director's control may supersede the producer's in certain areas and under certain circumstances. For example, the director's cutting rights may be superior to anybody else's; he may in fact have "final cut." But a "star" writer, for example, may have the equivalent of the Dramatists Guild covenant for his protection: "Thou shalt not touch my work." That's very rare, but there are a number of top writers who have the exclusive right to perform personally whatever changes are requested to be written and to be paid for doing so. In this fashion a writer is able to exert a degree of control over the progress and integrity of his work.

The actor is typically not in a much different position than a writer for hire or a director for hire from the standpoint of controls—he's going to be rendering certain services under the instruction of others. But there are certain rights that arise out of an actor's services that are unique to an actor. Merchandising happens to be one of them. The actor has a face. If it is well known, people will want to put it on box tops, sweatshirts, T-shirts, book jackets and record labels; it's a valuable right that has to be negotiated for an actor client with some attention. If he is an important actor, his image carries with it a certain dignity that must be preserved. This is also a valuable asset, which must not be merchandised in a denigrating fashion; controls must be negotiated and employed. It isn't just a matter of how much money will be paid and whether it will be in the form of a gross piece ("gross" in the merchandising business is generally represented by royalties) or a net-profit piece.

An important issue arises with respect to whether and when an actor may be paid off without his services having been used. What is the talent losing by not being used if he's been paid? The answer is credit and billing. Credit and billing have a value in this business far beyond just seeing a name on a marquee because credit and billing may get an actor his next job or the opportunity to advance his career in various ways. In the early days of the film business, if an actor was fired and a jury had to assess the damages that arose from the employer's breach of contract, they would often determine that the actor was guaranteed the salary he'd lost and should in fact be paid. But since he also wasn't used in the picture and didn't get the billing or the enhanced reputation that might have resulted, the jury placed another value on that. The sum might be astronomical because the issue is so amorphous. So the producers and financiers, in order to protect themselves, developed what is called the *pay or play clause,* which says in effect, "We can pay you off, we don't have to use you, and if we do that, all we owe you is your guaranteed compensation. Don't look to us for anything else; you don't have the right to." That has become the almost invariable custom of the business. But today the term *pay or play* is frequently used as a shorthand method for saying simply that there is a firm financial commitment under a deal. The distinction, however, is whether that firm financial commitment also requires the financier to use the person's services, which means it's more accurately pay *and* play (in which case the loss of credit may be compensable in damages) or whether it doesn't require the financier actually to use the actor's personal services, in which case it's pay *or* play, based on what has become the custom.

Producers' controls are many, and the negotiation of controls between a producer and a financing entity occupies a great deal of time because not only artistic and creative controls but financial controls as well have to be worked out between financier and producer in great detail. Attention also has to be paid to a producer's credit because he'll get not only the "produced by" credit but also in most cases some kind of entrepreneurial, proprietary, presentation or production credit. Questions arise: Does it go above the title or below the title? In what kinds of ads must it appear? These are negotiable points. When getting into directors', writers', or actors' credits, the lawyer can refer to the relevant guild contracts on the subject, which stipulate order and size. Beyond that (and this is particularly true of producers, whose credits are not regulated directly by guild contracts) credit is mostly a matter of prestige, stature and precedent, a question of who comes before whom. What is the placement? Must the credit appear in paid advertising? (*Paid advertising* is all advertising, as distinct from screen credit, issued by or under the control of the distributor—in print campaign ads, display ads, television ads, etc.) Most lawyers leave credit negotiations (except for producers) to the agent.

A director differs in one respect from an actor, writer or producer who renders services for hire. He has an interest in protecting the integrity of his work that goes slightly beyond the interest of the others because the director is generally considered to be responsible for the total creative rendition of the product, and his reputation rides on the overall result. The writer, for example, is responsible only for the literary contribution, as opposed to the visual and audio contribution made by the director. The actor is responsible for the performance of his role. The director is held responsible for all of the performances in the picture. Therefore, the director's cutting rights—the right to protect the integrity of his work—have to be negotiated with some care. Who gets to perform the television cut or supervise it? Who gets to do the foreign censorship cutting or supervise it? How many cuts does the director get? How many previews? When do they have to take place? Who has the right to select the preview sites (an important item, since audience reactions may differ widely from site to site)? Does a particularly important director's right to final cut depend on budgetary, marketing or other financial or commercial considerations?

In all of the foregoing matters the agent and lawyer ideally function as a team. But if a client is an entrepreneurial producer, the lawyer, and usually only the lawyer, will be involved from the very beginning, from the very moment the movie idea is conceived. First, there will be a rights question. If the entrepreneurial producer is planning to acquire a piece of published material, a title search is conducted. The lawyer should have one of the major organizations that perform this function do so at the copyright register in Washington, D.C., to find out whether the idea or material has been exploited before and, if so, to what extent and by whom. This has a definite bearing on how much should be paid for it. Second, it must be determined that the rights are indeed owned by the person who is planning to purvey them and, if they aren't, it is important to know who else owns any of them. Third, are there any further title problems? Can the title be used? Can it be used only in connection with this material? Can it be used in adaptations of this material? The lawyer will usually conduct the negotiation for the acquisition of the rights on behalf of the producer or entrepreneurial creator (which would include an entrepreneurial director, writer or, sometimes, actor who acts in a producing capacity).

When the rights have been secured, there is then the necessity (unless the client is also a writer) to have the material developed into screenplay form. A negotiation must take place with a representative of the writer who will develop the material if the producer client is going to finance the development himself. There may be still another negotiation, this one with a development-financing source, if it is not the producer client who will finance the development of the material.

There is one situation in which no negotiation takes place up front; that is where the client is an entrepreneurial writer who has conceived the idea and is also going to write the screenplay and ultimately sell it for financing or set it up for financing. In this case there won't be any negotiations until sometime later, but the rights-checking process still has to occur. The lawyer may have to learn whether the writer's idea is as original as his client thinks it is or whether he will perhaps be infringing on somebody else's rights if he develops this project in the way he wishes.

An early issue that arises is the establishment of price. From the standpoint of an entrepreneurial creator, what are his cost factors? What's his out-of-pocket expense, what is it for and what commitments have to be assumed by somebody else or discharged by him and then made whole with somebody else's money? That's bedrock. Next, what kind of speculative risk has to be taken? How much of the creator's time, energy and services are at risk for no money? What is the return to the financier likely to be if this project is exploited successfully? All these factors must be taken into account.

The price of services is influenced very greatly by the factor of sales strategy or positioning. How do you make something seem more than it is? How does synergism develop through the uniting of several different elements? These are intangibles. Selling is the function of an entrepreneurial producer or creator or of an agent; it is not a lawyer's function. But sometimes, when the entrepreneurial creator is engaging in packaging and the lawyer is assisting in this effort, the lawyer can usefully participate very directly in the politics necessary to get people together and then to create excitement around them. In this connection, is the entertainment lawyer acting as more than a lawyer? Is he allowed to do this? The California Labor Code and Business and Professions Code lay out the requirements for licensing on the part of those who solicit employment opportunities for people in the entertainment industry and the rules of professional conduct for lawyers. The codes, and the regulations adopted under them, as well as certain bar association rulings and opinions support the proposition that a lawyer (or an accountant) who performs such services incidental to his professional practice may do so without a separate license. Thus, it is not unethical or inappropriate for a lawyer to engage in the kind of activity that real estate brokers, loan brokers or insurance brokers engage in as long as it is incidental to the performance of his services in his professional practice. Certainly these functions extend beyond what most lawyers would consider traditional legal practice. On the other hand, there are lawyers practicing in real estate, in the insurance industry and in the financial community who perform functions above and beyond the traditional law practice. Entertainment lawyers are no different from those people, they are just more visible.

In any negotiation it's important to know the psychology of the buyer with whom one is dealing and to have a very real sense of the value of the elements within the package one is representing. Frequently the entrepreneurial creator who is in touch with the marketplace has a better sense of the situation than the negotiating lawyer himself. Therefore a lawyer negotiator has to stay in very close contact with his entrepreneurial client during negotiations. That may mean from one to twelve phone conversations during the course of any single day because frequently the lawyer will be negotiating with the business-affairs vice president, whose authority is limited to devising mechanical ways of shifting money to accomplish a certain result. This vice president doesn't have the authority to commit more than a certain amount without going back to his creative principal, who is usually the production head of the studio. Frequently that studio head and the entrepreneurial creator (the client) will be talking at the same time; thus the negotiation often proceeds simultaneously on two levels.

Once a transaction has resulted in a signed document, problems may arise in the administration or implementation of the deal. If it's a legal problem, then the lawyer had better become involved promptly. Frequently the problem will not at first be a legal problem, but rather a relationship problem. Perhaps communication has broken down between the creative people involved, and it's simply a battle of egos about whose creative views should prevail. Rather than expressing it in just that way, the creative people will often start looking for business reasons to further the dispute. This kind of problem can often be solved just by counseling and mediating. On the other hand, sometimes it genuinely is a legal problem. Perhaps something has come up that makes performance as originally contemplated more difficult, more expensive or in fact impossible. Circumstances may have changed; expenses may have exceeded what was originally estimated. In these situations lawyers obviously have to step in. Lawsuits are filed in some cases, but most are settled. There is more contention and less completed litigation in the entertainment business than in most other industries. However, there are some problems that cannot be solved by mediation or negotiation; arbitration and litigation are the only way to solve them.

There are a number of matters for the entertainment lawyer who is administering a career that are more financial than they are strictly legal in nature. For example, there is an important function to be performed in the area of tax planning and general financial planning and implementation. The matter of auditing net-profit participations or gross-receipts participations in connection with motion picture, home-video or television product is a very important function. It is imperative that the distributors who are responsible for the payment of these participations be policed, because they do make mistakes, both from an accounting and from a contract-interpretation standpoint. In all of these areas it is important that the lawyer and the business manager or accountant maintain an effective working relationship.

There is also the matter of organizing the personal affairs of a client. People in the entertainment business engage in politics; get married and divorced; have problems with lovers, spouses and children; pay taxes, buy and sell property and do all the things other people do, but often with more flair and a great deal more visibility. The entertainment lawyer frequently has to serve as a liaison between the client and the other partners or associates in his law firm who are performing the more traditional legal functions for that client, and often he must simultaneously monitor publicity (wanted or unwanted) that may be attendant to the legal matter at hand.

Finally the lawyer in many cases may be asked to perform a kind of psychological, rabbinical or personal-counseling function in addition to his business function. Advice sought is frequently personal advice about how to conduct not only one's career, but one's life. Lawyers in many cases may have neither the formal training nor the temperamental inclination for these kinds of counseling. It is always possible, however, for the lawyer to consult and involve others who do have the requisite background and either to use their advice in counseling the client or to involve them directly when the occasion requires it. One danger of the close working relationship between lawyer and client is the lawyer's temptation to assume the persona of the client. There is sometimes a seductive sense of power to be derived from those acts of guidance and decision making, both personal and professional, that shape a prominent client's life. As exhilarating as this can be, it can also take a toll on the practitioner's own sense of self. If one is sensitive to this danger, the practice of entertainment law can be fascinating, rewarding and satisfying.

# BUSINESS AFFAIRS

*by **STEPHEN M. KRAVIT**, who is a consultant and representative for entertainment industry clients and is based in Beverly Hills. Mr. Kravit was executive vice president of Kings Road Entertainment from 1989 to 1992, after serving as senior vice president, business affairs, of Twentieth Century Fox Film Corporation. He moved to Fox in 1978 as vice president, business affairs, following ten years at United Artists Corporation, where he was vice president, business-legal and earlier, head of the legal department. Mr. Kravit is a graduate of Columbia Law School, a member of the California and New York bars and is admitted to practice in the Supreme Court of the United States, the United States Court of Military Appeals and various other courts.*

***. . . top . . . talent receives a large cash payment as part of the negative cost, which is applied against, say, 10% of the gross from first dollar of film rental, escalating at higher levels. . . .***

Today's motion picture legal and business-affairs practices can be traced to the days, between the twenties and forties, when studios maintained their key creative talent under exclusive contract. Those agreements granted the employer the right to order activity and obligated the employee to perform as directed so as to get paid. Except for compensation, most contracts were standard, long-term and exclusive, and studio lawyers and outside law firms developed forms to speed up documentation.

By the 1950s the economics of the movie industry had changed as a result of government intervention (via antitrust consent decrees), heavy taxation, the growth of television, a decline in attendance and the development of profit-sharing arrangements with talent who used their increased bargaining power to free themselves from long-term exclusive arrangements. As a result, a new form of contract developed that provided creative talent with picture-by-picture, company-by-company relationships, instead of the previous long-term agreements with one studio employer.

Over the years the intricacy of these agreements has led to lawyers in entertainment firms becoming dealmakers as well as legal advisors, applying business, creative, legal and tax aspects in order to make the best possible deal for the client. (See "The Entertainment Lawyer" by Norman Garey, p. 195.) The role of the agent has changed as well. Whereas formerly they focused on getting their client the next job, agents today are dealmakers and packagers, in a climate where buyers compete for a limited and unique talent pool. From the buyer's point of view, whether studio or independent producer, deal-making expertise has evolved into a professional skill performed by business-affairs executives who generally possess a legal background. At a studio a business-affairs department is normally headed by a senior executive and is supported by administrative personnel and several negotiators.

Within a fully structured company, business affairs (or dealmaking) ranks as one of the five key management-level divisions, along with production, distribution (or sales), marketing and finance, all reporting to the chief operating officer.

Business affairs becomes involved as soon as production becomes interested enough in a project to pursue it. For purposes of illustration, let us assume we are describing the work of a studio financier-distributor. If a proposed project involves important creative talent and therefore significant money, the head of production, sometimes the board chairman and sometimes the president, gets involved, along with the head of business affairs, in setting financial parameters with the lawyers and agents representing the project. Deals with less complexity begin on other levels, but all deals involve a business-affairs representative as the point person of the company negotiating team, with input from the tax, finance, legal and production departments.

In making its recommendations on complex deals, business affairs will usually prepare and present to top management various pro-forma profit-and-loss calculations based upon expected and hypothetical levels of negative cost, marketing expenses and distribution revenues, factoring in various possible gross and net participations. These "numbers" are projection models inputed with the time-tested experience of company experts and will help to determine whether or not to make the deal, or "green-light" a production. However, since they are based on past performance, they will often not be accurate reflections of newly emerging or future trends, nor could they be, in light of the two-year average lead time between the decision to proceed and the release of a picture. More than one set of numbers has caused embarrassment when the public's response to a movie is unexpected and overwhelming, outperforming the rejecting studio's numbers model. The film business began as a gambler's business where one's "gut" reactions and passion for talent and material were the only guides. Since no one can accurately predict audience taste, this business—even with all of the financial prediction possibilities of a studio—will continue to remain a "gut" business.

The authorization to "make a deal" comes from the head of production. Then the head of business affairs judges the complexity of the deal and decides to what extent to become personally involved in the negotiation. Since scores of negotiations are going forward at the same time, delegation is important. In the press of daily work, getting the job done is more important than following a rigid command structure.

In dealmaking everything revolves around money. The negotiation between the studio and the talent must culminate in an agreement as to what duties the talent will perform in return for how much compensation (and other benefits). The rights, options or opportunities of each party—and their cost levels—as the project continues to develop, and, if the picture is made, the accounting and division of receipts from distribution, must all be resolved in this negotiation.

Once these basic terms are reached, the studio legal department (not the law firm for the other party) will generally draw up the contractual documentation covering the deal. This can range from an exchange of faxes to an 80–120-page formal agreement. Not all agreements are embodied in formal, signed documents. Oral agreements are not unusual and generally are equally as binding as those in tangible form. Naturally if oral agreements are questioned, they are more difficult to prove. As an alternative, and especially if there is an impending start date, the studio business-affairs department will prepare and secure the execution of a binding *deal memo,* a short-form agreement covering those essential money elements critical to the deal. In this case long-form formal contracts are not signed, the picture goes into production and any unresolved issues become moot with the passage of time.

The length and complexity of agreements in the movie business reflect a lack of trust between the parties, compounded by corporate instability. New managements, looking to enhance their relationships with talent and quickly build a production slate, often push for shortened agreements that can be signed quickly. Then, when a problem arises that is not specifically covered (inevitable when using short-form documents), management insists on expanding the document to assure protection on that detail in the future. Thus, the short form slowly balloons into the long form and the cycle is complete.

Most desirable talent are involved in developing several projects at once. However, since agreements are generally entered into on a per-project basis, the talent in effect selects the next project by making a deal. And since no one can force someone to make a deal, it can be said that creative freedom begins with this ability to choose.

Deals involving personal services (for directors, stars or writers, as examples) are generally direct employment agreements, though some are treated as loan-out agreements for tax purposes. For example, a star might create a company that has exclusive control over his or her services. That company, as the star's nominal employer, loans out the star's services to the real employer, the studio, and whatever tax advantages there might be would flow to the loan-out company, not the studio.

Writers' agreements can be in a variety of forms: a screenplay written-for-hire to be owned by the employer; or the purchase of an idea; or securing the rights to a play, published book or existing speculative screenplay; or agreements for rewrites or polishes. There are also agreements to be reached with the producer, director, executive producer, lead actors, cast and technical personnel, such as the production manager and cinematographer.

In addition there are agreements relating to shared financing. Tax-shelter deals were the most prominent of these. Although they have been virtually eliminated in the United States, they can be a spur to movie production elsewhere, as in Canada. A significant replacement for them has been the limited-partnership arrangement, notably Disney's Silver Screen investment series, which placed many millions of dollars into the production of specific movies, recouped on a certain basis. These deals reduced cash flow on the part of the studio (which saved on the cost of money) and guaranteed the principal back to the individual investors after a number of years. This form of "use of money" deal has recently been supplanted by Touchwood Pacific Partners, with a recoupment basis even more favorable to Disney. Another concept is joint venturing between an American studio, a producer and overseas distributors. In a recent example, production financing was fully pledged by the producer and his overseas distributors/venturers, with studio risk limited to guaranteeing a major domestic release and distribution in certain overseas territories, and the balance of territories remaining with the overseas distributors/venturers.

Another basic type of agreement is the *pickup deal,* covering pictures usually financed outside the studio system. Under this type of deal, subject to certain requirements, the studio will agree to pick up the picture for distribution and pay an amount upon delivery of the negative. This is useful to the producer, who uses the contract as collateral to borrow production financing and is attractive to the studio because it precludes the risk of cost overages (since the pickup payment is a fixed amount) and the risk of noncompletion, since that becomes the burden of the completion guarantor. (For the workings of a completion guarantee, see the article by Norman Rudman, p. 216.)

To secure the most desirable talent today, studios have reverted to the long-term, multiple-picture arrangement, known as the *overall deal.* From the studio's point of view this brings a sense of optimism in that their release schedule will be highlighted by the proven talent. Also, there is the hope for a breakout success "locomotive" picture that will carry along their other titles, especially overseas, where entire schedules of product are often sold together in a form of block booking. From the point of view of the talent, that person has a home, with the security, overhead, office, staff and perks that that implies. On the other hand, such talent cannot auction their services for the duration of the deal; they are taken off the market. In such deals the studio generally gives top talent early participation in gross. These gross participations accrue earlier in the revenue stream than traditional net participations and involve a reduced distribution fee to the studio. Typically these format in one of two modes. In the first mode the talent receives a large cash payment as part of the negative cost, which is applied against, say, 10% of the gross from first dollar of film rental, escalating at higher levels of film rental. In the second mode the talent takes a reduced cash payment (less than market price) as part of the negative cost, which is applied against an enhanced percentage (say, 12½% or 15%) of gross from first dollar of film rental. There are numerous permutations—sliding scales, increments of achieved gross, accruals/deferments versus payments as earned—that can be factored into these kinds of deals.

The *PFD agreement*, short for production, financing and distribution agreement, usually contains all the elements of the individual agreements outlined above. What follows is a generalized tour through its basic elements for a sample picture, simplified for the reader being introduced to the area.

The first section of a PFD agreement generally deals with development. It states the current form in which the project exists: as an outline, treatment or other form. Then it identifies the role of the talent tied to it. The producer involved is named, along with the terms of the producer's agreement. If there are writers attached, they are named, along with their terms and responsibilities. Are they writing a new screenplay? Are they rewriting? There are other issues. Has the project been budgeted? Is the producer seeking a director? What is the financial commitment of the studio and what payments are required as the project progresses? On what basis is it deemed to be a "go" picture? All of the answers are spelled out in the contract.

Because the development process is based on a series of steps, development deals within the PFD structure are called *step deals*. At each step a financial expenditure is made. For instance, when there's an existing script, the rewrite could be considered step one. Committing to a director before a second rewrite could be step two. Preparing a budget and production schedule could be step three. This is all spelled out in the development section, with each step having a cutoff to give the financing entity various "outs."

During this process the PFD agreement generally requires a producer or director to oversee the rewriting of the script, help in preparing the budget and shooting schedule and offer casting suggestions. *Suggestions* is a word of art that contractually means the producer must assist the studio in securing talent ready, willing and able to perform required services in the proposed picture, usually at a certain price.

Fees for the producer are normally also tied to steps. For responsibility over a rewrite only, that's one fee. For taking a project from an idea through various screenplay drafts and budgets, the producer would be entitled to more money because more services are being performed over an extended period of time. Writers' deals follow a similar pattern. Producers work on different projects at the same time and are considered nonexclusive. Writers may also work on different development projects, but are generally contracted to finish one specific project exclusively before beginning another. This is changing, though, to reflect the reality that writers may spend time setting up their next project while working on another one.

When the development process is completed, the financier-distributor must decide whether the picture will go forward. The time frame of development usually ends when there are in place those elements that would allow for making a decision on whether or not to proceed: generally a developed script, budget, director, producer and major casting. The theory of a time frame gives the producer a trigger to force a yes-or-no decision from the studio.

If the answer is no, *turnaround* usually kicks in. Turnaround is the effort, after abandonment, to set a project up elsewhere, repay prior investments and get the picture made. Many hit pictures have been in turnaround in one form or another. *Star Wars* was abandoned by Universal; *Home Alone* was abandoned by Warners; studios turned down *Platoon, Batman* and *Dick Tracy.* Turnaround can be affected by the *changed-element* clause. The abandoning studio does so on the basis of a specific screenplay, budget and talent. If an element changes, there is usually an obligation on the part of the producer to return to the abandoning studio and offer it with the revised element. For example, if a project is developed at Columbia without a star, then moves to Paramount with Kevin Costner attached, Columbia should have another chance to commit to the package with Costner on board. In the turnaround area most people act in good faith because it's a round world; they will have to deal with an abandoning studio on a new project in the future.

If the studio decides to proceed to production, the PFD agreement states how, what and when key talent will be paid. Usually, this means that salaries begin weekly when the cameras turn, upon "the start of principal photography." However, strong directors or stars may have a *pay or play* deal, whereupon total salary is due whether the picture is made or not, as long as the talent is ready, willing and able to perform during a specific period. When a producer, director or other talent receives salary amounts prior to principal photography, that is usually the result of bargaining leverage.

The production section also contains other requirements and obligations. There is language stating a "producer's obligation to produce the picture," which is standard, and states the producer's role. There are paragraphs stipulating the picture must be made on locations approved by the studio; the running time; where the picture is to be delivered; the picture's technical requirements, including the MPAA rating; studio consultation requirements as to music, casting and other matters; and business and creative approvals during production.

Generally the studio will allow the most experienced filmmakers to arrange for their chosen crew to be engaged, subject to union requirements, EEOC hiring and other government regulations. But the studio normally retains the right to approve the production manager and to select the production auditor because they deal directly with the money.

Since there are a mass of contracts involved in production, legal fees must be negotiated. Some studios allow the producer's own lawyer to handle this, rather than doing it in-house. Among the most prominent entertainment-law firms who handle the bulk of the documentation on major pictures, there have evolved certain mutually agreed-upon standard forms of legal boilerplate for contracts that can apply to different echelons of talent. Often these forms, with their permutations, are easily slotted in during documentation, in the legal back-and-forth.

The MPAA rating is stipulated in the production section, usually as "no more severe than R," or "NC-17." (See article by Jack Valenti, p. 396.) The producer, director and studio generally agree as to the target rating, but can't guarantee the picture will be so rated by the ratings board. The PFD agreement spells out what happens if the ratings board gives a disapproved rating.

Another area is "television coverage," which also applies to airlines and some overseas markets. The filmmakers making the theatrical picture must plan reshooting or redubbing to cover language or visuals unacceptable in these formats. This must be spelled out in terms of scheduling so that the parties know what to expect.

The right of the studio to view dailies can be a touchy area. Some directors object to studio executives watching their dailies; others are happy to show them. This comes down to the players, their sensibilities, relationships and bargaining power.

The contract also generally covers where the film is to be processed and what lab is to be used. A filmmaker may be more comfortable with one laboratory over another. In the old studio days this was a major issue, sometimes because of laboratory ownership by the studio and sometimes because of "kickbacks" paid by the laboratory to the producers and studio in return for obtaining the business. Today the sole criterion is technical skill, and the studio generally allows the director and producer to go to whichever competent lab they elect, but at rates no higher than what the studio itself would pay.

Pictures financed by studios are required to comply with industry union agreements. Studios are signatories to dozens of union agreements and therefore impose these obligations, as to salary rates, work rules and benefits, upon the picture-makers.

Another area to specify in the production section of the agreement is the rights of the parties. When a producer is producing, a director is directing and a studio is financing, who is ultimately in charge? The contract must spell out whose decision is final. The sensitive area of "final cut" comes into play. In most PFD agreements, the studio's decision is final, since its money is at work, and money speaks loudest. However, in practice a studio will rarely if ever intrude on a director's finished work, since that risks bad publicity and alienating talent.

Determining the start date is another crucial contractual element, especially for those pictures scheduled for summer or Christmas release. Working backward, it's important to allow for enough time in the production schedule to meet the goal. Too often a major production targeted as a big seasonal release can be rushed into production, placing everyone under stress. But all of the personnel agree to the schedule in advance, sometimes unrealistically.

Overhead costs must also be dealt with, and there are two types. One is overhead for purposes of budgeting a picture and represents a cash outlay to financier/investors. The other, which presents no cash risk to financier/investors, is overhead as it relates to recoupment in paying the various talent who are percentage participants. Major studios no longer include overhead in budgets for their own productions; they budget actual cash costs. However, for purposes of payouts to participants, the studios generally charge a 15% overhead fee on the negative cost as an item of recoupment in determining the level of receipts necessary for such payments.

Next there are questions about contract breaches, which can be complicated. What happens if the director gets drunk, doesn't show up, doesn't perform? What if a producer has completed much work before the picture starts shooting, but dies or breaches or does whatever producers do before going farther? The outcome is not necessarily that the picture doesn't get made or that the producer loses everything, but there must be a negotiated counterweight to discourage any breach. Settlements or payment vesting (if someone has breached) should be agreed upon as part of the employment contracts referred to in the PFD agreement.

That's a review of the production section, the P of a sample PFD agreement.

The financing section of the PFD agreement is set out in a few paragraphs, describing exactly how production money is to be allocated. This is often stated simply, noting that there will be a production schedule, production board and budget, and whatever costs are based on those items are to be paid.

The question of securing an outside completion guarantee for a studio picture is arguably a great scam, generally used by insecure executives to fool their bosses that their investment is covered. (For independent productions, however, it is often essential insurance that is required by banks to loan production financing using distribution agreements as collateral.) If a studio is financing a picture, it must do so until completion. In the vast majority of cases the outside completion guarantor will never be permitted to make good on that guarantee, since no studio wants a picture completed by bond-company executives or bankers rather than filmmakers. The classic example involved *Meteor*, which went millions of dollars over budget. The completion guarantor was perfectly willing to step in and finish the picture, but none of the financiers involved would dare accept a picture finished by a banker. Finally they had to put up more of their own money to complete it. The guarantor got his fee and did nothing, and the investors ended up carrying the overbudget costs for the picture.

In the distribution section of the PFD agreement, the first important area is credit, which can take the longest to negotiate of any topic in dealmaking. The agreement stipulates the precise credit to be given the director, producer, writer (and book author) and actors, both on the screen and in paid advertising. Details involve size, color, type, order and relationship to other credits. When negotiating credits for paid advertising, lawyers and agents spend hours discussing whose name should go first and on which line, if anyone else's name should be placed on their line, or whose name is larger. The result is that half a print ad can be taken up with names many of which have no meaning at the box office. Advertising space costs money. While distributors do not object to crediting the creators of a movie, or those well-known talent who help sell a movie, they can object to trying to sell a picture using a newspaper ad wherein, by contract, a name or likeness must be placed in a specific size and location, regardless of how these requirements may reduce the aesthetics or impact of the ad.

We now turn to money in relation to distribution, and this is the heart of the PFD agreement. It is here that *gross receipts* are defined, which is a calculation representing the "picture pot," into which picture revenues flow and out of which percentage participants are paid. For example, cash received from the theatres—not the box-office gross, but the amount the theatre turns over to the distributor (or studio financing entity) after making its own deductions is included in this pot. This amount is also known as *film rental* or *distributor's share of gross receipts*. Also included is income from home video (see article by Reg Childs, p. 328), pay- and free television, merchandising (see article by Stanford Blum, p. 407), soundtracks, book publishing (see article by Roberta Kent and Joel Gotler, p. 113), and music publishing. From this pot the distributor takes a distribution fee for each format as his cost of selling the movie in the United States. Generally the fee is territorial, at 30% for U.S. revenues and 40%–45% for overseas revenues. Sometimes this fee can be negotiated downward if a producing entity brings financing or print and ad expenditures to the table. Certain media and territories have different fees, which relate to the cost of selling in particular fields. Distribution fees are the lifeblood of the production/financing/distribution business.

Some fees may appear unfair to one party or another; competitive factors also exert a ceiling. There may be different fees in different countries for the same format because it may be more difficult to accomplish the sale in one country than another. For example, the theatrical-distribution fee for the United Kingdom can be 35% and for the rest of the world 40% because it is easier to distribute in the U.K. than in, say, Germany or France, because the latter require dubbing or subtitling, which are distribution expenses.

Sometimes a studio can make a "profit" on its fees, particularly with a big hit picture. Warners made so much money on *Batman* that the single picture virtually carried the entire distribution organization, and *Home Alone* did the same for Twentieth Century Fox. However, as others have noted in this book, the feature film game is one of high stakes and much risk. Even for a major studio, if the production and worldwide marketing costs of any given year's slate of films exceeds that studio's worldwide film rentals for the year (and the per-picture average is around $35-million, excluding out-of-pocket development and overhead not appearing on screen), no distribution fee is earned and the studio is in a loss position.

After deducting distribution fees, the remaining money is applied toward recouping advertising expenses, the cost of prints (today a broad domestic release can cost $3-million or more just for prints), shipping, insurance, inspection and delivery. Next in the PFD agreement is the studio's right to market the picture in any reasonable way, in two theatres or two thousand theatres, or anything in between. The studio's ultimate control over advertising follows, often with an obligation to consult with the producer, director and/or stars on the amount of the advertising budget and the conceptual advertising campaigns, which must comply with credit requirements.

Recoupment of negative cost and interest on the negative cost are covered in the next part of the distribution section. *Negative cost* is the cost of physically making the picture, before prints and advertising charges are incurred. The next step in the revenue flow is the recovery of the picture loan or investment, plus interest, from what remains of film rental.

Notice that, up to this point, all of the effort is being applied to simply recoup money that has already been paid out. Once the picture recovers its costs, it reaches the rarefied realm of breakeven or "net profits." Then, using our model, these profits are split, usually 50% to the financing entity (the studio) and 50% to the participants. As previously noted, special talent in superior bargaining positions, such as a top actor or top director, might receive percentages of proceeds at stages defined earlier than others in the revenue stream. For example, major stars like Arnold Schwarzenegger, Kevin Costner or Tom Cruise are likely to receive money ahead of other participants.

The accounting portion is also included in the distribution section of the PFD agreement, covering such details as when statements are due; who checks accounting statements; how much detail is provided in them; what their frequency is; what the auditing rights are; what effect an audit has; how much time there is to audit; how much time there is to object.

What follows is a series of relatively standard provisions covering the ownership of the picture, the right of the financing entity to copyright it; the right to distribute the picture in all media worldwide; the right to settle or file lawsuits and the right to settle with exhibitors and other licensees. This group of provisions is intended to state that the financing studio is the owner of the picture, not the filmmakers, although the filmmakers are entitled to certain financial interests. Insurance requirements are also stated, and if there is an insurance loss, how it is applied. For example, a studio would not take a distribution fee on an insurance recovery, but someone could suggest it and the producer could be naive enough to agree.

Another section covers theatrical sequel, remake and television-production rights. Normally people involved in a successful picture should have first refusal to do a remake, sequel or television production based on that picture. If they decline, there is some argument as to whether they are entitled to money from the derivative work for the act of making the original, even though the derivative work has its own creators and filmmakers. This is especially important today, with the multitude of sequels being made. After all, they arguably reduce risk, since the audience has approved of the prior, successful picture. Sequel, remake and TV rights, and the obligations of the studio and picture-makers to one another in these areas, can be intensely fought-over issues.

In reviewing the documentation for a picture, of course, all of the related agreements, such as those for the producer, director, stars and other key creative people, must mesh with the PFD agreement and its details. When the negotiations and documentation are finalized, the work of business affairs is completed. Once the picture is released, we can only hope we've protected the financier-distributor so that, if it's successful, there is enough revenue to satisfy the investors and to make more movies.

# THE FINISHING TOUCH: THE COMPLETION GUARANTEE

*by* ***NORMAN G. RUDMAN****, a member of the firm of Slaff, Mosk & Rudman, who has practiced law in Los Angeles since 1956. He was educated at UCLA and the Boalt Hall Law School (U.C. Berkeley). He has written and lectured on a variety of subjects including constitutional law, bankruptcy, divorce and the development, financing and production of feature films. Involved as counsel in many motion picture productions, he has also served as executive producer of several films.*

*... it can fairly be said that the cost of guaranteeing completion is the only cost of producing a movie that has actually shrunk, at least when measured as a percentage of budget, since the last edition of this book.*

For producers and their investors, the motion picture business is a risky, highly speculative business. The investors take risks on many factors: on the creative capabilities of the writer, producer, director and actors and on the ability to secure distribution and compete in the marketplace, to name a few of the more obvious. But the single risk that any investor will find intolerable is the risk that a picture will not be completed. In the production of any motion picture, whether a studio project or an independent venture, the need is always present to assure that what is started will be finished. That is where the issue of "completion" enters into the structuring of a motion picture deal.

If a studio-financed picture goes over budget, for whatever cause, there is reasonable certainty that the studio has the financial strength to cover the extra costs. For undertaking the risk of overbudget costs, the studio may add to the budget or retain a contractual right to recoup a cost item in the area of 5%–6%. Further, the studio may invoke contractual penalties against the producer and/or director for any excess costs unless such costs result from studio-approved enhancement (that is, content not found in the approved screenplay prior to shooting).

Assume a studio picture budgeted at $20-million. In addition to the producer's fee and overhead allowance, the producer's deal with the studio would typically give him a share of the net profits of the picture, say 50%, reducible for profits paid to other talent to a floor of perhaps 20% of 100% as defined in the studio's standard PFD (production, financing and distribution) agreement. (For details of a sample PFD agreement, see article by Stephen M. Kravit, p. 204.) Assume this is a generous studio that allows the producer a 10% contingency over the going-in budget so that he is not penalized if the picture comes in at a negative cost of up to $22-million. What happens if the cost reaches $23-million? One of the consequences may be that the producer's profits are, by contract, delayed while the studio recoups not only the $23-million (plus, of course, all distribution expenses, interest and distribution fees) but also double the overbudget costs, or an additional $1-million, before profits are deemed to have been reached. Or the producer may be required to agree that part of his fee be payable on a deferred basis and that, to the extent he goes over budget, such deferment is paid to the studio to cover the excess costs, rather than to him. Or the studio may take profit points away from the producer: the deal that gave him a floor of 20% of 100% of profits may begin to shrink on a formula whereby he may forfeit a point of profits for every, say, $200,000 of overbudget costs.

The specifics will, of course, depend on the producer's bargaining power vis-à-vis the studio, and what has been said with regard to the producer's deal might, in certain cases, apply to the director's as well. The point is that the studio will take the budget seriously; hence, if the costs run over, the studio is likely to believe either that the budget was false to begin with or that the producer did not manage the production competently. In either event it will likely want to spread the burdens of the extra costs to those most directly responsible.

In independent production the need for completion protection can be fulfilled in a variety of ways. A producer may be capable of meeting excess costs out of his own pocket; the deal with investors may permit an overcall; a standby commitment may be in place for overbudget financing on certain terms; or a producer may deal with a company whose business it is to provide a completion guarantee. Indeed, of late it has become a familiar practice even for studios to use outside completion guarantees.

A producer who is strong enough financially to sign his own guarantee of completion is generally strong enough on the line to assure that the picture comes in without invading his personal resources. This type of completion assurance can be as simple as the producer informally assuring his investors that he will complete the picture.

A second form of completion assurance is simply an overcall from investors. For example, if investors are organized into a limited partnership, the partnership agreement may permit going back to the investors for an extra 10% or 20% of their investment in order to meet overbudget costs.

A third approach might be a standby commitment to invest overbudget costs as called for. Assume an independent picture is budgeted at $8-million. The standby investor, for a negotiated cash fee or other consideration, might commit to provide an additional amount up to, say, $2-million. If called upon to put up any of that money, he might be entitled to take over production. (The completion guarantor, the fourth general category of completion assurance, is also likely to reserve takeover rights.) The principal distinction between the standby investor and the completion guarantor is that the standby investor will normally obtain a recoupment position prior to or at least equal to the people who put up the $8-million for the principal budget, and a profits interest in the picture besides. The profits interest is likely to be calculated at a better rate (say, double) than what the original investors would receive. (That profits interest would normally come out of the producer's share rather than the financiers' share of the profits.) If the original investors were receiving 50% of the profits for their $8-million, each would have one percentage point of profit for every $160,000 of money invested. If an investor put up $400,000 he would be entitled to 2½% of the net profits of the picture. But let's assume that $400,000 of the standby investor's money is used. On the standby deal hypothesized above, he would be entitled to five points of profit for that $400,000, but out of the producer's end, not that allocated to the principal financiers.

I distinguish between a standby investment and a completion guarantee because the standby investment dilutes or postpones the recoupment position of the original investors while the completion guarantee does not. Further, the completion guarantee as it has developed does not require the producer to give up any profits for the use of the guarantor's money. On the other hand, the standby investor may be willing to pay for the costs of enhancing a picture. The completion guarantor will not. The standby-investor format might provide acceptable completion assurance to equity investors. It would not satisfy a bank if a bank were putting up any of the production money, because the bank would not accept any position other than first recoupment. For the same reason, it would not satisfy most distributors putting up production money in the form of negative pickups of distribution licenses.

As another example, assume it actually cost $9-million to make the picture budgeted at $8-million, and that the standby investor advanced the $1-million overbudget costs. Assume further that the distributor has collected film rentals that, after deduction of distribution fees, print and advertising costs and other distribution expenses, leave $2-million of "producer's share" to be devoted to recoupment of production costs. If the picture had been brought in on budget, or as is explained below, if the overbudget costs had been paid for under a typical completion guarantee, that $2-million would be divided pro rata among the investors who funded the budget. Each would receive roughly one-fourth of his money at this point. But if a standby investor put up $1-million (because the picture actually cost $9-million to complete), under a deal calling for his recoupment prior to the original investors, the first $1-million of that $2-million goes to him. That leaves only $1-million to be divided among the original investors of $8-million, who would get back roughly 12½% instead of 25%. Similarly, if the standby investor's deal calls for recoupment pro rata with the principal investors, then the $2-million recoupment is divided among $9-million of investors, making a return of roughly 22% of investment. Thus the standby investment commitment is not a completion guarantee because *the essence of a completion guarantee is that the investors will get a finished picture for the budget amount they have financed.* If their recoupment is subordinated by the prior, or diluted by the concurrent (even if pro rata), recoupment of a standby investor, they have not had the benefit of their original bargain in that respect.

My fourth category of completion assurance is the only one properly entitled, in my view, to be called a completion guarantee. (It is common in the industry to refer to such a guarantee as a bond, and indeed, several companies in the completion-guarantee business use the word *bond* as part of their corporate names. For reasons of legal semantics this writer believes the word *guarantee* a more accurate description of the undertaking.)

The typical completion guarantee is a three-sided transaction among bank, producer and guarantor. The guarantor guarantees to the bank the producer's performance of the conditions of the loan agreement. It is not uncommon that completion guarantees are issued in two-sided (producer-guarantor) transactions, where the production financing comes not from bank loans but from equity investment or direct studio financing that is circumscribed by direct studio involvement in production. Here the guarantor may in effect be guaranteeing the performance of the producer to the producer himself or to a party so closely allied with the producer as to be chargeable with the producer's actions. The differences between three- and two-sided transactions are most clearly seen in the framing of issues in disputes over whether a guarantor is responsible for an overbudget cost.

To illustrate how a three-sided completion guarantee works, let's turn to an example of an independent picture budgeted at $10-million. To raise that money, assume the producer makes a pickup deal with an American distributor for theatrical, pay-TV and free-TV rights in the United States and Canada for $5-million, payable on delivery of the picture. A foreign-sales representative may sell distribution rights in twenty-five territories in the world and come back with as many contracts, some for $5,000, some for $100,000, making a total package of $3-million, all payable on delivery. And a video distributor might commit the remaining $2-million. Ignoring, for ease of illustration, such complicating factors as interest, discounts, fees, commissions, compensating balances and the like, these numbers round out the $10-million budget. With the exception of relatively nominal down payments, these contracts will usually be payable not on signing but only on or after the delivery of the specific picture. The contracts will not be collectible if the delivered picture does not meet specified conditions, such as that it must be based on a specific screenplay, directed by a certain director, have specific stars and running time, meet first-class technical requirements and be delivered by a specific outside date. The producer takes the contracts to a bank that deals in motion picture transactions and "discounts" them, that is, tenders them as security and the source of repayment for the $10-million production loan he needs.

What is the bank going to require in order to make the $10-million loan? It would be an unusual producer whose credit was good enough to support such a loan. All the bank can look to for payment is the fulfillment of the conditions set forth in the various (often numerous) distribution contracts. If it is assured that those conditions will be met, it may make the production loan. Furnishing such assurance is the function of the completion guarantee. Naturally it is the bank's burden to satisfy itself as to the creditworthiness of the paper against which it is lending. But the delivery conditions that must be fulfilled to make notes collectible, letters of credit draftable or contracts enforceable are what the completion guarantor guarantees.

As earlier observed, banks will make loans against distribution contracts only if they, and only they, have recourse to those contracts until their loans are repaid in full. For this reason the completion assurances of a so-called standby investor will not be acceptable to a bank if he has a recoupment position ahead of or even equal to the bank's. Professional completion guarantors, such as Film Finances, Inc., or the Completion Bond Company, have accepted this fixed truth. They offer a standard completion guarantee format involving subordinated recoupment that will very likely be acceptable, depending of course, on the bank's assessment of the specifics of the guarantor's contract and financial responsibility. Professional completion guarantors have learned to live with having recoupments of their advances, if any, subordinated to the bank or distributor recoupments or, if need be, to entire budgets.

The producer in search of a completion guarantee has several markets to explore. The guarantors are highly competitive, so much so that it can fairly be said that the cost of guaranteeing completion is the only cost of producing a movie that has actually shrunk, at least when measured as a percentage of budget, since the last edition of this book.

Historically, when the completion-guarantee business began and guarantors were conceived of as standby investors, the typical cost of a bond was 5% of direct-cost budget. When banks began to play a larger role in financing films, as described above, guarantors were compelled to accept the enlarged risks of subordinated recoupment. As a consequence, their fees increased to an average of 6% of budgeted direct costs. Quietly producers began to rebel, especially those producers of demonstrated competence (and good fortune) whose films usually finished on schedule and within budget. Guarantors then devised the "no-claim bonus" as an inducement to keep those producers coming back. Originating as a modest credit for a successful production that the producer could utilize against the cost of the guarantee on his next project, the "no-claim bonus" evolved over time into a refund of 50% of the guarantee fee paid on each successfully produced picture. The refund could be collected immediately upon timely delivery of the picture and release of any right to call on the guarantor for funds.

The current state of affairs is that guarantees are priced in the neighborhood of 2.5%–3% of a picture's budgeted direct costs, subject, however, to the condition that the guarantor will not be obligated actually to infuse funds into the production unless the producer first pays an additional amount, usually one calculated when combined with the initial fee to equal about 6% of budget. This formula works to deter producers from calling on the guarantor for production money in the absence of real need. The producer of a $10-million picture who has paid only $300,000 for the comfort of a completion guarantee to satisfy his bank is not likely to pay the guarantor another $300,000 in order to require the guarantor to cover his budget overruns unless he has no choice. If by cost-cutting measures, fresh capital, deferring supervisory fees and applying that extra $300,000 to production costs—instead of paying it to the completion guarantor—the producer can avoid calling on the guarantor for money, he is most likely to do so.

Of late, a new permutation of the completion-guarantee pricing formula has arisen. Over the years most, if not all, professional guarantors have relied on financial backing from insurance companies. Some, indeed, have been so closely allied with insurers, underwriting companies, general agents or brokerages as to be hard to distinguish from subsidiaries. Of course, every production needs a wide range of insurance coverages. The typical producer's package policy includes such coverages as cast, negative and faulty stock, workers compensation, liability and equipment; extraordinary conditions may suggest such coverages as foreign business interruption or adverse weather conditions; errors and omissions coverage will be a *sine qua non* for distribution and often for financing. Often this package is offered by the same insurance interests who have backed the guarantors. It is thus a natural development that completion guarantors have also become markets for the procurement of insurance coverage combined with completion guarantees. Thus far this development has generally been financially beneficial for producers. Combined rates have been less than the total of separately purchased completion guarantees and insurance. So long as the completion-guarantee business remains competitive, that condition should persist.

When a producer approaches a guarantor for a bond, the guarantor will commence an inquiry to satisfy itself that the picture qualifies and to determine the price it will quote. The inquiry begins with an examination of the screenplay, budget, shooting schedule, rights documents, principal employment and facilities agreements, and any and all supporting materials that bear upon the budget, production schedule and delivery requirements.

To a completion guarantor a screenplay is a legal document, analogous to what the plans and specifications for a building are to a construction firm. The guarantor regards the screenplay not merely as a story to be told on film but as the definitive description of the qualities and characteristics to be embodied in that film, including specific action, sets, props, locations, costumes and effects, among other elements. Like the screenplay, the budget, schedule, production board and delivery schedule are of vital importance. The guarantor will be concerned with the whole panoply of issues that may affect the bottom-line question whether the picture can be completed and delivered at the price, on the schedule and to meet the requirements of the distribution agreements being financed. The inquiry may range from the obvious (Is the cast budget large enough to pay the prices of the stars promised to the distributors?) to more subtle matters. For example, a guarantor once deemed it necessary to decline to guarantee a picture scheduled for exterior shooting on a school campus during summer vacations because the school refused to allow shooting to continue past opening of the fall term and weather-service records suggested that rain would probably delay shooting long enough to cast doubt on the producer's ability to finish filming at that location before the stop date. Few problems that emerge from such an inquiry require so drastic a response. Most lend themselves to cure by the negotiation of adjustments in budgeting, scheduling or staffing. If the distribution agreement calls for delivery of an interpositive and internegative, but the budget allowances for film and lab are too small, the solution is to find the extra money somewhere else. If the star has another commitment to follow this film, and therefore a stop date, reboarding to schedule his scenes to be shot with ample cushion for delays may cure the concern that the stop date might arrive before his services are completed. In addition to reviewing the documents, the guarantor in "vetting" the project may well want to meet with the producer, director, cinematographer or production manager to thrash out anticipated production problems and proposed solutions.

The guarantor's objective is to bring the picture in within the budget and on schedule, thereby fulfilling the delivery requirements of the various distribution contracts. Therefore it is the specific proposed picture disclosed by the screenplay and production plan that becomes the subject of the guarantee, just as it is the subject of the distribution contracts and the production loan. The difference between the guarantor's objectives and those of the bank and the distributors in this respect is that the latter normally have no responsibility for costs beyond their loan or pickup commitments. They can hardly be expected to object if the picture's cost zooms far beyond budget. The guarantor, however, has such responsibility if the cost of the specific guaranteed picture escalates. He will, therefore, resist being charged for cost escalations that are voluntarily or incompetently incurred by the producer. Changes in script, new locations, unscripted effects—all such things in the nature of enhancement of the picture will be outside the guarantee.

In reviewing a proposed picture to be guaranteed, the guarantor will want, among other requirements, to see the budget *proved.* He will be suspicious of round figures and will find the word *allowance* in a budget intolerable. The budget will be closely analyzed in relation to the screenplay, the shooting schedule and the professionalism of the principal creative, production and business personnel.

The completion guarantor will have to be satisfied with the director, cinematographer, production manager, location manager, production designer and stars, among others. Does the cinematographer take too long to light every shot? Does the director show up on the set without a shot list so that there is down time between setups? If the same person is going to star in and direct the picture and the director must be replaced during shooting, will the star show up on the set? Is the director married to the star?

Another issue the guarantor is wary of is a "sweetheart deal" whereby the producer is acquiring some essential service (such as postproduction facilities) on a better-than-prevailing-rate basis. If the "sweetheart" nature of the deal is what brings the budget down, then the completion-guarantee fee may be reduced; but the picture is almost certain to go over budget if the sweetheart deal falls out. All of these issues and more must be resolved to the satisfaction of the completion guarantor before the guarantee will be issued.

Aside from the fee, what does the producer "give up" to the guarantor? Guarantors normally require that they be vested with certain powers to police and oversee the course of production in order to keep it to schedule and budget. Included among such powers may be the right to cosign production-account checks; to observe on the set during shooting; to screen all the rushes; to sit in on all production meetings; to review the books; to receive on a timely basis copies of all production reports, such as camera, sound, production manager, and script supervisor reports; to demand and receive answers to questions, and to become coinsured under all the production's insurance. This way the completion guarantor can review the financial status of the picture on a daily or weekly basis, in partnership with the production company, to anticipate and correct potential overbudget incidents. In case of trouble, the completion guarantor can exercise the right to take over the production and, if need be, to fire personnel whose performance is below par (subject, of course, to the requirements of distribution, collective bargaining, and other relevant agreements). Despite the competitiveness of the completion-guarantee business, it has been next to impossible for any producer to convince a prospective guarantor to forgo any of these rights and powers during the negotiation process. Yet, though the guarantee documentation will almost inevitably contain all of these rights and powers, a guarantor once vested with them is unlikely to throw its weight around. It is only in that most rare case of extreme and otherwise unavoidable jeopardy that a guarantor is likely to exercise its takeover right, and so long as the producer, director and their cast and crew are responding to the exigencies of production with diligence and competence, the guarantor will content itself with keeping a watchful eye on things and offering counsel and assistance in solving problems as they arise. Most have ample qualifications and experience to bring to bear in doing so, and most prefer to be regarded as a resource to be called upon in bringing to fruition the picture the producer and director want to make and the distributor wants to distribute rather than as an assassin lurking in the shadows.

The typical completion-guarantee transaction contains many documents, including a complex one between the producer and the completion guarantor that details the completion guarantor's rights. The guarantee itself is likely to be a simple and straightforward document in favor of the bank, whereby the guarantor undertakes to assure completion of the picture. The guarantee may also provide that, in the event the guarantor is impelled to declare a production hopeless and abandoned, he will simply repay the bank its production loan. Abandonment is rarely invoked because it is simply too expensive.

Essentially the guarantee has both time and money aspects. Not only does the guarantor undertake to assure delivery of a specifically described film, the guarantor undertakes to deliver it by a prescribed date. Distributors who have committed to take the picture will ordinarily have conditioned their commitments on delivery by a specific date, because they will have begun the process of booking theatres and preparing their advertising and publicity campaigns while production proceeds. Moreover, production is a labor-intensive activity, hence the more time it takes, the more money it costs. It follows that an understanding of completion guarantees requires an understanding of why pictures sometimes run over budget. The reasons may vary:

- *Some budget elements may have been underestimated.* Although guarantors prefer to deal with budgets priced out in concrete terms based on actual deals or on generally prevailing rates, some budget categories are often not contracted until well after the guarantee has been issued. In the interim, prices may rise. This eventuality might affect the costs of services or facilities needed in postproduction or even during shooting. So simple an item as the cost of money may be a variable. Assume a $10-million budget and a picture to be shot on a foreign location. Assume that 40% of the picture's budget is to be expended in the local currency, for local cast, crew, facilities, housing and the like. If the producer has not purchased all of the local currency he expects to use at the exchange rates prevailing when the budget is approved and, later, those rates change 10% to the detriment of the dollar, the cost of the picture may rise several hundred thousand dollars. Yet, buying the foreign currency in advance might cost the producer the ability to save money if the value of the dollar rises. A completion guarantor may be expected to seek protection against such eventualities either by requiring the producer to make protective deals early or, if such exclusion does not cause the bank to object, by excluding such variations from its risk.

- *Unforeseen events may intrude.* Adverse weather or labor difficulties may crop up. Suppose the picture was budgeted to be shot nonunion. Even if the wage and salary scales were at union minimums or, perhaps, better, the producer did not budget to pay union overtime, fringes, vacation pay or penalties or to hire the full complement of people a union contract might require. But the producer hired certain key people who happen to be members of the union and, upon learning of the production, the union has told those people they may not work for a nonsignatory employer. The producer may find himself facing a Hobson's choice, between delaying production while he tries to hire replacements and negotiating a hasty adherence to the union's collective-bargaining agreement. The latter choice might increase the cost 30% or so in the affected labor categories; the former choice will also be expensive, though it is impossible to predict how much it will cost.
- *The pace of shooting may be slower than expected.* It has been said, perhaps apocryphally, that on the first day of principal photography a picture will usually be a week behind schedule. Inevitably, as the time required for shooting increases, so, too, does the negative cost. While some flat deals may be made, or "free" periods negotiated in connection with some engagements, most cast and crew hirings are done on a daily or weekly rate, and most stages, locations, facilities and equipment are rented on a time basis as well. The causes for delay are so many and varied

as to be impossible to exhaust or even classify. Temperamental stars, obsessed directors, poky camera crews, feuds between key personnel and a multitude of other such phenomena are all within experience. Although no completion guarantee was involved, one of the most celebrated of the industry's experiences with a production that could not be kept within bounds monetarily or temporally is chronicled in Steven Bach's book *Final Cut*. It details a sobering account of Murphy's law enforced with a vengeance. Other tales of similar woes abound. Perhaps it is human nature to dismiss from mind the overwhelming number of films that have been routinely and economically brought in by application of the competence and professionalism of their makers. Only the catastrophes grow legendary. We remember the *Titanic,* not all of the ships that did not collide with icebergs. Nevertheless the icebergs lurk in the shooting of every picture, and measures must be taken to avoid launching a new filmic *Titanic*. For example, in a picture familiar to me, one scene required forty-five takes. The scene, set in a pool hall over a table, required dialogue between the female and male leads while he made a difficult trick shot. Because the male lead was rather accomplished with a cue stick, the scene was not to be cheated with inserts. In the early takes he made the shot consistently, but she had trouble with her lines, sometimes misspeaking, sometimes stepping on his lines. After a break the dialogue went fine, but he was tired and kept missing the shot. The day's shooting extended to over twelve hours, and at the end of it only a fraction of the day's scheduled pages were in the can. That loss of time in itself posed no insurmountable problem. Had similar losses, for similar or different reasons, occurred again and again, however, the obvious cumulative impact might have been difficult to bear.

- *Illness, accidents and* force majeure *may intervene*. Cast insurance on most pictures covers only a few people. Unless special provisions are made and extra premiums paid, the cast insurance policy will cover only the top three or four stars and the director and producer. The theory is to cover those people whose performances on the film would be extraordinarily expensive to replicate or whose unavailability would paralyze the production, not merely delay it some. The completion guarantor bears the more comprehensive risk of delays caused by illness or injury to people not covered by cast insurance. Moreover, cast insurers will sometimes exclude from the risks they assume any losses that may result from the age of elderly actors, or known health defects that may afflict a person covered, or known insalubrious habits in such a person's history. If despite the cast insurer's warning the producer hires an elderly actor or one with a heart condition or one with a history of substance abuse, and the completion guarantor approves, the cast insurer will not bear or share in the extra cost if the risk posed by that condition becomes reality.

*Force majeure* events, such as war, civil disturbance, labor dispute, fire, flood and the like, are both impossible to predict and costly. Most distribution agreements will allow a modest and limited extension of the delivery date for an event of *force majeure,* hence it may be possible to collect on the distributor's cash commitment despite some tardiness in delivery if the delay was the result of such an event. But even where this is so, the additional costs incident to an event of *force majeure* are the guarantor's concern.

- *Enhancement.* It is not uncommon in the production of films that what is scripted is not shot and what is shot was not originally scripted. Although the script, budget, shooting schedule and other documents evidencing the production plan were supposed to be final when the guarantee was issued, subject only to minor variations responsive to the exigencies of production, a screenwriter may be kept on tap during the shoot. Ostensibly, when this happens, the purpose is to keep the dialogue fresh. Dialogue is inexpensive. What often happens, however, is that shooting discloses weaknesses in the plot line or characterizations, and some effort must be made to strengthen the weak spots. Hence one may find new scenes written, new characters introduced, new locations, sets, props, and action. These are not inexpensive. The producer may believe them essential to the making of a picture as good as he intended at the outset. In the eyes of the guarantor, however, the producer is welcome to them, but at his own cost, unless the producer can save the equivalent amounts in other areas.

New writing is only illustrative of the ways in which the scope of a production may be enlarged from what was guaranteed. Others include expansion of stunts or special effects, augmentation of the initially conceived music score (what director can resist the lure of a sound track crammed full of the sounds of hit records of the period being filmed?) or adding stars in cameo roles to the cast.

There are certain areas of controversy characteristic of completion guarantees. The most common issue in the area of claims is the difference between a legitimate overbudget cost, part of making the picture as contemplated and guaranteed, versus enhancement. In the three-sided transaction the guarantor usually has no choice but to finish the picture. If he has failed to police the production, and the producer has thereby succeeded in raising the cost of the picture by enhancing it, the guarantor is still obligated to the bank to make delivery notwithstanding overbudget costs. He may have to advance these costs and then bring an action against the producer to collect that portion of the extra costs occasioned by enhancement rather than by legitimate overbudget contingencies. In a two-sided transaction the issue may also arise in the context of a lawsuit by the producer (or a closely allied financier) against the guarantor, wherein the guarantor raises the defense that the costs in question are enhancement costs rather than legitimate overbudget costs.

Another source of dispute may be diversion of a picture's budgeted production funds to some purpose other than payment of the production's expenses. The possibilities are limitless. They include the charging of travel to the budget where the travel is actually related to another project or a personal frolic of the producer; concealment of personal debt repayment in checks to vendors ostensibly for materials or equipment for the production; and purchase of long-term capital assets out of the budget rather than renting them for the limited period of production. While the completion guarantor's watchword is always vigilance, such occurrences will have different ramifications in the two-sided as against the three-sided transaction. If, in the two-sided transaction, someone on the production team has succeeded in diverting funds from the production budget to some other purpose, resulting in a need for additional cash for completion, the guarantor may simply refuse to put it up. The guarantor could rely for a defense on the producer's own default by failing to devote the production budget solely and exclusively to the production of the picture. But this wouldn't be so if the guarantor's obligation is owed to a bank. Here the guarantor still must finish and deliver the picture and look to his own devices to attempt to retrieve any diverted funds from the producer.

The remarkable fact, considering the complexity and scope of feature film production, is that there are few such disputes. The guarantees of the established guarantors are rather routinely accepted, relied upon, and performed, and guarantors make payment of substantially more money to producers or investors in the form of "no-claim bonuses" or rebates than on overbudget claims.

The completion guarantee, as developed in the industry, is only one of several possible ways of assuring that the picture will be finished. It is probably the most satisfactory approach for independently financed pictures. The concept of the completion guarantee is very much in the minds of bankers considering entrance into motion picture work, securities brokers as they review movie financing through public issues and investment counselors as they give advice on private placements. Since even the most competently prepared budget is at best only an estimate of what it will take to bring a picture in, pictures do go over budget, notwithstanding the best intentions and efforts of all involved. For that exigency, the completion guarantee provides a significant measure of protection.

# THE TALENT AGENT

*by* ***BARRY WEITZ,*** *who joined the William Morris Agency after graduating from New York University in 1962. He was transferred to the Beverly Hills office of the agency and worked as a member of the motion picture department there until 1971, when he became an independent producer. His producing credits include the feature* The Seven Ups, *television series such as* Movin' On, *and television movies such as* Strike Force, Reunion, Time Bomb, Riviera, Ladykillers, Dead Reckoning *and* The Bride in Black.

*There is the danger of an agent becoming so deal-oriented that he or she may be more concerned about making a deal . . . than about the client's career. . . .*

The strength of a talented individual lies in his ability to continue to create. Whether it be a performance, a book, or a screenplay, creating is where the energies are. Often creators do not want to cope with business decisions or career guidance. An individual who must sell himself or negotiate terms in his own behalf might well be less successful than if another individual were taking care of these matters for him. Therefore the commission that is paid to an agent—10% or 15%—is often the best investment an individual can make in his career. An agent generally receives 10% of the gross fee that an actor receives. For instance, if an actor is to receive $100,000 for a role, the agent will receive 10% of that fee, or $10,000.

In the triangle of buyer, seller and agent, the agent performs two important functions. First, he sets up a very clear delineation between the buyer and the seller; second, he gives the buyer a professional person to deal with on a consistent basis and in a business atmosphere. Thus a freedom from emotion is established, allowing the parties to cut to the core of the negotiation and agree upon what each party considers essential. The presence of an agent is certainly a great aid to the buyer, since he does not have to become too closely involved with the artist during the negotiations. The buyer can deal with the artist on a creative level and with the agent on a purely business and career level. For these reasons the talent agency is a very real part of the entertainment business.

There are two ways for a talent agent to obtain clients. He can, through his connections or aggressiveness, bring the individual client to the agency on his own. He may also be assigned a specific client because of his experience, seniority, or a personality that complements the client's. Because human lives and livelihoods are so closely involved, the goal is to find the right kind of combination between agent and client. It is most important that the client be satisfied with his specific link to the business world, the individual who must communicate his thoughts and aspirations to the marketplace.

When an agency takes on a new client, he is required to sign an authorization contract. The contract calls for the agency to represent him in all areas for a stated period of years. Depending on the stature of the client, the agent will plan different approaches to potential buyers. If he is representing a relatively young, less-experienced actor, the agent may feel it necessary to have him meet casting directors and other individuals who are active on a daily basis in casting film and television roles. The agent has a fairly good idea of what his new client's abilities are, so he may decide to move him into the area of dramatic television for the purpose of getting a certain type of role that will show a young actor in a certain light. Of course, he may make the same decision merely in order to get the actor to work immediately. An agent may feel that it is best to move into TV at the outset, assuming that if the venture is successful, he and his client will be provided with the type of sample reel that can be shown to producers and directors in both the feature film and television areas.

Dealing with a new, young client requires a certain experience. The agent must take maximum advantage of the client's talents and expose them effectively in a fair atmosphere with a proper sense of direction. The agent's plan should be to move the actor along to various important roles while increasing his salary so that he is making steady advances in both areas. For instance, in television if an actor earns $1,500 a week in a supporting role for which he achieves proper acclaim and notice, the agent may then be able to move him farther along, perhaps to a costarring role where he can make as much as $35,000. If this level is successfully mastered, the agent may well have a young talent who will, as few actors can, move into major starring roles. Either because of a lack of opportunity, improper combination of role and actor, or various other indefinables, these ideal situations may never present themselves. In any event, assuming success along the line, the agent may have an actor who can move from $1,500 a week up to a very high six-figure category that also involves a percentage of the profits or a percentage of the gross on a given film.

An agent must be very cautious with an actor, however, and move properly to keep him working at good roles. Continuity of employment is an extremely important aspect of client development. The creative lives of many performers are generally not as long as those of other professionals, such as lawyers or doctors. The best earning years of their careers may be very few, and certainly during that time they have every right to try to make the most of their talent. The agent has the obligation to try to extend their creative life.

There is the danger of an agent becoming so deal-oriented that he or she may be more concerned about making a deal (and collecting a commission for the agency) than about the client's career. It's sometimes tempting to make, for example, a long-term television series deal for an unknown client that will effectively lock him up for years. The talent's only recourse is to renegotiate for higher salaries each year.

In an actor's career there may be a time when an agent has to decide whether to pursue employment or hold back from it. For instance, an actor may have completed a feature film in which his performance has a great impact. However, the picture is in the can and will not be released for six months. If he is immediately cast in a subsequent picture, his salary will be higher than for his yet-to-be released picture but will not reflect the heat and attention that will result when that picture is released. Should the agent make the deal or wait until the release of the big picture, with the offers and resulting bargaining power that are sure to come? One alternative is for the actor not to take the new deal unless the salary reflects what the agent believes he will be worth after the big picture opens. This kind of decision must be made on an individual basis, as part of the ongoing trust between client and agent.

Contract negotiations are probably the most time-consuming as well as important aspect of an agent's workday. Some agents like to think that there are certain secret methods that enhance their ability to negotiate, but there is no standardized formula or checklist that can be relied upon. Each negotiation is individual and unique. The aim of all negotiations is, of course, to secure the best possible deal for the client. Every agent, I am sure, has walked out of a negotiation assuming he has secured the absolute last penny only to discover later that the buyer was prepared to offer more than was finally paid.

There are times in an agent's life when he is working with an actor who is "hot"—one who is in demand throughout the industry because his popularity in earlier films indicates that his presence in a film will ensure a degree of success from the outset. The agent's responsibility in this situation is to exercise great care in analysis of screenplays that are being submitted to him. He must choose scripts that will have an important influence in the film business in the next year or so, thus exposing his client in a well-rounded fashion that will ensure greater career longevity. In these moments of great demand, the agent must take care to resist any tendency to overplay his position. Work and creative contribution are still the most important aspects in the performer's life.

If a client is not an "in-demand" actor, the agent must go out and strongly hustle for his sales. If the talent is there, and if various reviews have not overlooked that talent, then the agent has some sort of "hook" to use in selling his client. If, however, the performer has little motion picture ability—if he had the opportunity but has not been able to prove himself—then the agent must reevaluate his client's career. There are many clients, for example, who have not been able to succeed in features, but who have starred in their own television series. There are also clients who have not been able to succeed in either motion pictures or television but have become major stage performers. A great deal depends upon the talent finding his own medium. The agent certainly should assist in the search and give proper guidance to a client in this endeavor.

If, for any reason, the client should not be satisfied with the progress he is making in the agency, or if he is simply not happy with the way he is being represented, he can discharge the agency. The unions and guilds generally allow a certain number of days after he signs a contract, or after he has received his last offer, in which he is able to terminate his agency relationship. The client is always able to discharge his agency, but if he is not able to show cause, then he is obligated to pay any commission due the agency until the termination of the agreement.

Packaging has become an important part of the talent agent's work. Agencies such as International Creative Management (ICM), William Morris, Creative Artists (CAA), and Triad will assemble feature packages, usually consisting of the material, the writer, the director, perhaps the producer and perhaps the star, and present them to a financing source. In television, series are often created in such packages by agencies who receive a packaging fee in return.

The concept of stars wanting a degree of ownership and creative control over their product really goes back to the formation of United Artists in 1919, founded by Douglas Fairbanks, Mary Pickford, Charlie Chaplin and D. W. Griffith. Today stars are putting together their own packages, using material that is submitted to their own companies, and are being financed by major distributors. They give the studios the benefit of their talent, their box-office draw and an exclusive call on their pictures. In exchange the stars receive very high guaranteed compensation, low overheads, the most favorable distribution fee possible, the most favorable definition of gross participation possible, and a share in the ownership of the pictures. Stars such as Clint Eastwood, Robert Redford, Barbra Streisand, Kevin Costner, Eddie Murphy and Sally Field have enjoyed success as owners of as well as stars in their pictures. But along with the privilege of sharing profits or ownership of a picture comes the responsibility of sharing the risk of financing the picture in the event that it goes over budget.

Often business-oriented studio types do not concern themselves with a creative environment for the client. They may be more concerned with getting the project released or getting the return on their dollars. The agent's problem may very well start, then, after the deal has been made, when the buyer and seller start to "live together." Keeping the relationship alive and viable—and keeping the creative juices flowing—may be the most important aspect of the agent's work.

After he has closed the negotiation and the contracts have been signed, a different phase of the work begins. An agent must now live with his deals and not expect someone else to follow up. I think if an agent does not emerge from behind his desk, his relationships with his clients are going to fall apart, and his relationships with buyers, producers, directors and studios are going to deteriorate. The studio looks to the agent to be there at times to help solve his client's problems. The agent, of course, may have to walk a thin line between asserting himself on his client's behalf and meddling in matters that don't concern him. Generally, however, the studio is open and receptive to the agent. They feel his behavior and presence will, in the final analysis, be of help in easing difficulties.

One can easily see that a good agent will raise a business relationship to a human relationship, a circumstance that is difficult to find in any industry outside the entertainment business. The relationship can be a very personal, intimate one. An agent is a diplomat, negotiator, salesman, friend and a very real part of the performer's life. To review, the agent's responsibility includes careful shaping of the client's career, sound negotiating for the right kind of deal, awareness of what the trends in the market are, and the ability to anticipate what the market will be like after a film has been completed. It is difficult, however, for an agent to generalize to a client about this responsibility. He can put his finger on it only after he has found the right kind of project and has negotiated a deal offering a specific dollar amount for his client to perform in that project.

The agency will always be an integral part of the motion picture business because there will always be a need for the creative middleman—the buffer—who is in touch with the changing business and who can bring together various creative and business elements for the mutual satisfaction and reward of those who participate. Since many agencies continue to have training programs for qualified men and women, the level of professional expertise will be maintained.

# VI

# THE SHOOTING

# THE LINE PRODUCER

*by* **PAUL MASLANSKY,** *one of the most active line producers in the business, who has made motion pictures around the world. Before turning to producing, he served in production capacities for Columbia Pictures and United Artists in Europe. Notable among his producing credits are* Race with the Devil *for Twentieth Century Fox;* The Blue Bird, *a Fox coproduction with Lenfilm Studios, Moscow;* Damnation Alley *for Fox;* King, *the award-winning six-hour drama for NBC;* Hard Times *for Columbia;* When You Coming Back, Red Ryder? *for Melvin Simon Productions;* The Villain *for Rastar;* Hot Stuff *for Columbia;* Scavenger Hunt *for Mel Simon Productions;* Love Child *for the Ladd Company;* Return to Oz *for Disney;* Ski Patrol *and* For Better or for Worse *for Trans World Entertainment; the highly successful* Police Academy *series of six pictures at Warners; and* The Russia House *for MGM-Pathé.*

*The production period is very concentrated, with a high, intense energy level demanded at all times. Tremendous amounts of money are being spent in a short period, so it's important to be constant in terms of cost control.*

The line producer is the person hired by the financiers or producers of a picture to be directly responsible for the money, in charge of the day-to-day physical production of the picture, of planning and approving costs, of hiring and firing personnel, and of the many intangible requirements of maintaining a film crew on schedule. Although the screen credit may vary from executive producer, coproducer, producer or associate producer, it is the line producer who has the extensive background in physical production that makes him or her a key person on a picture, second only to the director and bridging the gap between management and the technical crew. If the director has creative control during shooting, the line producer has financial control.

A line producer can be hired by various parties: by a writer who wants to produce his screenplay and have the creative control, but who has no experience in physical production; by a new producer, bearing the title because he or she has acquired the most important element in a film deal—the property—but who may be a former agent, perhaps, with no production experience in the field; by an independent producer-financier, such as Morgan Creek, who will employ his services on a specific picture that they are financing; or by a studio that may be financing a picture on which they need someone to oversee their money, someone with whom the buck stops.

A project is already in the works, usually, when the line producer is hired. There is a completed screenplay, and there is financing (the line producer is a paid employee). When I join a picture, the first thing I do is read the script thoroughly and estimate how much it's going to cost. I start asking myself whether certain scenes should be shot on location or in a studio. Shooting on location can sometimes cost extra not only for travel but for per diem expenses for the cast and crew. But considering the fact that there are excellent local crews in cities in New York, Florida, North Carolina, Texas and Canada, certain location work can undercut the costs of constructing a specific set in a Hollywood studio. At a studio there's a five-day workweek; on location there's a six-day workweek. A six-day workweek on a seven-week picture on location is 42 days; a seven-week picture in a studio is 35 days. Time is of the essence, and you want to finish shooting at the earliest possible time without sacrificing quality.

If I am taking on the responsibility of bringing in a picture at a certain figure, I must approve the budget. This means I would take the existing budget, analyze it very closely along with the screenplay, discover the problem areas and make the appropriate changes.

A budget is divided into two main areas, above-the-line and below-the-line. The below-the-line is broken down into three sections: production period, postproduction period and "total other" charges. Different studios have different forms, and I would support an industry effort to settle on one standard form for all theatrical feature budgets. The top sheet is a summary of each area in the budget. (Please refer to the sample budget top sheet, p. 243.) The budget body can run from twenty to thirty or more pages, depending upon the complexity of the picture.

In analyzing a budget for production, the only figures that would be difficult to change are the above-the-line figures. *Above-the-line* costs refer to fixed costs of key creative commitments and personnel made before shooting starts, including the cost of story and other rights connected with the acquisition of the underlying literary material including screenplay costs; the producer's unit, covering salaries for the producer, coproducer, associate producer, plus their secretaries; direction, meaning the salaries for the director and his secretary; and cast, referring to the acting talent in the picture, including stunt people. For a location picture, add the traveling and daily living expenses (per diems) of all those people. Above-the-line costs may also include a finder's fee for someone who might have discovered the property and brought it to the attention of the producing entity, and maybe a production fee to the company that represents the original owner of the material (who was bought out).

Next, add the *fringes,* meaning the pension, health and welfare costs that are tacked onto the various guild members in the above-the-line area, such as those in SAG (Screen Actors Guild), DGA (Directors Guild of America) and WGA (Writers Guild of America).* Fringes add between 25% and 30% of labor costs.

The *below-the-line* accounts cover the costs incurred to physically

*Editor's Note: For information regarding current SAG minimums or other questions, contact the Screen Actors Guild, 7065 Hollywood Boulevard, Hollywood, CA 90028 (phone 213-465-4600). Inquiries as to the content of the current DGA Basic Agreement should be sent to the Directors Guild of America, Inc., at one of these addresses: 7920 Sunset Boulevard, Los Angeles, CA 90046 (310-289-2000); 110 West 57 Street, New York, NY 10019 (212-581-0370); or 520 North Michigan Avenue, Suite 1026, Chicago, IL 60611 (312-644-5050). For information on the Writers Guild, please see page 99.

make the picture. Whenever a budget has to be cut, it can usually be trimmed in the below-the-line area, where strategy must be applied to reduce the physical costs and cut shooting time. Below-the-line costs are divided into three areas. The production period refers to every cost directly related to making the picture during shooting. All the departments are represented here: set design (including art direction), set construction, special effects, property (props), wardrobe, makeup and hairdressing, lighting (electrical), camera, production sound, transportation, and film and lab costs. In each department the personnel costs include salaries, living expenses, travel and fringes. Add on the cost of the film, the processing, all the physical equipment used, catering, first aid, costumes—everything involved in the actual making of the picture during the production period.

Another part of the below-the-line costs is known as the postproduction or editing period. This area in the budget concerns itself with the editorial crew (an editor plus assistants) and any laboratory work directly related to the editing period. (The rushes and development of the film during shooting are budgeted in the shooting period.) This area also covers reprints, opticals, black-and-white dupes, and titles, right up to making the answer print. Add to this the cost of scoring the music and postproduction sound, such as looping and mixing the dialogue and sound effects. Fringes based on labor costs in this period are usually an added 30%–40%. If the editor is making $2,500 a week, his fringe account is charged perhaps 30% more, or an added $750. He will receive his $2,500 salary, and the $750 will be contributed to the editor's pension, health and welfare funds.

The third area of below-the-line costs, "total other" charges, can include publicity costs (during production only) and insurance costs, since the picture must be covered from the moment the first employee goes on payroll. Generally that person is the director. If monies have been spent but the director has to be replaced due to illness, insurance would compensate for those costs. It's also essential to insure the fact that you have the rights to the story. "Errors and omissions" insurance covers the possibility that the story's not original or that someone else owns it. It insures title, like title insurance for a house. Add the total above-the-line and below-the-line costs for the grand total of the budget.

If it is an independent production, I may have to arrange for a completion bond on the picture, which would mean making a deal with a financier, called a *completion guarantor,* who will insure that the picture is completed if it goes over budget. (See article by Norman G. Rudman, p. 216) Generally a completion guarantor gets 5%–6% of the budget for that service (with half rebated after shooting if the bond is not invaded), which becomes a budget item. A completion guarantee is written in such a way that the penalties are very severe;

| Acct # | Category Title | Page | | | Total |
|---|---|---|---|---|---|
| 11-00 | STORY, RIGHTS, CONTINUITY | | | | 0 |
| 12-00 | PRODUCERS UNIT | | | | 0 |
| 13-00 | DIRECTION | | | | 0 |
| 14-00 | CAST | | | | 0 |
| 15-00 | TRAVEL & LIVING | | | | 0 |
| 16-00 | PRODUCTION FEE | | | | 0 |
| | **ABOVE THE LINE TOTAL** | | | | **0** |
| 20-00 | PRODUCTION STAFF | | | | 0 |
| 21-00 | EXTRA TALENT | | | | 0 |
| 22-00 | SET DESIGN | | | | 0 |
| 23-00 | SET CONSTRUCTION | | | | 0 |
| 24-00 | SET STRIKING | | | | 0 |
| 25-00 | SET OPERATIONS | | | | 0 |
| 26-00 | SPECIAL EFFECTS | | | | 0 |
| 27-00 | SET DRESSING | | | | 0 |
| 28-00 | PROPERTY | | | | 0 |
| 29-00 | MEN'S WARDROBE | | | | 0 |
| 30-00 | WOMEN'S WARDROBE | | | | 0 |
| 31-00 | MAKEUP & HAIRDRESSING | | | | 0 |
| 32-00 | LIGHTING | | | | 0 |
| 33-00 | CAMERA | | | | 0 |
| 34-00 | PRODUCTION SOUND | | | | 0 |
| 35-00 | TRANSPORTATION | | | | 0 |
| 36-00 | LOCATION | | | | 0 |
| 37-00 | PRODUCTION FILM & LAB | | | | 0 |
| 39-00 | PROCESS | | | | 0 |
| 40-00 | SECOND UNIT | | | | 0 |
| 41-00 | TESTS | | | | 0 |
| 42-00 | MISC PRODUCTION EXPENSE | | | | 0 |
| | **PRODUCTION PERIOD TOTAL** | | | | **0** |
| 45-00 | FILM EDITING | | | | 0 |
| 46-00 | MUSIC | | | | 0 |
| 47-00 | POST PRODUCTION SOUND | | | | 0 |
| 48-00 | POST PROD. FILM & LAB | | | | 0 |
| 49-00 | MAIN & END TITLES | | | | 0 |
| 50-00 | 1ST DOMESTIC TRAILER | | | | 0 |
| | **TOTAL POST PRODUCTION** | | | | **0** |
| 65-00 | PUBLICITY | | | | 0 |
| 67-00 | INSURANCE | | | | 0 |
| 68-00 | GENERAL EXPENSES | | | | 0 |
| 69-00 | RETROACTIVE SALARIES | | | | 0 |
| | **TOTAL OTHER** | | | | **0** |
| | **Total Below-The-Line** | | | | **0** |
| | **Total Above & Below-The-Line** | | | | **0** |
| | **COMPLETION BOND** | | | | **0** |
| | **CONTINGENCY** | | | | **0** |
| | **OVERHEAD** | | | | **0** |
| | **INSURANCE** | | | | **0** |
| | **Total Fringes** | | | | **0** |
| | **Grand Total** | | | | **0** |

the guarantor can take physical control of the movie if it exceeds the budget plus a cushion. For example, on a $15-million movie, the guarantor will want a 10% contingency, or cushion, which would require him to start paying if production costs go over $16,500,000. There's rarely a picture that runs *under* budget; usually such a picture was overbudgeted and overscheduled in the first place. On a studio picture there is no completion bond. Instead there is overhead, which varies from studio to studio but averages about 20% and is part of the *negative cost* (literally the cost to deliver the negative, or production cost) of the picture.

Now, let's follow my procedures on a picture financed on an independent basis. Assuming the director has been chosen, the first person I would hire is the production manager. He's my right-hand man, along with the accountant. They're going to provide the fiscal control over the picture. The production manager will approve daily costs, and I will oversee this. The accountant feeds the daily costs, what each department is spending, to the production manager. It's important to select a production manager who has a good reputation from previous pictures.

Let's say we have a $15-million picture independently financed, and I have approved the budget. In order to plan the shooting schedule, I'll sit down with the production manager and break down the screenplay scene by scene. We'll specify which characters are involved in each scene, the description and location of each scene, whether it's day or night, interior or exterior. Each scene will be so described on a strip of cardboard about three-quarters of an inch wide and sixteen inches long. These strips will be juggled in various orders to determine the fastest and most economical sequence in which to shoot. Once this sequence is determined, the strips will be fit into a breakdown board and divided into shooting days. This becomes the road map for the shooting schedule. (For examples of production paperwork, see article by James Bridges on page 254.)

Part of this scheduling will be dictated by where the money is. Usually the most money is found in the cast. If I have Tom Cruise for four weeks and the picture's to be shot in eight, I may want to finish with him early, so that his expense is behind me. I would take all of Tom Cruise's scenes and place them in a sequence so that he's gone in four weeks. I also try to schedule all my exteriors early, because they involve the intangible of weather. If there are weather problems, it's better for them to affect us early in the shoot when we can move to an interior cover set. Later in shooting, waiting for clearing weather means losing time and money.

I will present the proposed shooting schedule to the director, and he will evaluate it, knowing how he works and making changes accordingly. The ideal would be to shoot a picture in continuity, that is, in order. But if the first scene of the screenplay is set in New York City, the second scene in Kansas City, Missouri, and the third scene back in New York, it's economical to shoot the New York scenes together, eliminating travel time back and forth.

Now the breakdown board and shooting schedule have been agreed upon, which gives us the number of shooting days. This helps us finalize the budget, since each shooting day represents an amount of dollars. What are some ways to save money? One way is to shoot seven pages of the screenplay on a given day rather than four, but that puts the company under tremendous pressure. A better way to save money is to cut out days by combining or even eliminating scenes, since time is money. Saving money usually means making compromises. Making movies generally is a series of compromises.

Saving time and therefore money is one benefit of applying computer software to scheduling and budgeting. In scheduling, when a health or weather emergency arises, we can swiftly analyze the dollar impact of alternative plans because choices are offered quickly. Let's say an artist becomes sick and we have to reschedule the next day's shooting. Instantly a computer prints out scenes in which the artist appears and then isolates alternative scenes at the current location. In budgeting, software enables us to track expenditures instantly. Every purchase order is posted into the system. At the push of a button, for example, I can learn that we have $65,000 more to spend on set construction for an interior of a ballroom set; before, I would have to run numbers with the production designer and construction coordinator and it was very time-consuming. Computers don't prevent pictures from going over budget, but the technology can warn about jeopardy sooner and also removes excuses.

With every scene now locked into a shooting sequence, I know which sets are required on which days, so I can hire a production designer. After consulting with the director, the production designer goes out to search for locations. Sometimes we'll send out letters to the various state film commissions with a list of locations needed and their descriptions. They will send back assorted photographs, since they all want us to come to their state and make the movie. This helps in location scouting and saves a lot of footwork.

Next there is other crew to hire, and each department is responsible for hiring some of their own. For example, the costume designer will hire the wardrobe master (or mistress), who actually takes care of the handling and cleaning of the garments.

On a six-week shoot, preproduction will last about six weeks. In the first week I've done the board and worked on the budget with the production manager and accountant. In the second week the production designer has left to search for locations, and casting is going on, since we've hired a casting director. The accountant has some payroll to contend with and sets up the books, the computer system and the picture's bank account. Next, the balance of the hiring is done, including the assistant directors, director of photography, other heads of departments and their staff.

Once we start shooting, the director runs the show. He has a department that includes the first assistant and second assistant or assistants. These people are responsible for running the floor logistically, keeping order, and handling the background action so that the director can concentrate on the principals in front of the camera. The second assistant generally does the paperwork, such as the *call sheet*, which announces the subsequent day's work schedule and is ready for distribution to the crew before the end of a shooting day.

The head of the camera department is the director of photography. His crew consists of the camera operator, who actually looks through the camera, and two assistants: one to follow focus and one to load the camera. The director of photography is responsible for the photographic look of the picture, so lighting falls in his domain. The chief electrician is called the *gaffer;* he's on the floor with the director of photography, helping set the lights. The chief electrician's assistant is called a *best boy.* There are two best boys on a picture, best boy to a gaffer and best boy for the chief grip. The chief grip is responsible for moving the camera to the place where the director wants to shoot and for moving the boards and panels that shade off lights. The construction crew is usually not around during shooting. They've been there, finished their work and gone to work on whatever construction is needed on the next location. The only construction person who remains with the shooting unit is there in case we have to float (or move) different walls or make set adjustments "to camera."

Very early in preproduction I ask each department head to provide a breakdown of their department budgets after I've made an initial budget estimate from the screenplay. This avoids a situation where I may have estimated wardrobe on a picture to cost $100,000 but the costume designer comes in very fancifully with wardrobe designs that cannot be delivered for less than $200,000. Or perhaps I don't know that the price of filigree has gone up 200% because Taiwan's having a political problem or that the cost of lumber in Houston is far more than in Oregon. Each department head confers with the director in preparing this breakdown figure, so they are expressly following the director's requirements. Generally a department head's figure and my estimate are close. If there's a real discrepancy, I make it known that I'm very distressed that a reasonable figure can't be met. If there's a lack of communication and that wardrobe figure, for example, is still not practical, I'd better find some extra money elsewhere in the budget or find another costume designer.

When the production designer returns from the location search, he will thoroughly sketch out his concept of the sets and supervise the drafting of the construction blueprints by a draftsman. The result is the production designer's vision of what the sets should look like, after collaborating with the director and producer, and guided by the screenplay. The production designer works with a set dresser, who is responsible for bringing in furniture, giving the set life. The set dresser has a lead man working with him as well as other laborers, called *swing gang*. The art department also includes the prop people (prop master and assistant prop master), who are responsible for the hand props used in the picture, as well as the cars and sometimes animals. If it's a western, livestock would be handled by a wrangler. Visual and mechanical effects also fall under the production designer's supervision. However, in elaborate special-effects pictures such as the *Back to the Future* series or *Who Framed Roger Rabbit?* the designer of the special effects is accorded a position equal to that of the production designer, and the two work together very closely.

The photography department consists of the D.P. (director of photography), camera operator, first and second assistants and still man, who takes all the still photographs used in publicity. The sound department is generally composed of two people: the *mixer,* who sits with earphones at his portable console, and the *boom man,* who holds the microphone. Sometimes a cable man is added to this crew. The mixer is responsible for proper sound levels and for clear sound; his quarter-inch audio tape runs at a speed exactly synchronized with the film in the camera. During preproduction the director, production manager and D.P. agree on what camera equipment to rent. This equipment is reserved, as is the raw film stock. The flow of raw stock is controlled by the camera department, and the production is not charged until stock is drawn from the source, whether Kodak, Agfa or Fuji. But film stock cannot be returned at full price, so estimating reserves, especially on location, is critical.

Wardrobe is headed by a costume designer who conceives of the garments, goes out with the wardrobe master and rents from wardrobe companies, or else sees that costumes are constructed. The wardrobe master cares for the cleaning of costumes and makes sure that each garment is tagged so that each artist knows on what day he wears what garment. They know that on day 15 Michelle Pfeiffer is going to be wearing the blue gown and on day 13 it's the green gown. In estimating budget items, the areas where one can go most wrong are the art department (covering set construction), wardrobe and transportation.

The transportation captain supervises the work of between eight and twenty drivers and other personnel, charged with the logistics of moving the entire company from hotel to set and back to hotel every day. A driver gets paid $1,000–$1,500 a week and is responsible for maintenance of his equipment. Teamsters work the longest hours of any department, except perhaps special effects, wardrobe and set dressing.

When we start shooting, there are two elements I have no control over: weather and health. There's no way to protect against bad weather except to have an interior cover set handy so that some shooting can be accomplished. In order to protect the health of the artists and key technicians, it makes sense to stay at a comfortable hotel so that they are mentally and physically prepared to put in minimum twelve-hour workdays and seventy-two-hour workweeks. Morale is an important intangible. Management and labor must have an open relationship on a movie because the crew becomes a family. It's pretty tough to get up at six every morning six days a week for twelve weeks and not slow down after a while. Motivation is not found only in the money: there has to be a good spirit created too. That's important for management, such as the production manager, line producer or financier, to understand.

Now we are entering the production period, the most intensive of the three stages, when money is being spent at the rate of $6,000 to $10,000 an hour. How is this organized on location? During preproduction a local bank will have been selected because of its reputation and its convenience to the location. The bank is usually delighted, because having a movie company's account is glamorous and means $5-million–$6-million worth of activity circulating through the bank over the weeks of shooting for our $15-million picture.

A checking account is opened through the use of signature cards after we have demonstrated to the bank the validity of our corporation. There are two signatures necessary on each numbered check; I generally sign on one line, and the accountant signs on the other. As backup, one of the financiers at home can sign in my place, and the production manager can sign in place of the accountant. We order books of numbered checks, and the accountant opens a master ledger in the production office that will account for each check in detail.

Checks are used to pay for physical equipment and wages. In the purchase of equipment, our company works like any small company. If a department member needs a piece of equipment, he will tell the department head, who will make out a purchase order (PO) for each item, approve it and get the production manager to approve it. When the bill comes in, the accountant will refer to a copy of that purchase order bearing the two approvals and make out a check for payment, which the production manager approves. One of them signs on one line of the check, and it's entered into the master ledger. Then I will review this check along with many other checks, study the backup paperwork and sign it on the other line before it is mailed out. If the item is a certain antique lamp to be rented for two weeks so that it can be in the scenes of a certain living room, the set decorator will have selected it from a dealer and negotiated the rental price approved by his department head. The set decorator will then hand in a purchase order for the rental of that lamp, approved by the art director and production manager. The company will use the lamp for two weeks and be billed accordingly.

To handle payroll, the production would generally hire a payroll-service company, an outside contractor that handles all the details of paying hourly employees, covering most people working on a motion picture. Each week the production manager and accountant cross-check employee time cards against the production reports. As an example, the production manager, reading a grip's time card, can verify that he did work twelve hours every day and six days that week, so he's entitled to be paid for a seventy-two-hour week. Call sheets and production reports can act as legal documents as to how much employees are to be paid. On location these approved time cards for a prior week are then sent to the payroll-service company on a Tuesday by Federal Express. The cards are analyzed, proper pension, health and welfare payments are added and payroll checks are cut and returned to the location for distribution on Thursday. Payroll services charge a fee based on the amount of money flowing through them, and certain monies are deposited with them in escrow in advance of production.

The only item that does not have all this formal backup paperwork is petty cash, used for many small purchases where checks are not practical. In order to minimize problems, I issue only $200–$1,000 of petty cash at a time, and each amount must be returned in the form of receipts. Each week I get an accounting of how much petty cash is out to whom. If one particular department is negligent in returning receipts, I will speak to those people. It's an incestuous business, and the crewmen who are hired have reputations to uphold. There are three aspects of reputation: skill, attitude and honor. If someone is very skillful but unwilling occasionally to give an extra twenty minutes or so necessary to complete a job without putting in for overtime, for instance, or if someone else is suspected of stealing, this affects their reputation in the business. After all, other productions will follow this one, and other line producers will hire based on the crew's previous reputation.

The production period is very concentrated, with a high, intense energy level demanded at all times. Tremendous amounts of money are being spent in a short period, so it's important to be constant in terms of cost control.

Each week during shooting a cost report is prepared for the financier. This compares the amount spent for each item from a budget top sheet against the original budgeted figure, gives a "cost to complete" figure and shows in one column to what extent these figures are over or under budget. The completion guarantor also receives this weekly cost report as well as the call sheets and production reports.

During production the daily paperwork is prepared carefully to memorialize the progress of each day's shooting. The script supervisor or continuity person sits by the camera during the entire picture. Every night he or she types up notes itemizing every take, recording script or dialogue changes, comments from the director, details such as what lens was used, what *f*-stop (for the lab to refer to later to match the shot), and what takes were to be printed. He also keeps a diary as to the times we start work, break for lunch and the time any accidents might have happened, as well as the time we finished work. This diary is given to the production secretary at the end of every shooting day, who fills out the production report from the data provided by the script supervisor and from other data from the second assistant director, who then verifies and completes the report.

On the set the first assistant director has his hands full with the actual functioning of the set; he also approves the call sheet and production report completed by the second assistant. The first assistant ultimately signs production reports (as does the production manager) and call sheets (the itemized schedule for the next day's shooting), but most of the paperwork is initiated by the second assistant and the production secretary.

By the last week of shooting I've given notice to most of the department heads. I don't need the photography or sound departments anymore, but I need people from wardrobe, props and set dressing to return articles that have been rented and to resell the items we've bought. Makeup and hair personnel are gone, unless there are some wigs to return. The continuity person or script supervisor stays on for at least a week to type up all the production notes. The production secretary, production manager and accountant remain until all invoices are in and all bills are paid. When bills are paid, all the time cards are in, the last payroll is done and adjustments are made for lost or damaged equipment, the production manager has completed his job and can prepare a final accounting. Generally this is two to six weeks after shooting ends.

It's mandatory to make a list at the start of preproduction of all items purchased outright by the department heads. Transportation may buy cars, generators, jacks, tires and tools. I'll take this inventory at the end of the picture and put the remaining items up for auction to the cast and crew. The production manager runs the auction, and the items go to the highest bidders as a cash sale. The money goes back into the production for use during the postproduction period.

According to the Directors Guild, the director has a minimum number of days to deliver his cut, roughly one day of cutting for every two days of shooting. Usually an editor is assembling the best takes during shooting, so that soon after the last day of production, the editor has a roughly assembled picture to show. This would be an assembly of the specific takes chosen by the director during the daily viewing of rushes and usually runs longer than our target length for delivery.

The picture is then turned over to the editor and the director for the director's cut. When the director delivers this cut, the producers, the studio or the financial entity can take over if they have the contractual right to do so.

A postproduction period can stretch out unless somebody places certain limits upon it. Pressure is brought to bear either by the studio or financier-distributor because of a contractual delivery date or an impending release date. Generally the first instincts of a director and editor in cutting a picture are close to what the final cut becomes, though it's tempting to refine a picture infinitely. Since interest continues to be charged on the money being used, the editorial process should be as efficient as possible.

During preproduction planning, certain bookings of facilities for postproduction have been made. For recording the musical score, a studio and an orchestra have to be scheduled; the composer may have just so much time before his next commitment. Rooms have to be rented for the recording of sound effects, looping and mixing. For the two to four weeks it takes to mix a picture, the mixing room must be reserved months in advance. For example, if this is August and I'm in preproduction on a picture, it would go into postproduction facilities in February. I must reserve the room now, because February is a rush period; pictures are being prepared for summer release then.

Let's assume we now have a final cut approved by the financing entity. Next we have ADR, computer-assisted automatic dialogue replacement, which is redoing some of the audio to improve the quality. For this we may call actors back to rerecord dialogue in a studio, matching in perfect sync with the dialogue on the screen.

During (and sometimes before) looping, sound editors start building the effects track. If a scene includes a motorboat and the audio during shooting sounds like a tiny engine rather than a powerful one, the sound editor will either search for the right effect in a sound-effects library or go out and reproduce the sound on location.

Since we've cut our picture to its final length, the picture is "locked." Now our composer and director view the cut and discuss where music is needed. The composer will create this music over a period of weeks and return to a recording studio to record the music after it is orchestrated, which requires the hiring of musicians and copyists.

The optical work to be done, such as titles, fades, dissolves or other effects, will be given to an optical house. Black-and-white prints, called *scratch prints,* are made for the sound, music and optical people to view while they're planning their work. At the same time the negative is being cut in the laboratory. All negatives bear numbers on the edges. By referring to these edge numbers, laboratory technicians conform the negative to the final cut, frame by frame, in a delicate process.

The sound track (including dialogue, music and effects on separate tracks) is then brought into a mixing room for the final mix. The chief mixer, who runs the dials that govern the dialogue tracks, is joined by a sound-effects mixer, who controls the pots and faders for the sound effects to be added to the picture. He may control ten different sounds at once: wind, cars, natural ambience (the "presence" of a room), sirens, footsteps. A third person controls music, with separate dials for the different recorded music tracks, such as percussion and strings; that person is a virtuoso in himself. The fourth person is the director, who makes sure the mixing achieves what he has in mind. This is a great process to watch, because at last the picture is coming together, joining final sound to picture. All sound is mixed in stereo, assigned to one of four channels (left, middle, right and surround) and prepared for enhancement in theatres through processes such as Dolby stereo or the Lucas THX sound system.

The preview process, contractual for directors and always instructive, brings the picture in near-final form to sample audiences. Previews can solve creative disputes and illuminate other decisions, since an audience is the final arbiter. Techniques range from recruiting an audience that knows what they will be viewing and afterward acts as a focus group, discussing questions with a professional moderator, to traveling to a city with average demographics and announcing a sneak preview but not the picture title. In either case the filmmakers are present, learn from audience feedback, return to the studio to fine-tune the picture and often arrange for another preview.

Once the preview process is complete, everything is turned over to the laboratory. Miraculously four or five days later there is an answer print, the first marriage of picture and sound on the same reel of film. Between the first answer print (or married print) and the first release print, there may be several efforts to correct color in various reels. The director of photography comes in and supervises the color correction of the picture scene by scene, involving the losing of color or infusing of color and contrast into different scenes.

Contractually my job on a picture is generally completed when the first release print is delivered to the financier-distributor; emotionally I follow a picture throughout its exploitable life worldwide.

Throughout the 1970s the overseas market was considered by most American distributors as a bonus added to domestic rentals. Today international revenue can be so significant that it can dwarf

American numbers. For example, the *Police Academy* movies performed more strongly overseas than in the United States from the second picture onward, particularly in Japan, Italy, France and Germany. This was especially significant in Japan, where American comedies historically did not do well. The overseas market has become so important to the *Police Academy* series that if we were to make a *Police Academy 7* today, the decision would be based on projections from overseas income, and all ancillary income, rather than relying on domestic theatrical income to return the lion's share of the cost of the picture.

# "THE CHINA SYNDROME:" SHOOTING PREPARATION AND PAPERWORK

*by **JAMES BRIDGES**, a writer-director who began his Hollywood career as an actor in over fifty television shows and several features. He then turned to writing such feature films as* The Appaloosa, Colossus: The Forbin Project, *and* Limbo *before becoming the writer-director of* The Baby Maker. *Bridges earned an Academy Award nomination for best screenplay adaptation for* The Paper Chase, *which he also directed. He has since written and directed* September 30, 1955, The China Syndrome, Urban Cowboy, Mike's Murder *and* Perfect; *directed* Bright Lights, Big City; and cowritten *White Hunter, Black Heart.*

***It would be nice if the actual shooting of a picture followed the shooting schedule exactly, but this is not the case. . . .***

During physical production huge amounts of money are spent daily for the one-time-only opportunity of capturing performers on film within the environment created by the artistic and technical members of the crew. Although the director is intimately involved in the meticulous preparation that precedes the first day of shooting, once shooting has begun, the crew basically takes over the details of physical production. This frees the director to concentrate on performances, the look of the picture and the telling of the story.

The director's immediate family during shooting includes the producer, production manager, first assistant director, second assistant, cameraman and script supervisor. The extended family includes the cast members, heads of departments and crew. Each film is different, with a unique set of problems, and each director develops a personal style in solving such problems. On *The China Syndrome* I was given great support by first assistant Kim Kurumada, associate producer/production manager Jim Nelson, executive producer Bruce Gilbert, actors Jack Lemmon, Michael Douglas (who also produced), and Jane Fonda (whose company coproduced), second assistant Barrie Osborne, cameraman James Crabe and script supervisor Marshall Schlom.

In order to trace preparation and paperwork, one scene has been isolated and will be followed from the pages of the screenplay, written by Mike Gray, T. S. Cook and me, through to the production report. All materials are furnished courtesy of Columbia Pictures.

## SCREENPLAY

The screenplay is the blueprint of the picture. Every bit of physical planning and each commitment of money flows from the screenplay. The screenplay form separates dialogue from description and is divided into scenes in a way that allows space for notations so that every department head can focus on specific needs of each scene. This example covers scenes 247–251, which are the last scenes of the picture. Reporter Kimberly Wells (Jane Fonda) and cameraman Richard Adams (Michael Douglas) are outside the nuclear power plant confronting coworkers of Jack Godell (Jack Lemmon), who has just been killed inside.

On the basis of the screenplay a preliminary budget is prepared, the shooting schedule is estimated, major casting commits to the project and all of this ideally leads to the producers securing a deal with a financier-distributor. *The China Syndrome* was financed and distributed by Columbia Pictures.

Early in preproduction there were two problems that were faced on *China Syndrome*. One was the set; enough lead time had to be afforded to production designer George Jenkins (who had won an Academy Award for *All the President's Men*) to construct the elaborate control room of the power plant. Second, the shooting schedule had to be planned to free Jack Lemmon to honor a prior commitment to the Broadway show *Tribute*. This meant that all of Jack Lemmon's work had to be scheduled first so that he could be released.

A lot of time was taken to assure that the technical aspects of the picture were accurate. The technical advisor, Greg Minor, was a scientist and the designer of nuclear power plant safety systems who had left the industry saying that such systems weren't working. He worked closely with the production designer and staff.

Informal meetings were held with certain heads of departments who had to reserve equipment for shooting. For example, the cinematographer, James Crabe, and I discussed the ratio we would shoot in, the kind of light the picture would use and the overall visual style of the picture. Then he reserved the necessary equipment, including dollies or cranes. He also worked closely with the production designer to consider lighting and physical movement on the set, such as which walls need to be wild (movable). The cameraman also met with the matte artist to plan how matte shots would be framed. In the case of our example, the basic exteriors of the power plant were matte shots, combining live action with the artist's rendering on glass.

Location selection and principal casting occurred very early in preproduction. When it was decided to use the outside of a power plant at Playa Del Rey as our nuclear plant (when we were close enough not to need a matte), I knew that I would want to have a big crane shot from above to establish all of the activity and news coverage outside, as well as an actual helicopter within frame to heighten the action. I was casually blocking the action of this scene when we scouted the location, and this would be carried over formally when it came to actual shooting.

CONTINUED

KIMBERLY
Did any of it get on the air?

MAC
Enough to make Godell look like a lunatic ---

Kimberly, Richard, and Churchill watch the TV monitor.

GIBSON
Gentlemen -- we'll have a prepared statement in just a few minutes -- again, I want to stress at this time that the public was in no danger of any kind -- the emotionally disturbed employee was humored only long enough for us to get the situation under control ---

REPORTER
Had he been drinking?

GIBSON
I've been told he had been drinking.

Kimberly, watching on the monitors, has had enough. She moves closer, looks off at MacCormack, DeYoung, Spindler etc., as they stand behind Gibson, listening.

GIBSON
Unfortunately he did cause some damage to the plant itself, but that damage was completely contained ---

REPORTERS
1. What kind of damage?
2. What happened to him?
3. Has he been arrested?
4. Were there any injuries?
5. How did you get into the control room?

Kimberly moves toward the group. DeYoung and Spindler are being brought forward by MacCormack and others.

248

GIBSON
Gentlemen -- this is Mr. DeYoung and Mr. Spindler -- Mr. DeYoung is in charge of this plant and Mr. Spindler was in charge of the operation through-out ---

CONTINUED

SPINDLER
It's over -- it's stable ---

Kimberly looks down at Godell. Richard is beside her.

KIMBERLY
He's dead.

DeYoung looks up at MacCormack who leans back from the window.

CUT TO

EXT. POWER PLANT - NIGHT - MATTE 24

The outside is all lit up. More reporters have arrive More television stations have moved in. Gibson moves of the plant with MacCormack, DeYound, Spindler and jo the reporters as Kimberly and Richard follow. The two at the crowd and start for the truck. Some cameramen shooting 16mm, others on tape. There is the gangbang of reporters clustered about the group. There are li obviously. The various vehicles are parked about. People rushing.

REPORTERS
1. What's happened?
2. What's going on?
3. Get out of the way!
Etc, etc.

GIBSON
If I can have your attention please.
I'd like to make an announcement.
(beat)
A few minutes ago, the situation inside was resolved. The Ventana Nuclear Plant is secure. I'd like to emphasize that at no time was the public in any danger.

REPORTERS
(overlapping, interrupting each other, shouting)
1. How was the situation resolved?
2. What happened in there?
3. What can you tell us about the armed man who siezed the plant?
4. Who's this guy Godell?

Kimberly and Richard arrive at the truck. Churchill

CONTINUED

The reporters start asking DeYoung and Spindler questions.

REPORTERS
1. How did you stop Godell?
2. What was he doing?
3. What happened to the television signal?

SPINDLER
We crossed some wires ---

Kimberly moves into the crush of reporters, elbowing and fighting like the rest. They are all shouting. DeYoung and Spindler are being interviewed as there is mass confusion, everyone talking.

Kimberly has managed to get herself up in front with her mike. Her cameraman right behind her, adjusting. The others turn and see her. Their cameras on her, their mikes being thrust forward at her.

KIMBERLY
Let me ask a question! A few minutes ago Jack Godell was shot to death in front of me. Why was he killed?

DeYOUNG
The man was emotionally disturbed -- he was ---

KIMBERLY
Who ordered the SWAT squad into the control room?

DeYOUNG
I told you, he was disturbed -- the man was dangerous ---

KIMBERLY
Mr. Spindler, do you believe Jack Godell was emotionally disturbed? Do you believe that?

Spindler looks at Gibson and at DeYoung.

CUT TO

INT. KXLA TRAILER - NIGHT 249

Mac is in the van with Richard, watching the monitors. Spindler is looking over at MacCormack and DeYoung. He hesitates.

CONTINUED

CONTINUED

KIMBERLY
Do you believe that? Do you believe he was disturbed?

RICHARD
Get him, Kimberly ---

CUT TO

EXT. PLANT - NIGHT 250

Kimberly questioning Spindler. The other reporters around, sensing that something very dramatic is about to happen.

SPINDLER
I don't know ---

KIMBERLY
How do you explain his behavior tonight? What did he say to you?

GIBSON
That's enough -- thank you. Mr. Spindler -- DeYoung ---

KIMBERLY
What did he say?

SPINDLER
He said he thought that this plant should be shut down ---

GIBSON
That's all -- thank you ---

Gibson takes Spindler by the arm and starts to lead him away. MacCormack and his men trying to stop the reporters. Kimberly in the crush, fighting to keep questioning, pushing after them.

KIMBERLY
Should it? Should it be shut down?

SPINDLER
It's not my place to say ---

KIMBERLY
If it's not your place, Mr. Spindler, whose is it?

Spindler stops and looks at her.

CONTINUED

CONTINUED

GIBSON
Let's go, Mr. Spindler -- we have to go ---

KIMBERLY
If there's nothing to hide, whose is it? Let him speak ---

SPINDLER
Wait a minute -- Jack Godell was my best friend -- these guys are trying to paint him as some kind of looney -- he wasn't a looney -- he was the sanest man I ever met ---

KIMBERLY
You believe he had reason to do what he did tonight? Was this plant unsafe?

SPINDLER
He shouldn't have done what he did if there wasn't something to it -- Jack Godell was not that kind of guy ---
(looking at Kimberly)
I don't know all the particulars -- he told me a few things -- but I'm sure there's going to be a big investigation -- the truth will come out -- and when it does, people will know that my friend, Jack Godell, was a hero -- he was a hero and not some kind of looney -- I've said too much already -- I -- got to go ---

CUT TO

INT. TV CONTROL ROOM - NIGHT 251

Kimberly can be seen on the monitor. ON AIR. She turns away from the crowd and to another camera. They are controlling what is being fed to the audience. Jacovich very moved by her reporting, standing back, watching, thinking. The technical staff are barking instructions.

KIMBERLY
Pete -- I met Jack Godell two days ago and I'm convinced that what happened tonight was not the act of a drunk or

CONTINUED

CONTINUED

KIMBERLY (cont'd)
a crazy man. Jack Godell was about to present evidence that he believed would show that this plant should be shut down. I'm sorry I'm not very objective. Let's hope it doesn't end here -- this is Kimberly Wells for Channel Three.

On the monitor we see Richard move up to Kimberly and they embrace as the technical staff cues Pete to appear.

PETE
Thank you, Kimberly, we'll be back with more on the Ventana Power Plant take over by one of it's employees right after this ---

DIRECTOR
Spectacular!

JACOVICH
Good job -- she did a hell of a good job -- I must say I'm not surprised.

The commercial continues in the ON AIR monitor. Richard and Kimberly move through the crowd. The music of the commercial rising and interestingly enough, it tells us to buy more electrical appliances.

FADE OUT



The casting director, Sally Dennison, worked closely with Jane, Michael, Bruce Gilbert and me to fill the supporting roles, and we would not cast anyone who Jack Lemmon thought might be wrong for the part. We had one reading with all of the actors together during preproduction. Out of that reading I went away and did some rewriting because I felt strongly that Jane's character didn't have enough to do in the piece. I spent two weeks rewriting in the mornings and working with the production people in the afternoons.

## BREAKDOWN SHEET

A budget figure is not meaningful until it conforms to the specific breakdown board and shooting schedule prepared by the production manager and first assistant director and approved by the producer, director and financier-distributor. Budget figures up to this point have been speculative; these are for real, prepared and approved by those people who are being paid to fulfill their own planning during physical production.

The whole process begins with breakdown sheets. The first assistant director takes the screenplay and isolates each scene or piece of complete action to a new breakdown sheet. The sheet lists all person-

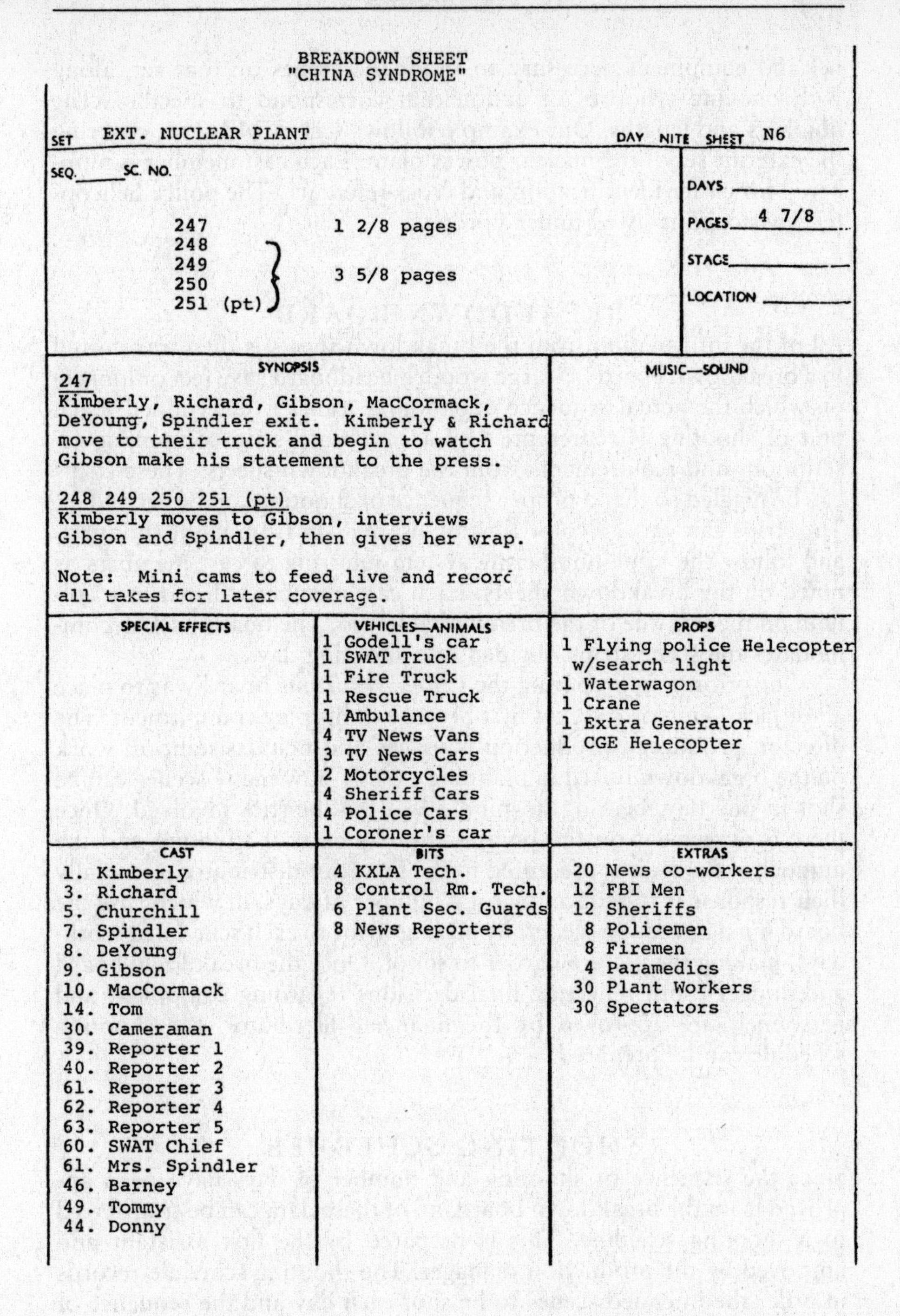

BREAKDOWN SHEET
"CHINA SYNDROME"

SET EXT. NUCLEAR PLANT — DAY NITE SHEET N6

SEQ.____ SC. NO.

| SC. NO. | | DAYS |
|---|---|---|
| 247 | 1 2/8 pages | PAGES 4 7/8 |
| 248, 249, 250, 251 (pt) | 3 5/8 pages | STAGE |
| | | LOCATION |

SYNOPSIS

247
Kimberly, Richard, Gibson, MacCormack, DeYoung, Spindler exit. Kimberly & Richard move to their truck and begin to watch Gibson make his statement to the press.

248 249 250 251 (pt)
Kimberly moves to Gibson, interviews Gibson and Spindler, then gives her wrap.

Note: Mini cams to feed live and record all takes for later coverage.

MUSIC—SOUND

SPECIAL EFFECTS

VEHICLES—ANIMALS
1 Godell's car
1 SWAT Truck
1 Fire Truck
1 Rescue Truck
1 Ambulance
4 TV News Vans
3 TV News Cars
2 Motorcycles
4 Sheriff Cars
4 Police Cars
1 Coroner's car

PROPS
1 Flying police Helecopter w/search light
1 Waterwagon
1 Crane
1 Extra Generator
1 CGE Helecopter

CAST
1. Kimberly
3. Richard
5. Churchill
7. Spindler
8. DeYoung
9. Gibson
10. MacCormack
14. Tom
30. Cameraman
39. Reporter 1
40. Reporter 2
61. Reporter 3
62. Reporter 4
63. Reporter 5
60. SWAT Chief
61. Mrs. Spindler
46. Barney
49. Tommy
44. Donny

BITS
4 KXLA Tech.
8 Control Rm. Tech.
6 Plant Sec. Guards
8 News Reporters

EXTRAS
20 News co-workers
12 FBI Men
12 Sheriffs
8 Policemen
8 Firemen
4 Paramedics
30 Plant Workers
30 Spectators

nel and equipment necessary to shoot the scenes on that set, along with one-line synopses of action that correspond to specific scene numbers and lengths. Our example follows scenes 247–251, covering the exterior set of the nuclear power plant. Each cast member is numbered for easier identification and cross-reference. The police helicopter envisioned is listed under "props."

## BREAKDOWN BOARD

All of the information from the breakdown sheets is then transferred to a breakdown board—a large wooden headboard (five feet or longer) on which the actual sequence of shooting scenes is determined. Each unit of shooting is represented by a cardboard strip containing descriptions and requirements from the breakdown sheets. These strips can be juggled so that a proper sequence of shooting can be arrived at. The strips can vary in color (distinguishing day from night shooting) and follow the same numbering system referring to cast members as noted on the breakdown sheets. Each cast member is listed in a column on the left side of the breakdown board. The board itself accommodates the scene strips, divided into shooting days.

One priority in preparing the *China Syndrome* board was to place all of Jack Lemmon's scenes first because of his play commitment. The director, producers, production manager and first assistant all work on the breakdown board in planning exactly how many scenes can be shot in one day, bearing in mind all of the logistics involved. Once there is agreement on the board—the sequence of shooting and the number of days—it is presented to the financier-distributor. Generally their response is to reduce the total number of days, in which case the board must be redone (generally adding strips to each scheduled workday), making the picture harder to shoot. Once the breakdown board and budget (both reflecting final decisions regarding equipment and personnel) are approved by the financier-distributor, the shooting schedule can be prepared.

## SHOOTING SCHEDULE

Since the sequence of shooting and number of days have been approved from the breakdown board, all of these data can be transferred to a shooting schedule. This is prepared by the first assistant and approved by the production manager. The shooting schedule records in order the intended scenes to be shot each day and the sequence of shooting days, along with the necessary cast members, equipment, personnel, props and sets to be used, again following the information set down in the breakdown sheets.

Because of the logistical requirements of *China Syndrome,* the

SHOOTING SCHEDULE
"China Syndrome"

| DAY & DATE | SETS-SCENES-DESCRIPTION | CAST | LOCATION |
|---|---|---|---|
| | END OF 42nd DAY | | |
| 43rd and<br>44th<br>shooting<br>days<br>Tue., Mar 16<br>and<br>Wed., Mar 17 | Ext. Nuclear Plant<br><br>Scene 147 -- Nite for Nite (N6)<br>1 2/8 pages<br><br>Kimberly, Richard, Godell, others exit. Gibson begins interview to the press.<br><br>Extras & Bits<br>Approx. 200 extras<br><br>Picture Cars<br>1 Godell's car<br>1 SWAT Truck<br>1 Fire Truck<br>1 Rescue Truck<br>1 Ambulance<br>4 TV News Vans<br>3 TV News Cars<br>2 Motorcycles<br>4 Sheriff Cars<br>4 Police Cars<br>1 Coroner's car<br><br>Props<br>Police guns<br>News Team Props, cameras, etc.<br>Hardhats<br>Work Tools<br><br>Note: Live mini-cams to work throughout the sequence. | 1. Kimberly<br>2. Richard<br>5. Churchill<br>7. Spindler<br>8. DeYoung<br>9. Gibson<br>10. MacCormack<br>14. Tom<br>30. Cameraman<br>29. Reporter 1<br>40. Reporter 2<br>41. Reporter 3<br>62. Reporter 4<br>63. Reporter 5<br>60. SWAT Chief<br>61. Mrs. Spindler<br>46. Barney<br>49. Tommy<br>44. Donny | Scattergood<br>Power Plant<br>12700 Vista<br>Del Mar<br>Playa Del Rey |
| | Scene 248 thru 251 (pt) -- Nite for Nite (N6) 3 5/8 pages<br><br>Kimberly interviews Spindler, gives wrap up.<br><br>Note: Film and video footage to be later used for Sc. 251 complete Int. News Room.<br><br>Trans., Props, Bits<br>Same as Sc. 147 | See Above | |
| | END OF 43rd & 44th DAY | | |

picture could not be shot in order. The scenes that ended the picture were actually shot around the middle of the schedule, on days 43 and 44, at night.

Once there is a shooting schedule, the producer, director and first assistant sit down with each department head to formally go over the requirements of each shooting day. For example, the transportation captain and I review the need for a SWAT truck, fire truck, ambulance and other vehicles for shooting outside the plant on days 43 and 44. The prop master will know that we'll need police guns, news-team equipment, hard hats and various work tools for the actors on that location. Naturally all of this is to review what had been discussed earlier in preproduction, when each department head prepared a budget figure and began renting necessary material or equipment. Now these elements can be secured for specific shooting days.

It would be nice if the actual shooting of a picture followed the shooting schedule exactly, but this is not the case. The reality is that pressures of time, money, weather, temperament and other contingencies generally throw a production company off the first shooting schedule, forcing them to revise constantly. This calls for real flexibility and imagination on the part of the staff, particularly the first assistant, production manager and producer. Restructuring the shooting schedule and planning around contingencies of weather and ill health are ongoing responsibilities of this collaborative mix.

## CALL SHEET

The call sheet—prepared by the second assistant director and then approved or changed by the first assistant and finally by the production manager—is issued each day to the cast and crew for the following day's shooting. Any departures from the shooting schedule or changes in equipment or personnel requirements will be listed in the call sheet. The call sheet lists the times the actors must report to makeup and what time they must be on the set. It also lists the reporting time of crew members, stand-ins, extras and certain vehicles.

This example of a call sheet was prepared for the 43rd day of shooting *The China Syndrome* at the Scattergood Power Plant location. Most crew members and vehicles were to be in place by 5:00 P.M., with extras and stand-ins in place by 5:30. Cast members were to be on the set by 6:30, and there is a provision for 215 dinners to be ready for cast and crew by 11:00 P.M. This call sheet covers the first of two nights at the location, which was the time allotted to shoot our sample pages.

Most of the daily paperwork during shooting is done by the assistant directors in order to free the director to focus on the actual shooting. Each director prepares for a shooting day in his or her own

**EYEWITNESS** LTD.
SUNSET-GOWER STUDIOS • 1438 No. Gower St.
Hollywood, Ca. 90028 • (213) 466-3295

PRODUCER: Michael Douglas
EXEC PROD: Bruce Gilbert
DIRECTOR: James Bridges

COMPANY WILL REPORT TO LOCATION, RD. TRIP: 52 MILES

**CALL SHEET**

Prod No. 132441

Day THURSDAY, MARCH 16, 1978
43RD Day out of 48+1HOLI.
Crew Call 5PM HAVING HAD
Shooting Call 6PM
Location SCATTERGOOD POWER PLANT
12700 VISTA DEL MAR - PLAYA DEL REY

| SET | D/N | SCENES | CAST | PGS | LOCATION |
|---|---|---|---|---|---|
| EXT. PLANT | N5 | 247,248,249pt. | | 3 | (see above) |
| EXT. PLANT | N5 | 250,251pt,253A, 254pt | | 2 5/8 | " " |
| INT. KXLA TRAILER | N5 | 249pt,251pt | | 5/8 | " " |
| NOTE: 5:00PM REHEARSAL SCENES: 247, 248, 249pt ! | | | | | TOTAL PAGES: 6 2/8 |

| CAST & DAY PLAYERS | PART OF | MAKE UP | SET | REMARKS |
|---|---|---|---|---|
| 1 Jane Fonda | Kimberly Wells | 5P | 630P | Courtesy PU 430P |
| 3 Michael Douglas | Richard Adams | 5P | 630P | Courtesy PU 430P |
| 4 Daniel Valdez | Hector Salas | HOLD | | |
| 5 James Karen | Mac Churchill | 5P | 630P | |
| 6 Peter Donat | Jacovich | HOLD | | |
| 7 Bill Brimley | Spindler | 5P | 630P | |
| 8 Scott Brady | DeYoung | 5P | 630P | |
| 9 James Hampton | Gibson | 5P | 630P | |
| 10 Richard Herd | MacCormack | 5P | 630P | |
| 13 Khalilah Ali | Marge | HOLD | | |
| 14 Jack Smith | Tom | 5P | 630P | |
| 15 Stan Bohrman | Pete | HOLD | | |
| 26 Reuben Collins | Sportscaster | HOLD | | |
| 39 Frank Cavesani | Reporter #1 | 430P | 630P | ND Meal |
| 40 Terry House | Reporter #2 | 430P | 630P | ND Meal |
| 61 Chris Woods | Reporter #3 | 430P | 630P | ND Meal |
| 62 Val Clenard | Reporter #4 | 430P | 630P | ND Meal |
| 63 Mark Carlton | Reporter #5 | 430P | 630P | ND Meal |
| 60 Clay Hodges | SWAT Chief | 5P | 630P | |
| Trudy Lane | Alma Spindler | 5P | 630P | Recalled for added scene |
| 46 Ron Lombard | Barney | 5P | 630P | " " " " |
| 49 Tom Eure | Tommy | 5P | 630P | " " " " |
| 44 Dan Lewkowitz | Donny | 5P | 630P | " " " " |

| ATMOSPHERE AND STANDINS | |
|---|---|
| 3 SI (Gary, Jerry, Jim) | 5PM |
| 4 SWAT | 530PM |
| 30 Plant Techs/15 F.B.I./4 Guards | 530PM |
| 4 Firemen/2 Fire Rescue/2 Medics | 530PM |
| 12 TV crew/4 Reporters | 530PM |
| 2 Coroners/10 Sheriffs | 530PM |
| 6 Police/2 MC Cops/Own Ward. | 530PM |
| 6 Wives | 530PM |

**SPECIAL INSTRUCTIONS**

PROPS/SET DRESSING: News mike & equipment, guns, SWAT, stretcher, recorders & notebooks, plant blueprints, hard hats, walkie talkies, plant cushman cart, tools, plastic explosives & blow torch, beer
FX: wet down exterior/practical red lites, all vehicles
POLICE: 2 @ 5PM
VTR: Playback in KXLA Trailer - scs. 249,251, 254
1 Security Cars 5PM

**CREW**

| | | | | | | | |
|---|---|---|---|---|---|---|---|
| 1 Cameraman | 5P | 4 Grips | 5P | 1 Mixer | 530P | 1 Make Up-Fonda | 448P |
| 3 Operator | 530P | 6 Electricians | 5P | 1 Boom Man | 5P | 1 Hair-Fonda | 448P |
| 3 1st Asst Cam | 5P | 1 Gen Op-Loc 40 | 5P | 1 Cable Man | 5P | 1 Costumer-Fonda | PU 415P |
| 1 2nd Asst Cam | 5P | 40 man | | 2 VTR Crew | 5P | 3 Make Up | 418P |
| Loader | | 2 Sp Effects | 5P | 1 Painter | 5P | Hair | |
| 1 Stillman | 530P | 3 Props | 5P | Greens | | 1 Men's Costumer | 418P |
| 1 Script | 5P | 1 Craft Service | 430P | X Set Dress | Per Art | 1 Women's Costumer | 5P |
| 1 1st Aid | 5P | | | | | 1 Xtra Costumer | 418P |

**TRANSPORTATION**

| PICTURE CARS/SPECIAL EQUIP | | | | MOTOR HOMES/CARS |
|---|---|---|---|---|
| 1 SWAT Truck | Catering: | | CamTk Loc | 1 Mot Hom Fonda Loc 430P |
| 2 Fire Tk & Fire Rescue | 30 Nd Meals Rdy | 515P | 1 Prod Van Loc 5P | 1 Mot Hom Douglas Loc 430P |
| 1 Ambulance | X Gal Coffee Rdy | 515P | Loc | Mot Hom Lemmon Loc |
| 4 Minicam Trucks | Doz Donuts Rdy | | Effx Trlr Loc | Mot Hom Bridges Loc |
| 1 Coroners Wagon | Lunches Rdy | | 1 Utility Tk Loc 5P | 1 Prod Car PU. Bridges 345P |
| 3 TV station wagons | 215 Dinners Rdy | 11P | 1 Prop Tk Loc 5P | Staff Car } Per Craig |
| 4 Sheriff & Police Cars | | | Loc | Key Car } Per Craig |
| 2 Motorcycles (Police) | Busses | | 1 M/U-Ward Trlr Loc 4P | Trans Car } Per Craig |
| 1 Police Helicopter | 1 Titan Crane & Arm | | 3 HoneyWgns 18 DR 4P | Fonda Car PU Dot 415P |
| 1 CG&E Helicopter | | | | Cast Car PU Douglas 430P |
| 1 Water Truck | | | | |

ADVANCE

Fri., Mar. 17 - complete above - total pages (6 2/8)

style, and every style is different. I get up every morning, make myself some coffee and go directly to the typewriter. I'll retype a scene sometimes to get it in my head. Then I'll type a diary of thoughts and images for myself and place this opposite the screenplay pages that I'll be shooting that day. I'll also make a list of shots needed for each scene and perhaps sketch out drawings of a scene, including camera placement and movement.

When I arrive at the set, I go to the first assistant and to the cameraman with my list of shots. We discuss the scene and review the shot list in terms of movement, style and content. The cameraman will cover the plans for the day's shooting with his staff, along with the lighting and sound crew. Then I am free to work with the actors.

## PRODUCTION REPORT

The production report is a document that states what has been accomplished during a day's shooting. It notes exactly what time the first shot was made; the number of scenes shot; the number of scenes to be shot in order to complete the picture; the status of the production (ahead or behind schedule); how many feet of film were shot, printed and wasted; the exact time players reported to the set, ate meals and were dismissed for the day; and any explanations of injuries or delays. The script supervisor, one of the key people around the camera, supplies the first assistant with the time of the first shot, time of camera wrap, number of scenes completed, number of setups and the total and daily timing of scenes. All other information, such as the company's status and how time was spent, is prepared by the first assistant. The first assistant confers with the production manager at the end of each day to determine how the picture is doing in terms of cost and what changes need to be dealt with in allotting additional time for shooting, thereby affecting scheduling, contract changes, and rental agreements. The information on the production report is prepared by the second assistant, checked and approved by the first assistant and approved as well by the production manager, who is responsible to the producer and to the financier-distributor for the cost of that day's shooting. The production report is a documented account of time spent, and therefore of money spent.

Today computer software can bring increased speed and flexibility to production paperwork. There are separate programs for screenplay formating, budgeting and scheduling. The results look just like these examples in this chapter, and variations can be generated at the push of a button. For example, using a budgeting program on a PC, any change during shooting involving cast, location or material can have its dollar impact reflected in the budget immediately and broken down in variations from cost-to-complete analysis to isolating pension, health and welfare benefit payments.

**EYEWITNESS** LTD.

Sunset-Gower Studios • 1438 No. Gower St.
Hollywood, California 90028 466-3295

| NO. OF DAYS ON PICTURE INCLUDING TODAY | | | | | |
|---|---|---|---|---|---|
| Holidays | Idle | Travel | Rehearsal | Work | Total |
| 1 | | | | 43 | 44 |

PRODUCER: Michael Douglas
EXEC PROD: Bruce Gilbert
DIRECTOR: James Bridges

PRODUCTION REPORT
PROD. 132441

DATE: THURSDAY, MARCH 16, 1978

Date Started 1/16/78 Est. Finish Date 4/4/78 No. Days Estimated 48 + 1 HOLIDAY Status 7 DAYS BEHIND

Location & Sets:
EXT. PLANT, FINALE
SCATTERGOOD POWER PLANT - 12700 VISTA DEL MAR - PLAYA DEL REY, CA.

Crew Call 5P Shooting Call 6P Lunch From 11P Til 1130P Dinner From - Til - Finished 440A
First Shot 855P First Shot After Lunch 1230A

| SCRIPT | SCENES | PAGES |
|---|---|---|
| Total in Script | 258 | 129 3/8 |
| Added Sc. (Credits) | | |
| Shooting Total | 258 | 129 3/8 |
| Taken Prev. | 157 | 84 1/8 |
| Taken Today | 1 | 1 2/8 |
| Total To Date | 158 | 85 3/8 |
| To Be Taken | 100 | 44 |

| MINUTES | | SETUPS | |
|---|---|---|---|
| Prev. | 81:15 | Prev. | 428 |
| Today | 1:10 | Today | 13 |
| Total | 82:25 | Total | 441 |

SCENE NOS. 247

Added Scenes/Retakes Other Pictures:

| | PICTURE NEGATIVE Exposed | Print. | No Good | Waste | SOUND Prod'n. Rolls | Footage |
|---|---|---|---|---|---|---|
| Shot Prev. | 149,010 | 83,630 | 65,000 | 16,130 | 75 | |
| Shot Today | 4,110 | 2,110 | 2,000 | 800 | 1 | |
| Total To Date | 153,120 | 85,740 | 67,000 | 16,930 | 76 | |

| FILM INVENTORY | |
|---|---|
| Ttl drwn | 209,310 |
| Rec'd. | 0 |
| Used | 170,050 |
| Bal. O.H. | 39,260 |

CAST — Contract and Day Players
Worked - W Rehearsal - R Finished - F
Started - S Hold - H Test - T
Travel - TR.

| No. | CAST | CHARACTER | W H S F R T TR | WORK TIME Report | Dismiss | MEALS 1st | 2nd | TRAVEL TIME Leave for Location | Arrive on Location | Leave Location | Arrive at Studio | REMARKS |
|---|---|---|---|---|---|---|---|---|---|---|---|---|
| 1 | Jane Fonda | K. Wells | W | 515P | 440A | 11P-1130P | | | | | | |
| 3 | M. Douglas | R. Adams | W | 535P | 440A | 11P-1130P | | | | | | |
| 4 | D. Valdez | H. Salas | H | | | | | | | | | |
| 5 | J. Karen | Churchill | W | 5P | 440A | 11P-1130P | | | | | | |
| 6 | P. Donat | Jacovitch | H | | | | | | | | | |
| 7 | B. Brimley | Spindler | W | 5P | 310A | 11P-1130P | | | | | | |
| 8 | S. Brady | DeYoung | W | 5P | 310A | 11P-1130P | | | | | | |
| 9 | J. Hampton | Gibson | W | 5P | 310A | 11P-1130P | | | | | | |
| 10 | R. Herd | MacCormack | W | 5P | 310A | 11P-1130P | | | | | | |
| 13 | Khalilah Ali | Marge | H | | | | | | | | | |
| 14 | J. Smith | Tom | W | 5P | 320A | 11P-1130P | | | | | | |
| 15 | S. Bohrman | Pete | H | | | | | | | | | |
| 26 | R. Collins | Sptscaster | H | | | | | | | | | |
| 30 | D. Arnsen | Cameraman | W | 430P | 315A | 11P-1130P | | | | | | ND Brkfst |
| 39 | F. Cavestani | Reporter #1 | W | 520P | 315A | 11P-1130P | | | | | | |
| 40 | T. House | Reporter #2 | W | 430P | 315A | 11P-1130P | | | | | | ND Brkfst |
| 61 | C. Woods | Reporter #3 | W | 430P | 315A | 11P-1130P | | | | | | ND Brkfst |
| 62 | V. Clenard | Reporter #4 | W | 430P | 315A | 11P-1130P | | | | | | ND Brkfst |
| 63 | M. Carlton | Reporter #5 | W | 430P | 315A | 11P-1130P | | | | | | ND Brkfst |
| 60 | C. Hodges | SWAT Chief | W | 5P | 315A | 11P-1130P | | | | | | |
| | T. Lane | Alma Spindler | W | 5P | 320A | 11P-1130P | | | | | | RECALLED FOR ADDED SCS. |
| 46 | R. Lombard | Barney | W | 5P | 310A | 11P-1130P | | | | | | RECALLED FOR ADDED SCS. |
| 49 | T. Eure | Tommy | W | 5P | 310A | 11P-1130P | | | | | | RECALLED FOR ADDED SCS. |
| 44 | D. Lewkowitz | Donny | W | 5P | 310A | 11P-1130P | | | | | | RECALLED FOR ADDED SCS. |

# EYEWITNESS

DATE: Thursday, March 16, 1978

EXTRA TALENT — MUSICIANS, ETC.

| No | Rate | Adj. To | O.T. | T.T. | Ward. | MPV | No. | Rate | Adj. To | O.T. | T.T. | Ward | MPV |
|---|---|---|---|---|---|---|---|---|---|---|---|---|---|
| 59 | 56 | | 1.4 | | | | 1 | 56 | | .3 | | | |
| 12 | 56 | 100 | 1.4 | | | | 2 | 56 | | .4 | | | |
| 7 | 56 | | 2.6 | | | | 1 | 56 | | .9 | | | |
| 6 | 56 | | 2.1 | | | | 3 | 56 | | 1.8 | | | |
| 5 | 56 | | 1.4 | | 11.00 | | 1 | 56 | | 1.9 | | | |
| 1 | 56 | | 3.2 | | | | 3 | 56 | 66 | 1.4 | | | |
| 1 | 56 | | 2.1 | | 11.00 | | 2 | 56 | 66 | .4 | | | |
| 1 | 56 | | .4 | | 11.00 | | 2 | 56 | 66 | 3.1 | | | |
| 1 | 56 | | | | 11.00 | | NOTE: | 10% NITE PREMIUM 8PM TO 1AM | | | | | |
| 1 | 56 | | | | | | | 20% NITE PREMIUM 1AM TO WRAP | | | | | |

**Staff and Crew**

1 Director
2 Asst. Directors
__ Asst. Dir. Trainee
1 Script Supervisor

1 Cameraman
3 Operator
5 Assistants
1 Still Man
1 Mixer
__ Recorder
1 Mike Boom Man
2 Cable Man
8 VTR Crew
1 Propmaster
2 Asst. Propmaster
1 Key Grip
1 2nd Co. Grip
1 Crab Dolly Grip
__ Crane Grip
2 Extra Co. Grips
1 Craft Service
__ Greensman
1 S.B Painter
1 Gaffer
1 Best Boy
6 Lamp Operators
1 Generator Man
__ 40 Man
1 Special Effects Man

4 Makeup Artist (1 Fonda)
1 Hair Stylist (Fonda)
__ Body Makeup Woman
2 Wardrobe Man
2 Wardrobe Woman (1 Fonda)
__ Wranglers, Trainers, Handlers
25 Drivers

**Miscellaneous Crew**

1 1st Aid
__ Fireman
2 Police
X Prod Designer + Dept.
__ Art Director
X Set Decorator + Crew
__ Swing Gang
X Construction crew
2 Watchmen: 1 @ 5P & 1 @ 5A
3 Water truck ops.
1 Helicopter Pilot

**Equipment**

__ Camera/Sound Truck
1 Prod Van—Grip/Elec/Gen
2 Crew Cab
1 FX Trailer
1 Utility Truck
1 Prop Truck
1 M/U Ward Trailer

3 Honey Wagon 19 DR
2 Motor Homes
2 Busses 40 passenger
1 Maxi Van
6 Station Wagon
__ Misc Cars
20 Picture Cars (see notes *)

3 Cameras 2 Panaflex, 1 Arri
1 Grab Dolly
__ Insert Car
1 Crane Titan
X Lunches 235 Dinners

REMARKS & EXPLANATION OF DELAYS:

* Vehicles: 1 SWAT Truck, 1 Fire Truck, 1 Fire Rescue, 1 Ambulance, 4 Minicam trucks, 1 Coroner's wagon, 3 TV station wagons, 4 Sheriff & Police cars, 2 motorcycles, 1 helicopter, 1 BMW

10% Nite Premium 8PM to 1AM & 20% Nite Premium 1AM to Wrap

Assoc Prod-Production Manager Jim Nelson    Assistant Director Kim Kurumada    2nd Asst. Director Barrie Osborne

What I find exciting about making a movie is that every day I hold the entire picture in my head, constantly arranging, rearranging and working as it takes on a life of its own. *The China Syndrome* posed a specific cinematic problem, and I was determined that it would be resolved. With the support of the creative and technical team, the picture was completely realized. Good directing, to me, involves hiring the very best people in all departments and creating an atmosphere wherein they all can contribute to the end product. With the help of my staff on *The China Syndrome*—including first assistant Kim Kurumada and executive secretary Debbie Getlin, who helped gather these sample pages—I was free to concentrate on the dramatic telling of the story. Making *The China Syndrome* was a rich and satisfying experience.

VII
THE
SELLING

# DISTRIBUTION AND EXHIBITION: AN OVERVIEW

*by* **A. D. MURPHY,** *who was financial editor, film critic and news reporter for* Daily Variety *and weekly* Variety *from 1965 to 1978. He continues to write for those publications in a special-correspondent status, having shifted to a new full-time teaching career in the School of Cinema-Television, University of Southern California. He was the founder-director of a unique two-year graduate program in motion picture producing and management and remains on the faculty of that program.*

*This article is from the journalism of Mr. Murphy that originally appeared in the* Daily Variety *issues of January 10 and February 14, 1977, and February 1, June 14, June 28, August 3 and October 26, 1978; it is included here through the kind permission of Syd Silverman, chairman, Variety Inc., as updated and revised by Mr. Murphy expressly for this volume.*

*. . . one might . . . multiply rentals by 2.5 or 3 to get an idea of the box-office. But that's a largely meaningless exercise, more suitable for cheap puffery . . . than for really understanding what makes the film business tick. . . .*

In the dawn of the industry, say, before 1910, the earliest films (one-reelers) were sold outright to exhibitors at so much per foot. In time public demand and selectivity forced production costs higher and led to the introduction of *percentage rentals* to justify the investment risk.

## EARLY PERCENTAGES

Film distributors originally licensed films to exhibitors for rentals that were less than 20% of the box-office receipts. By the 1920s the terms were under 25%. By the mid-1960s a theatre circuit comprising a mix of first-run, key neighborhood and lesser theatres wound up at the end of the year paying out to distributors in the aggregate no more than about one-third of the related box-office. This overall one-third cost of renting film derived from hundreds, even thousands, of different film play dates, some at high percentages (50%–60%), some at low (35%–40%), some at flat (or fixed) rentals. Circa 1990 the statistical payout is near 45%.

The realization that an exhibitor's overall film costs are composed of various percentage splits of the box-office with the distributors on thousands of different runs of hundreds of different films is the beginning of wisdom in understanding the film business.

In key first-run dates the distributor takes 90%. But, as in all such things, one must always ascertain the percentage of *what?* In this case it is not of the total box-office, but 90% of the box-office minus an allowance for what is often held to be the theatre's operating costs (or "nut").

## FIGURING THE NUT

A major New York or Los Angeles first-run house, circa 1990, can claim a house expense of, say, $20,000 per week; this, like so much else in the industry, is a *negotiated figure*. It is not arrived at by examination of actual records; the exhibitor announces his claimed costs, and the distributor accepts or rejects. They finally agree.

Assume for the moment, at least, that house nuts approximate reality. (The old distributor claim that there was a lot of "water" or "air" in house nuts—in other words, a built-in profit—may be set aside by the realization that house allowances often do not change for years, though everyone is aware of inflation.) An example or two may be of benefit.

Say that a certain key N.Y.–L.A. first-run theatre does, in the first week of a film's premiere run, a box-office of $50,000. Take out the $20,000 house allowance, and the distributor's share is 90% of $30,000 (the adjusted box-office figure), or $27,000. That $27,000 in film rental is, in actuality, 54% of the public's money, not 90%. If the raw box-office figure had been $70,000, the adjusted figure would be $50,000, of which the distrib's $45,000 rental would be 64% of the public's ticket money. The bigger the box-office gross, the bigger the rental percentage; but at bigger numbers *nobody complains.*

But let's say the film is a dud, and the week's box-office is $25,000. After taking out the house allowance, the distrib would *seem* to get 90% of $5,000, or $4,500, which is 18% of the raw figure, unless floor figures are operative (see below).

Looking at the 90/10 deal another way, the exhibitor's share can be considered a guaranteed covering of costs plus a guaranteed 10% profit. Cost plus 10%—it doesn't sound nearly so bad that way. A lot of businesses would be glad to know that their expenses would be covered and they would make a 10% profit as well.

## FLOORS

But now what about these floor figures? They came into being in the late sixties as a countermove by distributors who questioned the validity of house nuts and who no longer could afford to absorb most of the losses incurred by a failed film. The *floor*-percentage innovation meant that, however the 90/10 arithmetic went, in no case would the distributor take less than that floor percentage of the raw or actual box-office figure.

Floor figures typically are 70% for the first few weeks of a film's run, 60% for the next leg of the run, then 50%, and then 35%–40%. Once again, when business is torrid, nobody complains; when it is sluggish or bad, the floor figures, ranging from 70% on down, are operative. Floor figures, however, only apply to film-ticket revenues, not total theatre receipts, which include concessions and, nowadays, pinball machines and so on.

There has always been grousing over film licensing terms. An exhibitor instinctively objects to relatively high percentage takeouts by the distributor, but the film supplier points out that high-profit concession income is not divided at all. The distributor can further claim that he has assumed the total up-front risk in making the film, a process that may have taken as long as eighteen months before opening day, and has assumed most of the promotional burden as well.

Aha, you say, the distributor has assumed *all* the up-front production risk? What about those (nonrefundable) guarantees and (refundable) advances that exhibitors these days sometimes put up before getting the film delivered? Good question, even though the record is filled with cases of exhibitors offering front money *even when it wasn't asked for* in order to get a film away from a competitor's house. (On that point 90/10 deals are often requested by an exhibitor, not always imposed by a distributor.) Let's explore this guarantee-advance (front money) situation.

The bulk of the front money pledged by exhibitors is not due until a couple of weeks before opening day; thus, while the nonrefundable (guarantee) portions of it can be numerically compared to production-investment risk, it is not in the distrib's hand until shortly before opening, although the filming may have begun a year or more earlier. And then, after opening, the exhibitor keeps every cent of the box-office money—*all of it*—until the guarantee is earned. With the blockbuster films of recent years, front money has been earned in the first days or week of a run, or at least very early on.

The point is, the exhibitor has not posted collateral until just before opening; he keeps all the box-office money until the front money is earned, and if he has one of those big pictures, he is inconvenienced for a relatively brief period.

To be sure, some films don't earn their guarantees (though the advances are refundable), and that is certainly a potential burden on an exhibitor who guessed wrong, especially when he made his commitment without even seeing the picture, through blind bidding. (Contrary to uninformed opinion, you can't find any responsible people who actually defend blind bidding as a positive market force; some state legislatures have outlawed the practice, under pressure from exhibitors.) Of course the distributor who financed the film guessed wrong, too, so sympathy must be allocated to both parties.

In the early days Marcus Loew and William Fox owned theatres, and, then as now, a theatre owner would go berserk not knowing what film to put on the screen when the film showing had ended its run. For Loew and Fox the solution was to go into direct production to assure enough films for their own houses; since neither had theatres everywhere, they automatically went into de facto distribution to market their films after their own first-run preemptions had occurred.

## SELF-PROTECTION

Also in that era Adolph Zukor, a producer and then a producer-distributor, was faced with rising talent costs (spell that Mary Pickford) and decided that if he couldn't dominate the industry through talent, he'd protect himself by owning enough theatres at least to guarantee first-run exposure for Paramount pictures. Thus, from opposite directions, the "vertical integration" emerged, with production, distribution and exhibition functions consolidated under what became five major corporate roofs.

These companies had it made in the shade, but they were too aggressive. Nonaffiliated theatre owners, not permitted to participate in various cozy reciprocal product poolings, began to agitate. It took nearly forty years, but finally the government acted. After a key Supreme Court opinion, the studio-owned theatre chains were divested; however, the producer-distrib partnerships were left legitimately linked.

The loss of guaranteed key first runs on films, the competition of the new TV medium and a general expansion of what is now known as the *leisure-time* sector all worked havoc on the industry. As previously documented by *Variety,* the real end of this postwar decline came in 1971, fully twenty-five years of industry erosion after the peak year 1946. Since 1971 (which climaxed a disastrous industry depression escaped by no major company), there has been a steady improvement in industry health.

It is true that major producer-distribs dominate 85%–90% of the market. It is also true that such concentration occurs in most industries; the definition of a "major" anything presumes the impact. Not fully appreciated is the fact that the film industry's traditional source of production money is not commercial banks; the largest single source is recycled film rentals paid to the major distribution companies. To be sure, the majors have lines of credit with the big banks, but they properly represent standby money, not unlike one's use of bank credit cards. When many large companies forgot their self-financing responsibilities and drew down all their credit to the limit, the result was the 1969–71 Hollywood depression.

A lot of people—not just exhibitors and their new-found political and populist allies—are wondering more and more why it is that, given the generally improving conditions of the theatrical film industry since 1971, the major American film companies are making fewer and fewer pictures? One answer is that within the current economic realities of the business, they cannot extend production investment too much without repeating the mistakes of overproduction that helped cause the disastrous Hollywood depression of 1969–71.

In the following analysis, which attempts to explain the proffered answer to the question, a few preliminary ideas are worth stating.

- Neither a film producer-distributor nor an exhibitor can claim some Divine Right of staying in business; staying alive in business is the perpetual problem of every individual company.
- A major diversified film production-distribution company—active in music, recreation parks, telefilming, whatever—cannot be expected to lose money on theatrical filmmaking (in many cases the primary business of such companies) and make it up on the diversification.
- The last preliminary idea to get is the speciousness of citing exhibitor "demand" as a reason for making more films. Retailers of all kinds in all industries always clamor for "more" product (more "successful" product is always meant, but is rarely explicit). This is specious because the only safe "demand" to which a supplier can respond is that of the ultimate consumer—the public.

Thus you have to start with the prevailing public demand (as measured in box-office dollars) and then work back up the system, through the film suppliers' share of that box-office (measured in film rentals); through the additional license fees from posttheatrical markets; through current and anticipated costs of marketing (prints and advertising); through outlays in creative talent gross and net participations; through actual production costs and actual overhead (that is, people overhead, not the percentage kind); through project development (the industry's R & D expenses); through the annual interest on any outside production borrowings; and so on all the way to a final dollar figure that you sincerely hope is a positive number.

But what of the film projects that, before release, are often largely covered by a television license deal, a soundtrack or book deal, a merchandising deal, or some other deal? That's fine on a case-by-case basis. However, not every film lends itself to such lucrative peripheral and secondary exploitation. In other words, when you find these angles, great; but when you *need* them up front, to justify the financial risk, you're already in trouble.

Moving now into some numbers, assume one of the major companies early in 1991 is trying to get a reasonable fix on a feature production program outlay (including pickups from independent producers) that will go to market in 1992–3; this assumption allows for the long lead time in production.

## BUDGETING MODEL

By 1989 domestic box-office revenues were near $5.0-billion, from which major American distributors generated something near $1.9-billion in film *rentals*.

Foreign markets in this period returned about $1.1-billion in rentals, some of which was of course left in those markets for business purposes. In round total, say that the annual average global rental from theatres at that time came to approximately $3.0-billion. Factor in the exploding posttheatrical markets of home video, pay-TV, free-TV, and so on, and the all-media rental total by 1989 was $10.0-billion or more.

This industry-wide rental figure could reasonably support *direct* production investment in the $4.5-billion range. This ballpark figure is determined by taking out the distribution fees (but then applying the excess of such fee income over actual fixed distribution organization costs to new investment and recoupment); subtracting the costs of prints, promotion and participations; and charging off abandoned development costs, interest, and so on.

Next divide this hypothetically "safe" industry-wide production figure by eight majors. The resulting "safe average" of production investment for each major studio is therefore in the range of $560-million.

Finally, for each of the majors, divide that studio figure by the prevailing $24–$25-million average cost per film—the "event" pictures plus the costs of all the rest—and you wind up with the reasonable number (about twenty to twenty-two, give or take) of films that figure can yield, including acquisitions, negative pickups, and so on. In other words, the proposed future production investment planning has got to be based on the current and recent public theatrical and posttheatrical market, using as a point of departure an "average" expected market share.

All right, then, assume that you budget on the averages, enjoy the greater market share if it happens and don't hurt too badly if another company's greater market share squeezes your own (as it truly will). But now comes the question, why can't a studio get more films for that $560-million annual outlay? It would be nice if it could; it's safe to say that if a studio could get thirty films for the figure, it would make them. However, notwithstanding the explosion in posttheatrical markets, a film's economic value is set in the initial theatre market, and that market is flat at 1.1-billion annual domestic ticket sales.

## OUTSIDE MONEY

Next question: Why not then augment the studio's investment with outside money—overseas tax shelters and so forth? That's being done, too, but remember that an outside partner isn't working for nothing either; the more partners—fiscal as well as creative—the more outside payments to them from gross and net.

Well, then, what about these big new independent production companies that are fully financing the production of their films? A complex question; for one thing, they are competing with major studios for the same talent, which means talent prices won't be going down, all other things being equal. On the other hand, these self-sufficient production companies at least provide more places for film ideas to flourish and develop. Also, ultimate distribution through a major company will enhance the number of released pictures (unless of course the distributor simply allocates more acquisition money at the expense of house production coin, in which case there is no increase in the number of films available to exhibitors from that source).

But isn't there a paradox lurking here? If in a given year there isn't that much of an increase in the public market, won't a significantly larger number of films being released result in more pix fighting for essentially the same business? Yes, that's true. Then won't somebody, or somebody's films, get squeezed in the crush? You'd better believe it. The surviving suppliers will be those who have the broadest and most diversified base, as well as the least fiscal recklessness.

## THE LAST MYTH

The last myth—and maybe the biggest one of all, among the unsophisticated—is worth discussing. There are those who with apparent sincerity believe that if the current major film production-distribution companies weren't so firmly established in their ways, and if filmmakers and exhibitors dealt directly in an Adam Smith–type competitive open marketplace of many suppliers and many buyers, there would be peace and prosperity. "Eliminate the middleman," as the saying goes; "Plain pipe racks, direct to you."

If that happened—and it did happen once, in the dawn of the film business, when the pioneer producer sold his product direct to his exhibitor customers—there would be no strong industry. Filmmakers would be living hand-to-mouth, waiting for the last film's payoff to begin the next one; exhibitors would be scrambling for product when the promised film was delayed; and so on. But in time there would evolve an idea: a person or company would undertake the risk of financing volume production in some orderly fashion and then assume the added risk of marketing to make it available to theatres concurrent with periods of maximum patronage, and licensing the films to world media.

The decisions of this entity or force would frequently become controversial; filmmakers and exhibitors alike, for common and different reasons, would grouse, often bitterly. But the force would be with them, and when business conditions stabilized again, there would be perhaps five to ten of them. Their collective name: *major distributors*.

## ROADRUNNER VS. WILE E. COYOTE

If one wanted to make an animated picture out of the film business, the characters of distribution and exhibition already exist: the Roadrunner for distribution and Wile E. Coyote for exhibition. Those two Warner Bros. cartoon personalities fit the situation like a glove, especially Coyote, the clever but frustrated adversary whose best-laid plans always blow up in his face.

The soap-opera relationship between distribution and exhibition began more than half a century ago, perhaps from the day that theatres could no longer buy film for so many cents per foot. Whatever the initial incident, the drama has been running ever since. The usual episodic climax finds exhibition sandbagged, just like Wile E. Coyote.

A few of the more dramatic episodes come to mind. When Paramount's Adolph Zukor decided that he couldn't keep paying Mary Pickford more money and revised his strategy from controlling the industry through stars to control via ownership of key first-run theatres, a major theatre-acquisition spree ensued. It was marked by the delicacy often associated with waterfront labor practices. Some key regional exhibitors decided to band together to resist. Their brainchild: First National Pictures, which they bankrolled in return for regional first-run preemption and, presumably, lower film costs. But to get established, First National immediately paid Pickford and Charles Chaplin more money than they had ever extracted from anyone else. (So much for lower film costs and rentals.)

Then came the outcry in the twenties from exhibitors who, shut out of the various theatre-circuit product-pooling arrangements, complained about restraint of trade. The United States government accumulated a lot of evidence, and many indictments were expected. But along came an unexpected event—the Depression—which diverted attention to more pressing national concerns.

In 1938, when the national recovery was stumbling along, exhibitor complaints again got the ball rolling, and the government finally took action against these distributor miscreants via a *civil* complaint, not a criminal one. The usual legal jockeying took place again, only to be interrupted by another event—World War II.

Then, by the late forties, when the handwriting was on the wall for theatre-circuit separation from major integrated companies, the inroads of television and an enlarged leisure spectrum on theatrical box-office made such separation most attractive for the distributors. Exhibition is a cash business and also a cash drain (the majors, which went into receivership in the Depression, did so because of their theatre subsids); the lights must burn every day, the staff must be paid every week and so on.

Distributors by 1950 were not unwilling to unload what was in fact a major cash hemorrhage in their overall operations. In the process, by no longer having to supply their own and affiliated houses every week, the studios lopped off their creative contract talent as well.

## PADDING THE NUT?

Come the fifties and sixties and distributors were getting higher percentages to license films to exhibitors. In the 90/10 licensing deal, the distributor gets 90% of the box-office *after* a reduction for the exhibitor's house expense. But those house-expense figures were (and still are) negotiated numbers, and there was the lingering suspicion that some of the house nuts had a little built-in profit to the theatre's benefit.

Thus the next countermove was a minimum floor figure, a percentage of the *actual* box-office total that the distributor would get regardless of the 90/10 house-nut formula.

All this time, mind you, the haggling over settlement of played-off films resulted in many tens of millions of overdue film rentals, held by exhibitors pending settlement of the play-date contracts. This situation has become less severe in recent years through a combination of factors: the simple march of computer accounting technology; and a new ploy by distribution, the obtaining of guarantees (nonrefundable) and advances (refundable in certain circumstances) on major films.

It should be apparent by now that every trade practice breeds its own countermeasure, which in turn leads to a new gambit.

## SHAKY COUNTERMEASURE

But there was one exhibition countermeasure that, in its own small, shaky and unstable way, occasionally worked out: the *split,* a prearranged local formula by which the local exhibs divided up the product. Splits are very shaky since a couple of bad pictures can get an exhibitor so hungry for a hit that he will torpedo the local detente. Distributors look forward to such an event since, with more than one potential buyer, the terms can easily improve to the distributor's benefit. When most local exhibitors became tired of losing money on overbid films, a split often resumed.

Exhibitors, seeking government help, turned up the heat on distributor licensing practices. And what happens? By the perfect timing that exhibition has always managed to achieve, the government seizes upon and declares illegal the one trade practice that, in some places and at some times, alleviates a perceived burden on exhibition—the split.

The lessons of history are obvious. In a reverse of the situation that obtains in the TV and radio industries, where legislators (or their families, or their law firms) are cozy with broadcasters, exhibition (which pretends to grass-roots consciousness and under certain circumstances has it) lacks sufficient friends at the source of the magic federal power it wishes to turn against its natural enemy.

Will Hayes, Eric Johnston and now Jack Valenti, as successive heads of the Motion Picture Association of America, have a long-standing relationship with Washington. Maybe exhibition can get its act together and rapidly develop a strong lobbying pressure there. But considering past history, maybe this latest situation is just another example of Wile E. Coyote assembling his Acme-brand Roadrunner kit only to find it's another boomerang. In fact exhibitors flopped so badly at the federal level that they are lobbying at the local level, with much better success.

## BOX-OFFICE VS. FILM RENTAL

Perennial confusion, both in the film trade and in consumer media, centers on the terms *box-office* and *film rental.* Feature magazine writers interchange the terms with abandon, and filmmakers don't help the situation with their careless intermixture of figures. Confusion arises from the premise that, since the paying public buys tickets, the theatre box-office total is therefore the best measure of a film's economic performance. Right? Wrong!

Every film booking every week generates some box-office coin. But that money is divided between the exhibitor and the distributor according to the terms of the licensing agreement (90/10 with floors, 50%, sliding scale, etc.). The exhibitor's share remains in the theatre to pay the mortgage, staff and so on and sustains the brick and mortar of the exhibition arm of the industry.

The distributor's share of the box-office is what is known as *film rental* (or distributor's gross; the terms are reasonably synonymous). Film rental is the crucial figure because film rentals pay for the production, promotion and prints of the film as well as for participations in gross when the creative talent is in the superstar echelon.

In short, film rentals are what recycles the industry—the pool of money that is continually reinvested in new production, in addition to sustaining the distribution organization and being allocated to marketing new product.

If this is the case, why do people keep talking about box-office figures? Because, when a new film opens, the box-office results in the first few hundred play dates give an indication of the film's potential strength. There's another reason that, for devious reasons, box-office figures are flung about: Being the larger numbers, they make a conversation more exciting.

In percentage licensing engagements, distributors do in fact get a box-office report that can also yield the number of admissions and other data. But in tail-end play dates (where the exhibitor pays a fixed amount to play an older film), no box-office report is required. Thus for many films nobody really knows what the box-office total was.

Finally, it's very difficult to extrapolate from rentals to box-office, since a film generates its box-office (and hence rentals) from many different price scales and licensing contract terms.

On a statistical basis (that is, throwing dozens of films of different appeal and performance into one pot for analysis), one might *with care* multiply rentals by 2.5 or 3 to get an idea of the box-office. But that's a largely meaningless exercise, more suitable for cheap puffery and aggrandizement than for really understanding what makes the film business tick.

## DISTRIBUTION FEE

There is also some confusion surrounding the concept of the distribution fee.

The distribution fee is a percentage charge on incoming film rentals from theatres, TV, pay-TV systems and so forth. It is simply a service fee (not unlike the doctor's office-call or house-call fee) designed *solely* to support the existence of the sales organization; it does *not* apply to the recovery of any expenses directly related to releasing a film (such as prints, advertising, etc.). Deductions for direct costs of marketing a film are made after imposition of the fee.

For U.S.-Canadian theatrical distribution the fee is normally 30%; for the U.K., 35%; elsewhere, 40% or even higher. When doing arithmetic, it's often easy to use a global average of 33⅓%. The fee pays the fixed costs of the marketing organization: salary, office leases and so on. For a major American production-distribution company operating its own sales arm directly on a global basis, these annual expenses now approach the $30-million mark, slightly less than half being the domestic-market component. Since one can use (statistically) an average worldwide distribution fee of approximately one-third, this means that when, in a given year, theatrical rentals pass $100-million (depending on the company), the sales department is off its nut. Beyond that incoming rental figure, additional distribution-fee income represents . . . what? Call it what you will, except don't call it *profit*, because it isn't, in the sense of free-and-clear money available for stockholder dividends and such.

What this excess-fee income (over actual costs of maintaining the distribution organization) does is accumulate a pool of money that is often absolutely essential for use in recovering direct and actual marketing and production costs otherwise unrecouped; this pool of money also gets recycled into new film production.

Film companies have large lines of credit with banks, but these are more or less standby credits for seasonal and occasional use. The idea that film companies are forever running to the bank for more money is ludicrous; there wouldn't even be a film business if that were so.

Excess distribution-fee income, and any later distribution profits that it helps augment (or create), is quite literally the blood pressure in the circulatory system that keeps pumping money into new production, thereby keeping the system alive. Reducing this financial equivalent of blood pressure to a minimal life-support level can have the same effect on a company that it would on an individual.

In describing the chronic anemia of the pre-1951 United Artists Corporation to an interviewer a few years ago, Arthur Krim (who certainly knows) put his finger on the matter. He pointed out that the concept of a passive distribution organization—lacking sufficient financial resources for direct and continuing production investment, relying on privately financed producer affiliates to deliver on time and hoping to make money but nervously praying simply to break even—is a blueprint for instability and financial disaster.

To the comment that there are many current and potential new markets for feature films besides theatres, one can agree that other sources of revenue and profit exist. But the economic value of a film is more or less dependent on its theatrical-market performance.

Assume a major American distributor can figure an average of something near $375-million per year in global theatrical film rentals. (The fee income from this is therefore in the $125-million range; after covering the fixed costs of the sales department, this leaves excess fee income of about $95–$100-million.) After deducting the fee from the annual worldwide rental take, this leaves something like $250-million available to pay for four items conveniently remembered as all starting with the letter *p*: prints, promotion, production and participation costs.

This figure is slightly less than half of the production budget model of $560-million noted earlier. Where does the remainder come from? From incoming revenues on *older* films moving through posttheatrical markets.

## RESTRUCTURED EXHIBITION

Resurgent American film-industry prosperity over much of this past decade derives in part from the obvious factors of rising attendance levels in combination with rising ticket prices. Less apparent, but no less important, is the significant transformation in the exhibition branch of the industry: there has occurred a major increase in the number of theatres that enjoy key or subkey status in the types of runs. This has had the effect of keeping major films at a higher level of release, during which admission prices are higher, for longer periods of time.

From a ten-year analysis of the *Variety* key U.S. city box-office data, it is estimated that the proportion of the nation's theatres that regularly enjoy key or subkey run status has doubled, from less than a quarter of all houses to a level that is steadily approaching two-thirds or more.

This upgrading of theatre run status has been effected by both exhibition and distribution. Exhibition has responded to the related factors of urban decay and suburban sprawl by expanding into shopping centers with multiscreen theatres and subdividing older picture palaces; distribution has played its part by radically changing release patterns so as to bestow key and subkey status to thousands of these smaller houses.

A brief look back may make the current situation clearer. For decades there was a sort of three-tier exhibition industry. At the highest level were the key first-run presentation houses, charging the top ticket prices and seating thousands; among them were world-famous local landmarks. The next level comprised both very important neighborhood (more precisely, regional) theatres and many more early subsequent-run houses, where the ticket scales were lower. At the bottom were those late-run, small, low-priced screens that are accurately, though perhaps crudely, known in the trade as *dumps*.

A feature film in former times would move down through this market structure like clockwork. In fact, before major circuit exhibition was divorced from production-distribution, it *was* a clock, with play-off intervals separated by fixed clearances of fourteen days, twenty-eight days, forty-two days and so on between successive play-off periods. Thus a major and popular film would play in first-run theatres to audiences willing to pay top prices; next would come the subrun fanning out to less-prestigious (but also lower-priced) theatres for those audiences unwilling or unable to pay top scales; finally would come late-run exposure at the cheapest price with a lower-income audience to support that tail-end market. The merchandising pattern was the same as that of other consumer goods—the exclusive shops, then the general department store and finally the close-out sale.

## ENTER TELEVISION

But over the past forty years, major social and business upheavals have occurred. The introduction of TV rapidly dried up most of the tail-end theatres (though some shifted to foreign films and thereby became the entering wedge for new film styles from abroad; this important causal influence on the art-house subindustry is rarely discussed). Also, the transformation of metropolitan areas from a wheel-spoke configuration to that of a doughnut with an empty core killed off many of the key presentation houses. At the same time, this change had the positive effect of gradually upgrading the run status of those surviving middle-tier secondary regional and neighborhood theatres, augmented by the postwar drive-in boom and the shopping-center phenomenon in these newly affluent middle-income suburbs.

The twenty-five-year decline (1946–71) in film audiences naturally made many of the older middle-tier theatres into a new bottom rung—the nouveau poor of exhibition. Look at it this way: A lake dries up from the edges, and a consumer market dries up from the fringe; theatres that once were in moderately deep or shallow economic water wound up on the beach.

## SEVENTIES TURNAROUND

This brings us up to the early seventies and the turnaround in attendance. Rising production costs and, more significantly, soaring costs in advertising and promotion combined to influence distributor marketing decisions so as to launch films from a wider base than the old single-screen exclusive run. Hence there are the minimultiples that involve not one theatre but six or eight or ten strategic houses that fairly well ring, but do not saturate, a metropolitan area. Suddenly many older suburban houses and many brand-new outlying theatres have become first-run situations.

Following the launching of new major films comes the first general release, which automatically elevates many more suburban screens into subkey status. The net result of this transformation has been to upgrade the middle-tier theatre into, as often as not, day-date, first-run, and regionally exclusive status for films that play their preem runs longer in smaller houses but at top ticket prices.

Thus, (1) rising numbers of patrons (2) willing to pay the highest prices at (3) conveniently accessible theatres (4) for popular films (5) with longer runs at those locations, along with (6) ticket-price inflation throughout the entire exhibition spectrum constitute the more comprehensive explanation for the latter-day revenue boom. That distributors are getting a higher proportion of film rentals from this higher box-office gross is an effect of both stronger licensing terms and the fact of life that newer product commands a better price than last-run bookings.

That old three-tier play-off of films is now largely obsolete; it's a two-tier marketplace at best—first-run, leftover-run and not much in between.

The standardization of wide-release film distribution was both cause and effect of an explosion in the number of theatre screens—from 14,000 in 1970 to 24,000 or more in 1989. Yet the annual screen total is a *net* number—new openings minus closings. Exhibition is always in a moulting process: Newer shopping districts and residential areas draw newer screens, and these supplant and replace older screens in areas no longer fashionable. Over any 10-to-20-year period, the bottom one-third of screens are driven out of business by newer screen construction.

The enlarging market array for feature films—theatres, pay-TV, free-TV, home-entertainment systems and so on—accelerates and energizes the upgrading of theatres.

Also, the population demographics have changed—that onetime post–World War II "baby boom" that became the college crowd in the sixties has moved into adulthood and is still turned on to films; through the end of the century those people will be a big factor in both theatre attendance and home-entertainment markets. *More* people are watching *more* films in *more* media *more* often than ever before. And despite alarmist fears, the box-office is still the key market because it establishes the value and reputation of a film.

The question now arises, Who is going to control these film houses? Who is going to be listed among the twenty-five to thirty circuits that will dominate the theatre business in a decade? General Cinema Corp. and United Artists Theatre Circuit are already very large chains; their continued presence in exhibition seems ensured by their current moves.

But in a decade the exhibition end of the business would be far more stable if there were half a dozen to a dozen key circuits operating in all or most areas of the country. The long-term costs of current and competing expansion by regional circuits is enormous.

One answer to this potential situation is the consolidation of some major regional theatre chains into larger, more financially stable, nationwide operations. Exhibition is filled with many successful entrepreneurs who have built themselves, over the past thirty years or so, into mighty regional businessmen. They are properly proud, properly powerful and perhaps understandably reluctant to surrender their positions of success.

But that's what has happened over the past decades in the production and distribution ends of the business; the older pioneers and entrepreneurs eventually yielded, sometimes under the force of new business circumstances, to a more contemporary era of economics and management. It may be approaching the time when these regional exhibitor powers ought to begin considering the merits of orderly advance planning and combinations of units into larger chains. The perilous alternatives include localized overbuilding and outbidding, which amount to a deadly game of chicken in which some people do get hurt very badly.

Given the temperaments of some of these regional exhib titans, quite a lot of diplomacy will be needed to effect a combination of business operations. However, a proper long-range look seems to demand that some thinking and action begin to take place now. Since 1986 some film distributors have resumed theatre ownership, which enhances the financial stability of exhibition. Fears of the old trade-practice abuses through theatre ownership by distributors are nullified by many factors. The most important one is the multiscreen nature of exhibition, which means that every theatre needs the films of *all* distributors in order to fill the many screens located at the theatre site. No major distributor can make enough films to satisfy the concurrent needs of all screens.

# MOTION PICTURE MARKETING

*by* ***ROBERT G. FRIEDMAN,*** *president of Warner Bros. worldwide theatrical advertising and publicity. Mr. Friedman began his film career in the Warner Bros. mailroom in 1970 and rose through the ranks to jobs of increasing responsibility. He has been instrumental in heading the advertising, publicity and promotion campaigns for many Warners pictures, including these popular successes:* Superman, Chariots of Fire, The Killing Fields, Risky Business, Gremlins, Lethal Weapon 2, *the* National Lampoon Vacation *films,* Driving Miss Daisy, Robin Hood: Prince of Thieves *and* Batman.

*In determining the media buys, spending levels are set based on how much the picture is expected to gross. . . . The warning becomes, Don't outspend your revenues, but don't underspend your potential. . . .*

Originally movies were driven by publicity generated on the set by studio publicity mills, created to constantly churn out provocative stories in order to keep star names and picture titles before the public. Throughout the 1960s, movies enjoyed a longevity in the marketplace, remaining in theatres for months, in contrast to the make-or-break intense competition of today. Advertising dollars were limited to newspapers and radio.

In the early 1970s the process began to change. First a company named Sunn Classics made a strong impact by distributing family-oriented films regionally, supporting them with saturation network-television advertising, the first time such an approach was methodically applied to movies. Warners refined this strategy for the national-saturation release of *Billy Jack* in 1971, spending unprecedented television advertising dollars to achieve an unusually high level of gross rating points, or GRPs. (*Gross rating points* are a measure of the level of audience tune-in to television commercials, a way of converting advertising dollars into impressions.) *Billy Jack* was highly successful, and movie marketers realized that national-television advertising was the wave of the future.

As a result, release patterns changed to make the ad expenditures more efficient. The national-saturation release quickly replaced roadshow exclusive runs and limited national runs as the preferred release pattern. A picture that previously opened in 500 theatres could now open in 1,000 or more and be cost-efficient, measured against the reach of network-TV buys. By the 1980s the saturation release was expanded to 2,000 or more prints, and exhibition continued building multiplexes to meet the demand.

Today competition is very intense, with two or more new movies opening most weekends. And the audience is unforgiving: there are first-choice movies, and the rest are also-rans, a list that changes every week. The work of movie marketing is a painstaking and very expensive effort to position the audience for that critical opening weekend. Our goal is to make our movie the first-choice movie for its market target; if it doesn't succeed, the picture is usually forced out by new product coming in behind it.

At Warner Bros. we separate the areas of sales and marketing, although some studios combine them. As president of worldwide advertising and publicity, I am across the hall from my colleague Barry Reardon, president of distribution. (See article by Mr. Reardon, p. 309.) We work hand-in-glove; there is constant interaction.

Advertising and publicity becomes involved once production green-lights a movie. Under the ad-pub umbrella are these disciplines: creative advertising, publicity and promotion, market research, media and international. Briefly stated, creative advertising oversees the creative strategy and execution of advertising materials, which are done in-house or through outside vendors. The publicity and promotion staff works to convey the same message in public relations, starting with unit publicists and photographers on the set, through release publicity. Market research interfaces with the creative process through surveys of advertising materials; with production in fine-tuning the film in the preview process; and with sales through marketing-opportunity studies, tracking studies and competition studies. Media is the division that controls spending, physically places the advertising through ad agencies and subcontractors, and is responsible for the strategic design and execution of the media plan. International covers these areas outside the United States and Canada. There are also administrative support groups, which assist these divisions.

The marketing of a picture is like a race, in that each discipline may start at different times, but all finish together, at the target, opening weekend. Creative advertising starts as early as publicity, in preproduction; market research begins simultaneously; media starts just prior to release, and international, which has been active throughout, really kicks in after the domestic release (except for those pictures that open overseas prior to domestic).

Let me track the involvement of these specialties through the life of a movie.

## CREATIVE ADVERTISING

Early strategic thinking is the same on any picture, regardless of size. Generally during preproduction, a marketing strategy has been devised, which undergoes constant revision. As this process proceeds, we will review the screenplay and production schedule to learn if there are opportunities to shoot advertising concepts during production, since it's difficult to call everyone back for a photo session after the fact.

As marketing people we have an idea of the potential of the project at this stage, from the casting of the movie (both in front and behind the camera) and the story. During production we learn even more from the look of the picture. At that point we start thinking about which projects might have break-out success (like *Batman* or *Lethal Weapon*), although it's impossible to predict this. Also, we can sense the gemstones, the smaller movies that might break out with support from critics, from awards and ultimately from the public (such as *Stand and Deliver, Hamlet, Roger and Me, Driving Miss Daisy* and *Chariots of Fire*). This type of movie relies more on its own quality, and must be carefully handled.

Once the picture has completed principal photography, a select number of advertising and publicity people are allowed to see a rough cut in order to start working on the trailers. Occasionally we may look at dailies to get a feel of the tone and visual style of the picture. After viewing the rough cut, marketing really kicks in. The first focus is on preparing the trailer, and we take care that it reflects the important moments that the movie delivers. This becomes the backbone of all broadcast materials. Perhaps we will first create a ninety-second (:90) teaser trailer, offering just a glimpse of the movie to audiences in theatres as far as six months in advance of release, followed by a trailer (2:00) with actual scenes. Because the theatrical trailer is the first impression the movie audience has, this is some of the most important, persuasive work we do.

## PROMOTION AND PUBLICITY

Once the picture is green-lighted, the publicity department meets with the filmmakers to select the unit publicist, the on-scene public relations expert during filming. The unit publicist writes the production notes, which is a history of the filming, the biographies of key personnel, and communicates to us everything that is going on during shooting, including media requests that will be fulfilled or not. Also, there is a stills photographer who chronicles production from the point of view of the motion picture camera, to supply stills of the on-camera (and off-camera) action as it proceeds. In most cases the stills photographer will shoot at the same time that the motion picture camera is capturing performance; however, some actors prefer an alternative, and stay in place after shooting a scene, posing for such stills, as if they were filming. Shooting stills during actual filming is of course preferable.

Other promotional tools, such as behind-the-scenes programs and video press kits, are being compiled during shooting. This is vital if certain actors will be unavailable afterward, especially for international use, since most actors can't spend the time to travel around the world for publicity. For a sense of costs, an hour-long behind-the-scenes TV program can range from $75,000 to $150,000; a short featurette, three to seven minutes, can cost less than $30,000. A combination can find a featurette packaged for TV stations with a series of :60 or :90 news wraps, self-contained stories that can be "wrapped around" by local newscasters, plus a selection of location footage and film excerpts that can be built into a bigger piece locally, all offered with written material about the movie. Such packages can cost from $35,000 to $125,000 to produce.

## MARKET RESEARCH

In market research, advertising material is tested via intercept, where a pedestrian may be asked to respond to certain concepts, stars, images and advertising copy. Concept testing is the first measuring we do, for a sense of how the audience relates to a specific movie idea. Then we will turn to a title and star measurement (without the concept), then to the print campaign. Assume the research company has been asked by our head of market research to seek out a certain target audience, whether male or female, in a given age group, and of a certain moviegoing frequency. People who meet the criteria are recruited through intercept, usually at a shopping mall, where a booth is set up for them to watch and react to material, in exchange for a premium.

An interview is one-on-one. A subject will be shown the ads; then the interviewer will ask what the ad says; whether the ad is appealing; whether the person is interested in seeing that movie; and if so, how interested. Valuations depend on which research company is at work. A sample system might range from "definitely want to see," "probably want to see," "might," "might not," "not very interested" to "definitely not interested." These interviews are an additional series of voices that either confirm or call into question the overall approach. If a less-than-terrific response comes back, then refinement of the print campaign is necessary, by fine-tuning the image or copy. In the rare case where conceptually we are not communicating what we intend to, we may have to start from scratch.

Market research also helps us refine through testing our broadcast campaign, both trailer and television. Since the trailer is created earlier than the TV commercials, it is tested earlier. How does this testing work? Small audiences are recruited demographically, shown the trailer or TV commercials, asked questions, and the data come back to us for assessment. There are also adjective profiles, where a subject will choose from a list of adjectives to measure what the material is communicating. Separately a subject would be asked to assess the material itself. Earlier testing levels are compared with later ones, as a gauge of how well we are improving (or decreasing) interest, in addition to researching our target audiences with the intended message. Market research is also involved in previewing.

## PREVIEWING

Previews can be divided into two types, production or marketing, although they often overlap. A preview audience can be a recruited audience, whose demographic makeup is preselected. At a production preview the production executives work closely with the filmmakers to creatively fine-tune the movie. At a marketing preview we are studying audience reaction in connection with what we've been preparing in creative advertising as to the style and message of the campaign. And just as we are constantly reassessing release strategies, we are also reviewing advertising strategies, along with expenditures commensurate with those evolving strategies.

The preview process can alter release strategies. For example, a movie that had been identified for release as a slower, single-exclusive, review-driven picture may be changed to a wide, national release, for one of two reasons: It previews either much better or much worse than expected. If it previews better, we have the chance to reach a wider audience faster. If it previews worse, it can't be released slowly, because bad word-of-mouth will destroy it. But if it's conceptually advertisable and might achieve grosses for perhaps the first two weekends, it could be worth releasing the picture into the market nationally to "steal a couple of weeks" before word-of-mouth spreads. Now consider a picture that was always planned as a national release and turns out even better than expected. This could call for an opening two weeks earlier in very few cities, to allow for strong reviews to push the word-of-mouth and maximize the national opening.

Sitting through a preview, one learns a lot from audience reaction and from reviewing preview cards that viewers have filled out. Preview cards are always instructive. They express not only audience likes and dislikes but also details about the demographic makeup of those who most appreciate the film. For example, if our advertising is working for younger people but the film is also playing well to older ones, that presents an opportunity to broaden the audience base. We must then go back to create advertising materials that will appeal to the older audience as well. This kind of revision can apply to male and female audience segments too. As an example *Lethal Weapon* was clearly an action picture that strongly appealed to men; but in the preview process we learned that Mel Gibson's warmth, charm and good looks overcame the genre, creating broad interest among women. With that in mind, we reviewed our material, and it became a date movie.

Are there better cities for previews than others? This depends on the movie. San Diego and Sacramento are terrific preview cities overall; Seattle is a good market to preview a sophisticated movie. Today most previewing is done in Los Angeles, where it's possible to recruit any type of audience, from sophisticated to blue-collar. Consider how much activity this represents. At Warners alone we release on average twenty-five movies a year and preview each at least twice. That's at least fifty screenings, one a week, handled by market research. Most are recruited screenings, since it's best for an audience to know in advance what they are seeing. Another approach to previews is focus-group screenings, which feature a question-and-answer session with a moderator after a preview, as the filmmakers in the audience listen to the feedback.

At the preview stage ad-pub meets with sales to review release strategies and screening policies, to reassess the broadcast advertising budget and to decide whether the picture will be helped or hurt by publicity. (For a sense of timing, a summer picture opening in June is previewed in March–April.) There also may be elements that one loves in the broadcast trailer or print campaign but that the test audience (recruited by market research) doesn't respond to, calling for revising the materials. What is unlikely to change from the preview is the print campaign. By that time, market research has honed in on the images that are effective and on the copy that is working to appeal to the target audience.

## MEDIA

After the preview process, which helps us determine who the target audience is for the picture, we begin to devise the media strategy, around eight weeks before opening. As noted, if we learn we can broaden that audience base, we will reassess the media plan to accommodate this.

In determining the media buys, spending levels are set based on how much the picture is expected to gross: costs are driven by projected performance. The warning becomes, Don't outspend your revenues, but don't underspend your potential.

In the media buy, television is the greatest expense (broken down into network, spot and cable), followed by newspapers, magazines, radio and outdoor (billboards, subways, bus shelters, sides of buses). The levels of advertising in different media are known as weight levels: television weight, newspaper weight.

Let's consider television advertising. In what is known as the prebuying season (June and July for the following year), Warners has already garnered a large amount of up-front network commitments, paid for in advance. This gives us the security blanket of prebought time at fixed rates and is generally less than 50% of our total network buys. Specific television spot advertising is committed to, three to four weeks prior to opening weekend. Recognize, however, that the TV marketplace changes constantly. Buying a network spot on a Wednesday night at 9:00 for a certain program's demographics might instead deliver a substitute program or a repeat. The buyer must be aware of that and adjust accordingly. The buyer must also have enough flexibility so that if a release date shifts, entire media schedules can be changed or reassigned to other pictures.

To buy our spot and network-television advertising, we use a national agency, Grey Advertising, which receives a commission. They get their directions from our vice president of media, who provides an execution strategy and must approve the dollar amount to be spent. For an example of the high-end cost of network spot advertising, a recent Super Bowl :30 cost over $800,000. By comparison a :30 for *Cheers* was in the $350–$400,000 range; for *The Bill Cosby Show* and *The Simpsons*, $300–$350,000. For a :30 during one of those TV series on the local level, the New York market cost $40–$50,000.

In planning a media buy, formulas are used that vary from movie to movie. This is because a certain price delivers a certain size of audience, and a theoretical number of impressions upon each audience member is needed to achieve results. An equation is applied involving reach versus frequency. *Reach* is how many people see your spot; *frequency* is how many times they see it. How many impressions are required? No studies have been done relating to movies, since the product is so changeable, but in the packaged-goods world, involving soap or cars, the goal is three or more impressions.

In order to achieve the most impressions for the money, the buyer invokes reach vs. frequency in devising a spot-buying campaign. A television buy for a sample movie might find spots placed in certain important network prime-time programs, surrounded by lesser network programs, perhaps nonprime time, and placement in many more less-expensive spot market programming, perhaps in the local fringe time slots, both early (six to eight P.M.) and late-night (after local news or eleven P.M.). Another part of the buying equation is *cost-per-thousand,* a measurement of efficiency, stated on a station's rate card. With this computation the buyer learns whether too much money is being spent to reach a targeted number of people. If the same number can be reached at a lower cost-per-thousand, that would call for changing to a different, more cost-efficient programming mix. For example, the cost-per-thousand to reach females aged eighteen to thirty-four for program X may be cheaper than buying program Y to reach the same people, and therefore more economical.

A given media buy can be translated into gross-rating-points estimates for each program purchased, based on station rate cards. Gross rating points, GRPs, are a media-world measurement of the reach and frequency of a particular program. GRPs can break down into two measurements, household points and target points. *Household points,* the widely publicized TV rating figure, measures the number of American households tuning in. But each program has a specific demographic profile, and within each household there are certain target groups. *Target points* measure these groups: young adults, old adults, male, female, children, and so on. Since we know through experience what levels of GRPs are required to reach the goal number and makeup of impressions at the desired efficiency, one then works backward and makes the proper buys.

The Warners head of media and the people at Grey know all these permutations. With their budget they plan the strategy as to whom to buy, when to buy and how to buy, so that the result reaches the target audience for each movie with the greatest economy. That ties in with the earliest work on each movie, as noted, about identifying the primary and secondary audiences.

As for buying media other than TV, strategy depends on the movie, and each is an individual case. Generally, though, I use outdoor sparingly, in certain cities such as Los Angeles, and certainly on New York subways. In radio I believe that talk, all-news and easy-listening is more effective than youth radio, since adults are less-frequent station-changers. Youth radio can be effective on certain movies, however, such as *New Jack City*. MTV is also a good vehicle to target young people, as is Arsenio Hall; cable stations such as Arts & Entertainment and wrestling channels are very targetable in that they reach quite specific audiences.

## HOW MUCH?

The first question in preparing the media buy is how much to spend, which depends on what the prospective gross is. That judgment can be influenced by the season, the competition, and that figure is constantly evolving. The decision is locked in from six weeks to two months before opening weekend. *Robin Hood: Prince of Thieves* started earlier than that. If magazines are included, that prolongs the lead time to three months.

The next question is what the media mix should be, utilizing that money. The average MPAA figure for advertising expenditures is just over $11-million, and that includes prints.

When we prepare our advertising budgets, we include our creative costs, such as internal costs in creating the advertising campaign and the publicity campaign (much like overhead), the physical expenses of creating the various campaigns, spots and trailers, the cost of sending the filmmakers on tour, the cost of reproducing press kits and other costs. Per picture this can range from $750,000 to $3-million or more. For purposes of the following examples I'll exclude the creative costs.

Media buys differ considerably between a national release and a limited release. The following figures apply to media only, throughout their theatrical life. A low-end campaign for a national release can range from $5–$6-million in media costs; for a bigger picture the figure can be $10–$20-million. Media costs on *Batman* were finally just under $20-million. Because of its high awareness level I actually reduced the media costs on that picture. The first *Lethal Weapon*, released in March 1987, cost around $9-million for media; *Lethal Weapon 2*, in 1989, cost around $15-million for media, higher because it was released in the more competitive summer environment, as well as inflationary factors.

For a limited or exclusive release of a picture (in New York, Los Angeles and Toronto), in anticipation of a later wide release, media buys can range from $300,000 to $400,000, but this is with very little television, and most money going to newspapers and outdoors. The limited release range of media buys can vary from $200,000 or less if no further rollout is expected, to a more lavish opening, generating a national reputation, at $1.5-million or higher. The early, exclusive run of *Reversal of Fortune*, which included television, cost around $600,000 for media buys. This established a presence and generated reviews. The picture slowly widened to a national release and ultimately cost $5.6-million for media.

## CLOSER TO RELEASE

As the release date approaches, the earliest concern is placement for magazine deadlines, which can be as long as four months for some, or six weeks for *Rolling Stone*, as an example. Publicity has taken the written material from the production stage, polished it and put it into the form of a press kit. The press kit also includes stills that have now been selected from the unit photography (eight to fifteen in black-and-white, and an equal number in color) and that capture the spirit of the movie. These press kits are serviced to some two thousand journalists around the country, generally six to eight weeks in advance of opening. They are used to familiarize the press with the movie, or as references when they review, and to help supply feature stories for smaller papers that may not have an arts editor. From the creative standpoint the written press kits get much more use than video press kits.

At the same time, publicity is charting artists' availabilities with an eye to setting a series of interviews with key press. This depends on how keenly an actor may want to promote the film; you can't force anyone to do an interview. Scheduling is one side of the equation; pitching stories to the media is another. All Warner publicists are busy pitching all of our product, as coordinated by the head of publicity. For example, a publicist based in New York will try to position *Rolling Stone* for a story on, say, Michael J. Fox, because his movie's demographics are similar to those of the magazine. The publicist will call the contact person, usually the arts editor or an assistant, present suggestions and determine interest. Based on that interest, conversations follow as to whether a cover is appropriate, and the kind of story content. This becomes a two-sided negotiation, between the studio publicist and the magazine, and between the publicist and the artist. Diplomacy is required, and the studio is in the middle. This process is repeated with magazines, newspapers and television stations across the country up to the first day of release, and sometimes through the second week. For instance, in the second week of *Out for Justice* with Steven Seagal, the movie had clearly dug in, and there were opportunities for more interviews in order to maintain it. Steven Seagal agreed to add them. We had to be particularly inventive on *Robin Hood* because Kevin Costner had just enjoyed every type of publicity as a result of his sweep of the Academy Awards for *Dances With Wolves*.

Six to twelve weeks before release, on the promotion side, local market radio promotions are being discussed; local tie-ins with merchants and giveaways are going forward; and toy stores are stocking merchandising toys, such as action figures, if applicable. The promotion pieces of the puzzle are all coming together much like the advertising pieces, all focused to one target, the release date.

## OPENING WEEKEND

On opening weekend sales and marketing executives stay as close as possible to theatre grosses. In addition, our market-research people are doing exit interviews, to learn how the audience is responding.

The three possibilities are that the picture opens bigger than expected, around where expected or smaller than expected. If it opens bigger, naturally support it, but that support must depend on how much bigger the numbers are than expected. This reevaluation must be based on the picture's performance, exit interviews and competition. If the picture is performing exceptionally, it may even be prudent to reduce spending and save that money for later in the run, when it will be more useful. That decision will also depend on the competitive environment and is made by the sales and marketing executives. Every picture that opens big and has the potential to run, one should chase.

If a picture opens at around expectations, the follow-through plan continues. If those expectations were low, it becomes a self-fulfilling prophecy, and nothing more than intended is spent. Such a picture may hang on for the second or third weekend, depending on what's following it into theatres, but then it may be gone. For a national release, movies have become a three-day business; if the opening grosses are not strong, the picture will not survive for an extended run.

If a picture is not performing as expected, it's virtually impossible to rescue. Once $4- to $6-million has been spent to convey a specific message to the audience, that cannot be changed. On the first weekend we know, by tracking grosses from exit polling, just how our picture will perform. Let's assume we, as marketing executives, know we have a good movie, stemming from the recruited screenings. And on opening weekend nine out of ten exit interviews confirm such research-testing screenings. If the picture opens weakly, the only element that can turn things around is reviews.

The foregoing were examples of national releases. A picture opens in a limited release if it is of high quality, does not possess the marquee or concept value to open powerfully and will probably get good reviews. If such a picture opens weakly but still gets good reviews, it might be saved, upon adjusting the message of the ad campaign. In this case there is more flexibility (although rescue depends upon the quality of the movie) because a limited amount of ad money has been spent, and the picture hasn't been exposed to a large number of people.

As noted, exit interviews during a picture's opening will tend to coincide with our marketing previews. This confirms that we are on target. But it becomes interesting when this is not the case, and ingenuity must be applied. *Goodfellas* is an example of a sophisticated movie with exit interviews that ran counter to market testing. It tested below average in research screenings, largely due to violence and length. Then many critics dubbed it one of the best pictures of the year. With that stamp of approval, exit-interview ratings were double the market-research screenings. The picture opened big, we supported it and it kept growing.

Now let's take a case of a good movie with strong exit interviews and good reviews that opens weakly, perhaps due to heavy competition or other market conditions. It might be rescued with an ad campaign highlighting review excerpts, along with more spending than expected in order to support it and enhance word-of-mouth. *Defending Your Life* is an example. Reviews were strong, exits were good, but it opened below expectations. We supported it with a solid review campaign and more expenditures than planned. The second week dropoff was only 7%, indicating solid holding power, and the third week held similarly. To sum up, if it's not a good movie, gets poor reviews and opens poorly, it can't be saved. If it's a good movie, gets good reviews and opens poorly, it might be saved. If it's not a review-driven movie, such as an action or teenage movie, and opens poorly, it probably can't be saved.

In marketing a motion picture in its theatrical release, the percentage of failure is much greater than the percentage of success, which makes it a highly expensive business. It is analogous to speeding down a mountain on a train, heading toward the release date. Nothing can push the train back up the mountain; there's no second chance, all expenditures have been made. A recent survey placed the figure of pictures not recouping prints and ad costs during domestic theatrical release at 80%.

## CONSERVING COSTS

Marketing costs keep escalating. While there is no absolute solution to this, there are ways to try to conserve.

The primary expense in movie marketing is television, either national or spot. The network audience base is decreasing, so for the foreseeable future, one is paying more to get less. How can this become cost-effective? One solution is to buy off-network, or during fringe (rather than prime time) or on cable or on independent TV stations.

As an example of off-network, buy the Fox Broadcasting Network, which reaches moviegoers more efficiently than ABC, CBS or NBC. Their demographics often dovetails with target movie audiences, and they are less expensive than the other three.

Another solution is to buy without regard to the competition. If a certain level of gross rating points can achieve a given reach and frequency, don't be distracted by what a competitor is doing at higher levels. If the creative materials are strong enough to break through with your message, don't overspend. The heightened competition to be number one on a given weekend and at the end of the year has helped drive movie marketing costs up. If someone else is spending 700 gross rating points in TV to open, one may be fearful of spending only 500 gross rating points, even though one knows that level is enough. It comes down to remaining confident and not allowing the size of the competition's dollars to influence decision making.

In addition make sure the advertising materials are delivering on the message, since it's cheaper to improve the materials than trying to spend to achieve certain awareness and want-to-see levels with inadequate materials. This again returns to the beginning of the marketing process, when it's not enough to believe you have strong materials, either broadcast or print; that faith must be proven in market testing.

Yet another way to save money in TV is to experiment in buying strategies. For instance, use more :15 commercials, a very effective medium. If :15s can be made that deliver the same message as :30s, 50% is saved. Use them at the end of your flight, closer to the release date, once the strategic message has been delivered.

A general caution in television buying is to always reevaluate the impact of network versus spot, because costs are always changing. One season it may be more efficient to buy more network than spot, and the next season, based on shifting costs, this may be reversed.

Another area where spending is high is in newspapers. One way to save money is not buying full-page ads; three-quarters of a page commands the page. If the competition buys a full page, resist the vanity temptation to copy, even though artists may urge you to do so. The public does not evaluate whether an ad is three-quarters or a full page; they are simply responding (or not responding) to the message.

It is also useful to study historical data about cities where particular genres perform. Warner Bros. has a vast data base tracking grosses on every picture's performance. This is valuable research in arriving at marketing and sales decisions. If a sophisticated picture is opening in a small town where such pictures have historically not performed, be careful. Support it, don't make it a self-fulfilling prophecy; but don't chase it.

Other ways to save money can be found in the creation of the materials. For example, create more in-house; do less; if you go outside, use a single vendor rather than two or three. Vendors are creative sources, outside companies or individuals who can be trailer and TV-spot makers, print and one-sheet makers. Their services are extensions of our in-house creative people. The Warners head of creative advertising sets the strategy and works with an outside vendor to execute that strategy. How are vendors chosen? They are cast, much the way one casts a writer or a director of photography. Each vendor has specific strengths in his or her creative execution that lend themselves to one project over another.

## INTERNATIONAL

Philosophically the world is the same as it relates to the strategic marketing of a movie. It's not totally homogeneous, but there are reliable generalities. Some movie genres work better in certain countries than in others. For instance most people will like an action picture worldwide, whereas some comedies, more endemic to the United States, may not travel well. This changes where there may be peculiar nationalistic tendencies, or where the media mix is different, or where viewing habits are different.

Once that is said, there are ways, in certain territories, to sell a movie differently; but these are the exceptions, not the rule. Generally, advertising mixes differ on a country-by-country basis. In western Europe, for example, since most television was government-controlled, TV advertising was difficult to purchase; but with privatization, this is now changing. In other countries it's just too expensive to buy TV in the way it is done in America, which calls for change in fulfilling marketing strategies on a local basis, as overseen by the international division.

Pictures are opening overseas earlier than previously, to take advantage of the wide reach of American publicity. Since media coverage is immediate and worldwide, the impact of an American release can generate huge revenues overseas, as with *Dances With Wolves* and *Robin Hood*. When deciding when best to open abroad, the question involves a combination of circumstances, including how long it takes for excitement to travel, local traditions and moviegoing habits.

Both domestic and international distribution are constantly reviewing their release dates and plans. In ad-pub we are preparing materials for the earliest opening, wherever that is. Of course, where a picture must be subtitled or dubbed, that creates an inherent delay in non-English-speaking release dates. In Paris they insist on subtitles, but outside the city they prefer dubbing. In certain countries such as in the Far East or in Latin America, it's too expensive to dub, so a picture is subtitled.

To review the divisions in relation to international, both creative and publicity work the same until a picture is completed and release schedules are being mapped out. Then international personnel take over. In each major country there are Warner advertising, publicity, promotion and media-buying people, with levels of hierarchy that report back to the studio.

Here's a sampling of some major overseas markets, as they relate to ad-pub issues. Japan, the second-largest-grossing country, is very expensive in terms of opening costs and television advertising; newspaper costs are small. As it relates to movies, it is a young-female-driven society, since women decide what movies to see on dates. The United Kingdom is a strong market, strategically analogous to America, as is Australia. France is unique in the world in that it is cinema-driven; movies are deeply ingrained in the culture. Although one can't buy television commercials due to government regulations, that is changing. Italy is also a strong market, but people don't go to the movies in the summer because few theatres are air-conditioned, although this is being improved. Also, most people are away on vacation during August, and much of the country closes down. Theatres in Germany had been in need of repair, and now there are plans for new construction. Eastern Europe represents a whole new potential revenue source. (For more details on the global market, see the article by Peter Dekom, p. 417.)

Movies are the second largest export of the United States, and advertising-publicity positions and helps generate that popularity. But what we do is movie-driven. Every movie is unique and requires different levels of expenditure, in terms of creating materials (whether they are easy to make or not), in media spending (whether it's a concept that's easy to communicate or not); or for different levels of success (whether it's working or not working). Systems can always be improved. When new media become available, they must be evaluated. If television becomes a factor in countries where it wasn't earlier, how does that change the media mix? The way to improve the business is to continue to learn, to constantly push the envelope. Don't make an average campaign, make an above-average campaign. Because the business is changing so quickly, one can't sit back. There's no time for sorrow or elation, because next week's another movie.

VIII
THE
DISTRIBUTORS

# THE STUDIO DISTRIBUTOR

*by **D. BARRY REARDON,** president of Warner Bros. Distributing Corporation, based in Burbank, who has served on both sides of the distribution-exhibition equation. A graduate of Holy Cross College (B.A.) and Trinity College (M.B.A.), Mr. Reardon began his career as assistant to the president of the Eureka/Carlisle division of Litton Industries. In 1968 he started a seven-year tenure as vice president and assistant to the president of Paramount Pictures before shifting in 1975 to one of the largest theatre chains in the United States, General Cinema Corporation, where he served as senior vice president until 1978. That year Mr. Reardon was named executive vice president and general sales manager of Warner Bros. Distributing, and was elevated to president in 1982.*

*When . . .* [Batman] *opened . . . on 2,194 screens, grossing $40-million the first weekend and $70-million the first week, we knew we had a phenomenon on our hands. . . .*

Warner Bros. Distributing Corporation, charged with licensing pictures to the theatrical market, works hand-in-hand with Warner Bros. Pictures, the financing-producing-marketing entity, in an interlocking team effort under the Time Warner umbrella.

The Warner Bros. Distributing Corporation blankets North America with four divisions of sales executives in the United States and one in Canada. The eastern division includes offices in New York City (covering Philadelphia); Boston (New England, Cleveland and Cincinnati); Washington, D.C. (and Pittsburgh). The southern-division offices are in Dallas (with New Orleans, Memphis and all of Oklahoma) and Atlanta (Charlotte and all of Florida). The midwest-division office is in Chicago (including Milwaukee, Minneapolis, Indianapolis, Des Moines, Omaha and Detroit), and the western-division office is in Encino (covering Los Angeles; Denver/Salt Lake; Kansas City/St. Louis; San Francisco and Seattle/Portland). The head office in Canada is in Toronto (covering eastern Canada), with additional offices in Montreal (including Quebec province) and Calgary (Vancouver and the west). Each branch office is supervised by a branch manager and staffed by sales people, bookers and a cashier.

In our home office in Burbank there is a support group of about twenty, including print control and business personnel, along with auditing, one-sheet (poster) and trailer staffers. Print-control people order and handle prints, overseeing quality control and transportation to theatres. The business staff tracks grosses and analyzes and circulates figures generated by our in-house computer; the auditing people watch over receivables, collections and theatre accounts to determine which theatres to check in order to protect our grosses.

In a good example of synergy, we distribute all our own trailers and one-sheets to theatres through WEA (Warner Elektra Asylum), using the same apparatus as the record/tape division. This unit alone can generate from 2,500 one-sheets and trailers for smaller pictures, to

10–12,000 for a huge picture like *Batman*. Standees and instant marquees are also distributed by WEA. The operation is highly efficient through the use of computers. The day after we contract to show a film in a Kenosha, Wisconsin, theatre calling for two trailers and two one-sheets, that material is automatically shipped via WEA.

We track all Warners projects in development, but the real focus occurs when production green-lights a project to go forward and begin preparation for shooting. During budgeting and casting we will read the script and think about how the picture might slot into a tentative release schedule.

Another source of product is the negative pickup, when a completed or nearly completed picture financed and produced outside the studio is seeking distribution and we acquire it. Examples include *Chariots of Fire, Stand and Deliver* and *Roger and Me.* A pickup deal would usually be structured with a guarantee against a percentage of the gross after recoupment of costs. Our staff is always searching at film festivals and through other sources for pictures available for either domestic or overseas theatrical and home-video release, and we market pickups exactly the way we handle Warner-developed pictures.

Although the business is no longer seasonal, the pictures with the biggest market potential will be targeted for summer or Christmas, such as *Batman 2* and *Lethal Weapon 3*. More upscale or esoteric pictures will perform better in the fall and will also be positioned for the Academy and Golden Globe (foreign press) awards. But major grosses can be generated anytime by the right pictures. *Oh, God!* and *Private Benjamin* were two very strong September releases. Simply put, there's money to be made any time of the year, with early November and early May (the two lead-in periods for Christmas and summer) as the last remaining soft spots.

Once a picture is in production, the advertising-publicity kicks in. My office is next to Rob Friedman's, the president of worldwide advertising and publicity, and we are in and out of each other's areas ten times a day as concepts for spots and print campaigns are conceived and evaluated. Each movie has its own personality, and calls for subtlety in developing its release and marketing strategy.

When there is an answer print, we are out testing our movies with various audiences for fine-tuning and to confirm our planning as to the potential audience and release date. Competition is a crucial issue in targeting a release. Will we be up against a movie with similar audience-segment appeal? Sometimes this leads to moving dates up or back. During Christmas and summer there is much jockeying for position as the majors shift dates against each other's product. For example, in 1989 *Lethal Weapon 2* was scheduled to open against James Bond, on the same day. Since both pictures would be drawing from the same audience, we moved our date one week earlier. The picture opened with strong momentum, which carried it to an ultimate gross of $150-million.

A picture is locked into a release period perhaps four months in advance. Let's use Christmas as an example. We may not have set an exact release date, but we know it will fall between Thanksgiving and Christmas Day. Trailers and one-sheet posters will be in theatres by September. The teaser trailer for *Robin Hood: Prince of Thieves,* for instance, a summer release, was in theatres in January.

Now two important decisions are made concurrently. The picture must be shown to exhibitors to lock in theatres in a release pattern, and an advertising budget must be decided upon for money spent between now and opening day.

In determining release patterns, there are three basic choices: a wide (or broad, or general, or saturation) release, swiftly, in 800 to 1,800 theatres if the picture has wide commercial appeal; or a platform release, slowly, opening in perhaps three theatres (New York, Los Angeles, Toronto) exclusively and fanning out afterward; or a limited (or intermediate) release, in 600–800 theatres, with plans to spread wider one month later, depending on performance. (There is also a regional release, but that does not have much commercial appeal anymore.)

The wide release is selected for a picture that can play in every town in the country and must be supported by high media spending and national publicity. This can be economical in terms of advertising, since you are purchasing mass-audience saturation buys.

The best example of course is *Batman,* which became an event as the fastest-grossing picture in history. The campaign began during Christmas 1988, with a teaser trailer in theatres emphasizing the logo and including some short film clips. We were hoping for strong numbers, but there were some nay-sayers who criticized the campaign and ever-presence of the *Batman* logo. When the picture opened on June 23, 1989, on 2,194 screens, grossing $40-million the first weekend and $70-million the first week, we knew we had a phenomenon on our hands. For the second weekend, June 30, we increased the run to 2,201 screens and were grossing over $122-million. In thirty-one days we had broken the $200-million mark, before slightly reducing screens to 2,072 the following week, August 4, and reaching $218-million. By week 10, August 25, the picture was up to $235-million (1,302 screens); by week 15, September 29, $246-million (783 screens). At week 20 the picture passed the $250-million milestone (628 screens) and gradually began playing itself out until the end of the run, early December 1989, after twenty-five weeks, for a cumulative domestic gross of $251,188,924.

A platform or slow release may launch a picture in single, exclusive openings in New York and Los Angeles, where it sits while garnering reviews and word-of-mouth before spreading out to perhaps 150 screens, slowly. Examples include *The Color Purple, Driving Miss Daisy* and *Hamlet. Driving Miss Daisy* opened so strongly in four theatres in New York and Los Angeles that first weekend, mid-December 1989, that more theatres were immediately added in those cities through the Christmas holiday period. Because the picture continued to perform so strongly, all of our decisions were in support of its popularity. By Martin Luther King Day in January we were up to 277 screens in perhaps 100 markets; by January 31 we were up to 895 screens, in anticipation of Academy Award nominations the following Wednesday. There was an intermediate step to 1,302 screens by February 14, 1990, increasing to 1,432 two weeks later, up to 1,668 by the end of March, following the Academy Awards, when *Daisy* won four Oscars including Best Actress and Best Picture. By June 20, after twenty-eight weeks, the picture was playing on 775 screens and slowly played itself out by August 15, week 36, for a cumulative total U.S. theatrical gross of $106,593,296.

The third release pattern, the limited release, might open a picture in a few theatres in New York, Los Angeles and Toronto, then skip the platforms and move right to a wide release. *Accidental Tourist* is an example, which opened exclusively at Christmas 1988, took advantage of good reviews and jumped to over 800 screens the first weekend in January.

Although bidding is on the decline, in some markets where theatres are so competitive that they request it, we will provide for bidding. In other markets we may send out bid requests, get no response and end up negotiating with those theatres we think will be best for the picture. In either case we will set up a common date and invite exhibitors to screen the movie in some twenty-eight cities. If we are planning to bid a picture for broad release, the day before the screenings we send out bid solicitations to most theatres in the country. We will ask them to respond up to seven days after the screening, and those interested will either offer a bid or be willing to negotiate. Other studios may act differently; Fox doesn't bid at all; Paramount bids in some cities and not in others.

In analyzing bids in a market, the local Warners branch manager will judge which bid is best for Warners' interests, down to specific screens being offered in multiplexes and clearances; the division manager will approve and relay it to the home office in Burbank, where a top official will approve it and award the bid. The theatre is notified and sent a confirmation stating the terms. Since every exhibitor signs a formal Warners contract, which remains on file containing our standard terms and conditions, the paperwork for each picture need only state the specific terms of that run and refer to the long-form agreement on file.

The following two pages trace some of the paperwork on a sample playdate of *Batman*.

The first page reproduces the top sheet of the Bid Invitation Letter, dated February 23, 1989, which I sent out to all division and district managers shortly after the national exhibitor screenings of *Batman*. It sets down the terms for all theatres who are about to be solicited to play first-run engagements and is generally attached to a synopsis of the film and its major selling points. The terms for *Batman*, "90/10 over an approved house allowance with minimums," means that Warners would receive, on a weekly basis, the greater of two calculations: 90% of the box-office gross after deducting an agreed-upon house allowance, or a straight minimum percentage (called "floors"), expressed in three choices: either eight weeks with percentage floors of three weeks at 70%, two weeks at 60%, two at 50%, one at 40% and the balance of 35% (with a suggested guarantee); or six weeks (from 70% down to 35% with a guarantee); or four weeks (with a guarantee). In the earlier, high-grossing weeks of a run, the 90/10 calculation prevails, while in later weeks the percentage minimums are invoked. (For an extended review of a variety of exhibition calculations, see the article by A. Alan Friedberg, p. 341.) A "no-pass policy" prohibits the use of group or discount tickets until later in the run.

After the district managers receive this letter, they rewrite it and send similar letters to each exhibitor in their territory, expressing the requested terms and including the picture synopsis. By this time the theatre executives have seen the picture, and the transaction continues between the Warners district (and division) managers and their client theatres.

This bid response was accepted, and *Batman* did play successfully, as indicated by the second illustration, from our computer system, of the trial balance (our current, ongoing accounting of all recent pictures) of our sample theatre engagement. This excerpt (generated on March 16, 1991) covers the box-office gross and billing to Warners for weeks 5–8, from July 21 to August 17, 1989, plus certain subsequent entries. For purposes of this article, isolate the line items *flash bill* for each week. ("Flash bill" is our internal projection of weekly film rental.) For example, for week 5, the "flash bill" box-office gross was $48,232.00. To verify which of the two calculations applied for this week—90/10 over house allowance or straight percentage minimum—run the numbers for each. Under 90/10 begin with the gross for that week, $48,232. Deduct from that a $7,750 agreed-upon house allowance (which the theatre retains), and the remainder is $40,482. 90% of that is $36,433.80, the figure that would be returned to Warners if this formula prevailed. Under the minimum percentage formula, since this is week 5, 60% of the box-office gross (without deducting a house allowance) would apply. 60% of $48,232 is $28,939.20, which is

Inter-Office Memo

To BRANCH AND SALES MANAGERS-U.S. From D. BARRY REARDON

Subject Bid Invitation Letter - BATMAN

Date February 23, 1989 Copies to

Enclosed is the bid invitation letter and market plans for BATMAN. You will be notified shortly when to bid the picture.

Terms for BATMAN are to be 90/10 over an approved house allowance with minimums as follows:

| 8 Weeks | 6 Weeks | 4 Weeks |
|---|---|---|
| 70 | 70 | 70 |
| 70 | 70 | 60 |
| 70 | 60 | 60 |
| 60 | 60 | 50 |
| 60 | 50 | If held-40 |
| 50 | 40 | Bal.--35 |
| 50 | Bal.--35 | |
| 40 | | |
| Bal.--35 | | |

Suggested guarantees:

There will be a no pass policy. Your attention is directed to the capitalized paragraph in the bid letter, which reminds exhibitors of our 2% limitation on passes and discount tickets.

As usual, you will need to discuss with your division or district manager the appropriate minimum playing time.

Sincerely,

Barry

D. BARRY REARDON

cv
Enclosures

RPT. NO. DMM725R1 WARNER BROS. DISTRIBUTING CORP. BRANCH 29 SAN FRANCISCO WEEK ENDING 03/16/91 PAGE

| RELEASE | PLAYDATE | ENG | GUARANTEE | TERMS | GROSS | REFERENCE | TRANSACTION | DEBITS | CREDITS | NET DUE |
|---|---|---|---|---|---|---|---|---|---|---|
| | | | | 90.0 | 59,611.50 | 09/23/89 | 33 FINAL ADJ. | | 1,077.75 | |
| | | | | 90.0 | 59,611.50 | 10/21/89 | 33 FINAL ADJ. | | 2,615.40 | |
| | | | | 90.0 | 59,609.00 | 07/22/89 | 35 FLASH BILL | 46,673.10 | | |
| | | | | | | 845-10/21/89 | 51 A/R APP. | | 42,079.95 | |
| | | | | | | 859-12/09/89 | 51 A/R APP. | | 900.00 | |
| | | | | | | 845-10/21/89 | 52 CO-OP DED. | 1,014.63 | | |
| | | | | | | 849-11/11/89 | 52 CO-OP DED. | | 311.66 | |
| | | | | | | 002-11/18/89 | 82 APPROV CO-OP | | 702.97 | |
| | 07/21-07/27/89 | 05 | | 90.0 | | 04/15/89 | 02 PCTG BILLING | | | |
| | | | | 90.0 | 48,229.50 | 09/23/89 | 33 FINAL ADJ. | | 1,082.25 | |
| | | | | 90.0 | 48,229.50 | 10/21/89 | 33 FINAL ADJ. | | 2,615.40 | |
| | | | | 90.0 | 48,232.00 | 07/29/89 | 35 FLASH BILL | 36,433.80 | | |
| | | | | | | 845-10/21/89 | 51 A/R APP. | | 31,836.15 | |
| | | | | | | 859-12/09/89 | 51 A/R APP. | | 900.00 | |
| | | | | | | 845-10/21/89 | 52 CO-OP DED. | 769.81 | | |
| | | | | | | 002-10/21/89 | 82 APPROV CO-OP | | 769.81 | |
| | 07/28-08/03/89 | 06 | | 90.0 | | 04/15/89 | 02 PCTG BILLING | | | |
| | | | | 90.0 | 36,467.50 | 09/23/89 | 33 FINAL ADJ. | 1.35 | | |
| | | | | 90.0 | 36,467.50 | 10/21/89 | 33 FINAL ADJ. | | 2,615.40 | |
| | | | | 90.0 | 36,466.00 | 08/05/89 | 35 FLASH BILL | 25,844.40 | | |
| | | | | | | 845-10/21/89 | 51 A/R APP. | | 22,330.35 | |
| | | | | | | 859-12/09/89 | 51 A/R APP. | 180.00 | | |
| | | | | | | 845-10/21/89 | 52 CO-OP DED. | 531.55 | | |
| | | | | | | 002-10/21/89 | 82 APPROV CO-OP | | 531.55 | 1,080.00 |
| | 08/04-08/10/89 | 07 | | 90.0 | | 04/15/89 | 02 PCTG BILLING | | | |
| | | | | 90.0 | 26,903.00 | 09/23/89 | 33 FINAL ADJ. | | | |
| | | | | 90.0 | 26,903.00 | 08/12/89 | 35 FLASH BILL | 17,237.70 | | |
| | | | | | | 845-10/21/89 | 51 A/R APP. | | 17,237.70 | |
| | | | | | | 845-10/21/89 | 52 CO-OP DED. | 516.13 | | |
| | | | | | | 002-10/21/89 | 82 APPROV CO-OP | | 516.13 | |
| | 08/11-08/17/89 | 08 | | 90.0 | | 04/15/89 | 02 PCTG BILLING | | | |
| | | | | 90.0 | 20,104.00 | 10/14/89 | 33 FINAL ADJ. | .90 | | |
| | | | | 90.0 | 20,103.00 | 08/19/89 | 35 FLASH BILL | 11,117.70 | | |
| | | | | | | 845-10/21/89 | 51 A/R APP. | | 11,118.60 | |
| | | | | | | 845-10/21/89 | 52 CO-OP DED. | 296.50 | | |
| | | | | | | 002-12/02/89 | 82 APPROV CO-OP | | 296.50 | |

$7,494.60 less than the $36,433.80 Warners would receive under 90/10. Therefore, of the two, the 90/10 formula prevailed, and $36,433.80 was billed, to be remitted to Warners. That figure is found in the debits column; the balance of the box-office gross remains with the theatre. Similar calculations can be run for each week.

To review, in a negotiating market the interested exhibitors respond with their willingness to discuss terms, and our salespeople contact them and select the most favorable deal for us. New York City is a negotiated market; Los Angeles contains pockets of bidding and negotiated areas (such as Westwood). Certain negotiated deals could include guarantees and terms, others just terms, depending on conditions. Choosing the proper theatre calls upon the special expertise of sales executives, who know intimate details of theatres in their area, including ambience, numbers of screens and seats and grossing history. The apparently strongest theatre proposal may not be the best deal for the picture, and awarding a picture is often a complex and difficult decision.

At the same time that theatres are being locked in, a marketing budget is drawn up as to the cost to open the picture. (See article by Robert G. Friedman, p. 291.) This involves deciding upon dollar commitments for advertising in newspapers and on television, broken down among network, local spot and national and local cable. Cooperative advertising is shared with theatres in the top fifty markets. (For a list of the top fifty markets, in order of approximate revenue potential, see the last page of this article.)

Say we intend to open in 1,500 theatres. Our trailers, one-sheets and standees are poised at each theatre. Publicity has been working full-throttle. The filmmakers are involved in these decisions from the trailer stage forward, and we welcome their input, since they've been intimate with the movie.

All of this energy targets the critical opening weekend. Our sophisticated data base and in-house computer tracking system kicks in, and executives are at their desks Saturday morning analyzing the grosses as they are received from EDI (Entertainment Data Inc.), a service that reports on every theatre opening our picture. Since our data base contains the grossing history of every theatre in the country over ten years, we instantly know where we stand and can make reliable projections. Built into these projections are weather issues (heavy snow in Chicago), TV competition (a hot miniseries), current events (international unrest) or other influences.

What follows is one of three scenarios. If the picture opens strongly, everyone is happy and there is little to do but monitor the progress. If it opens moderately or poorly, I am on the phone Saturday morning discussing grosses with top marketing executives and deciding whether to change the campaign, add more television or radio advertising, or alter the newspaper approach, since backup campaigns are ready. If the picture opens weakly to poor reviews, we will try every alternative to support it, since we strongly believe in our pictures, and look to the second and third weekends for the results of our efforts. Because of the high cost of media, however, it's unwise to go beyond the fourth weekend on a failing movie. If a picture is doing well, during subsequent weeks we are always reviewing advertising expenditures and theatre grosses, fine-tuning expenditures in order to support audience response with proven levels of TV commercials and newspaper ads. This in turn will keep the picture at a high-grossing level.

Showmanship and audience targeting are important factors in selling pictures. For example, we screened *Hamlet* with Mel Gibson to a wide number of teachers around the country. Then we devised a mailing to schools and universities including a brochure and study guide (and offering a half-hour video) that invited students to take advantage of group sales through an 800 number or reply cards. By January 1991 we had heard from some 13,000 schools. Then, with the cooperation of theatres, we devised a group-sales coupon for students only, priced at $3.50, and mailed out over 3-million of them.

It is nice to report that relations with exhibitors have improved dramatically. Ten or so years ago there was an adversarial relationship between distributors and theatre owners; today we both know that we need each other's support, that we are mutually reliant. As a result of the theatre acquisitions by studios in the 1980s, Paramount and Warners each owns 50% of Cinamerica Theatres, which includes Mann and Festival theatres. But we in distribution regard them as simply another exhibitor; there is no favoritism. We continue to try to place our pictures in those theatres that will maximize our return.

One sensitive issue between distributors and exhibitors is the playing of commercial advertisements on screens. At Warners we are strongly against it. It's an insult to patrons, who can see plenty of advertising at home on television and are going out to see a movie, not a commercial. This practice diminishes the moviegoing experience, and I find it offensive. Disney was the first company to take a stand, contractually prohibiting ads in theatres showing their films. We've asked theatres not to show commercials with a Warners picture, and most have agreed. Another issue is ticket pricing, which is set by the theatres. At this writing the top ticket price is $7.50, and we have no control over the prices that individual theatres charge.

Although the business seems healthy, having reached the $5-billion annual gross plateau in the United States, attendance is flat, at the 1-billion level. Every executive would like to attract more people to the movies, but that's not easy. Theatres have done their part, undergoing substantial building and upgrading, with new inviting multiplexes featuring top-quality picture and sound. Theatre sound technology such as Dolby is entrenched, and digital sound is just around the corner. Picture technology promising some kind of video delivery as an improvement upon film on a large screen is in the talking stage and would have to be quite extraordinary to satisfy our demanding audience, which responds to the subtle nuances of beautiful photography. At Warners our challenge is to create, market and sell product that is continually more appealing to the audience.

### The Top 50 Markets—United States and Canada (Listed in Approximate Order of Revenue Potential)

1. New York Metropolitan Area
2. Los Angeles, CA
3. Washington, DC
4. Philadelphia, PA
5. San Francisco Bay Area, CA
6. Chicago, IL
7. Toronto, Ontario
8. San Diego, CA
9. Orange County, CA
10. Detroit, MI
11. Atlanta, GA
12. Dallas, TX
13. Miami, FL
14. Houston, TX
15. Boston, MA
16. Seattle, WA
17. Phoenix, AZ
18. Montreal, Quebec
19. San Jose Area, CA
20. Denver, CO
21. Baltimore, MD
22. Kansas City, MO
23. St. Louis, MO
24. Vancouver, British Columbia
25. Cleveland, OH
26. Orlando, FL
27. Minneapolis, MN
28. Fort Worth, TX
29. Portland, OR
30. Buffalo, NY
31. Sacramento, CA
32. San Antonio, TX
33. Austin, TX
34. Columbus, OH
35. Long Beach, CA
36. Pittsburgh, PA
37. Honolulu, HI
38. Norfolk, VA
39. Rochester, NY
40. Milwaukee, WI
41. Indianapolis, IN
42. Las Vegas, NV
43. Ottawa, Ontario
44. Memphis, TN
45. Calgary, Alberta
46. Edmonton, Alberta
47. Cincinnati, OH
48. Ventura County, CA
49. New Orleans, LA
50. Tucson, AZ

# INDEPENDENT DISTRIBUTION AND MARKETING

*by* ***IRA DEUTCHMAN,*** *president of Fine Line Features and senior vice president of parent company New Line Cinema, based in New York. Among Fine Line's releases have been Jane Campion's* An Angel at My Table *and Gus van Sant's* My Own Private Idaho. *As president of the Deutchman Company, he provided marketing consulting services for such films as* sex, lies, and videotape *for Miramax,* To Sleep With Anger *for the Samuel Goldwyn Company and* Metropolitan *for New Line Cinema. Mr. Deutchman was one of the founding partners and president, marketing and distribution, for Cinecom Entertainment Group, Inc.* (A Room With a View, Stop Making Sense). *An adjunct professor in film at Columbia University, Mr. Deutchman serves on the boards of the Independent Feature Project/West, the Collective for Living Cinema and on the advisory boards of the U.S. Film Festival and the Sundance Institute. His screen credits include associate producer of* Matewan *and executive producer of* Swimming to Cambodia, Miles From Home *and* Straight Out of Brooklyn.

***The movie business is a business where there are no rules, and the minute you think you've learned the rules, they change on you. . . .***

What is an independent? I get this question all the time, and it amazes me that people are so hung up on definitions. It has become my recent opinion that the word *independent* is completely useless in that it defines us by what we aren't instead of what we are. It has nothing to do with size or scale, as witnessed by the fact that several so-called independents are currently in much better shape than certain so-called majors. And it has little to do with potential, since clearly a film like *Teenage Mutant Ninja Turtles* can glide over the $100-million mark with the best of the studio product. So what is the difference?

I think it is more accurate to call us "niche" companies. The primary difference is in marketing styles. Studios, by needing to feed an enormous overhead machine, are forced by their very nature to swing for the fences every time out. This requires a marketing style that works on the macro—reaching out for large numbers. We, and our competitors, however, are practitioners of micro-marketing—sticking our fingers into little pockets of business and scaling down the economics accordingly. The first part of the formula is to identify a market that is currently underserved by the studios. The second part is realistically sizing up the potential of that market and weighing the costs inherent in trying to reach it. The final part is being flexible enough to react to the marketplace—we can change things around a lot quicker than a larger organization. Compared with the majors, the profit margin is smaller, but the risk is smaller as well.

None of this is new. Most real innovation in the movie business has originated in the independent sector. The modern horror genre, the bread-and-butter market for the majors for much of the late seventies and eighties, was pioneered in the fifties and sixties by an independent, American International Pictures, and taken up in the early seventies by Roger Corman's New World. Even such a studio signature as the "wide-release" was in fact created by an independent, Sunn Classics. Whenever an independent breaks through in a new area, the majors are quick to move into that area with a vengeance, and the "indies" by necessity move on.

Another area where there has been much success by independents is in foreign-language and more specialized films. Through the fifties and into the sixties the rather small niche was left to a few smaller companies such as Cinema V, Libra Films and New Yorker Films to mine. By the late seventies and early eighties, with the breakthrough of a number of smaller films to larger audiences, the studios began a short-lived trend toward forming "classics" divisions in order to jump on the bandwagon. These were quickly abandoned when it became clear that prestige was higher than profits and that there wasn't enough quality product around to feed them all.

During the time that the majors were fighting over Truffaut and Fellini, a new breed of independents was already looking ahead to the next trend—using the niche marketing methods previously used for foreign-language product and applying it to smaller American films. Thus the success of such companies as Cinecom (*A Room With a View, The Brother From Another Planet, Stop Making Sense*), and Island/Alive (*Trip to Bountiful, Kiss of the Spider Woman*). By this time the majors were out of the art business and firmly entrenched in wide releases of teen-oriented films, leaving the field wide open for the adult audience.

Perhaps the biggest boon that has ever occurred in the independent sector was the explosion of home video in the early eighties. It was a voracious market for anything with sprocket holes, and even the major studios couldn't provide enough product to satisfy the demand. All of a sudden there was enormous capital available to independent theatrical distributors as advances against the home-video rights. Not only was all this money being used to acquire films, it also fueled the entrance of many independents into production.

And then bust! There were far too many films in the marketplace and not enough audience to support them. Also, rather than staying with the kind of cutting-edge theatrical marketing efforts that had established them, the video-driven money put the pressure on to achieve broader and broader audiences. This put them, by default, into competition with the studios. Finally the rush to satiate the video market created a rush toward quantity rather than quality, and many bad films were made. Thus the late eighties saw such companies as New World (years after Corman sold it), Cannon, Vestron, Atlantic, Cinecom, Skouras, Island and Alive (now separate) in decline, out of business, or in a different business. When the smoke cleared, New Line Cinema, the Samuel Goldwyn Company and Miramax were the survivors.

Other than market niche, the indies operate much like their studio brethren, but on a slightly smaller scale. One difference is that the studios rely almost entirely on their own productions, where the indies are a good deal more acquisition-based. Finished films are acquired for distribution from independent filmmakers in the U.S. and from foreign production companies that are seeking distribution in the U.S. Therefore heavy emphasis is placed on attendance at major film festivals and markets throughout the world. It's a very competitive market, and distributors are always on the lookout for a breakthrough movie.

A filmmaker-producer team with a completed picture seeking distribution must be very careful in this atmosphere to develop a strategy before approaching distributors. All festivals have dangers as well as assets, and one has to know how to use the tools available to be able to spark interest in his or her film. After all, buyers are being enticed in a million directions at once by sellers of hundreds of available movies at a typical festival or market. In my opinion the best festivals to find independent American buyers are New York, Sundance, Cannes and Toronto. Filmmakers seeking information on festivals should contact the Association of Independent Video and Filmmakers in New York, which publishes *The Independent* magazine as well as a yearly festival guide containing addresses, phone numbers, application deadlines and procedures.

The reason that festivals are so important in getting interest from potential buyers is that it's the only way to get buyers to see the picture in an exhibition context where they can observe how the audience and press respond. This is the most persuasive way to demonstrate to a distributor that a picture is worthwhile. Once a film is accepted for a festival, the filmmakers have to try to get a representative from the various distributors interested in seeing the film. This is where it gets tricky, and where it helps to have someone involved who has some experience. There are producers' reps, lawyers, agents and marketing consultants who specialize in helping to work the festivals and in making the deals. Depending on how much help is needed and what kinds of services are required, these professionals work on either hourly charges or for a percentage of the deal. Many reps, lawyers, and so on have developed reputations that immediately help in attracting buyers to the product.

Most of the time independent films are sold to U.S. distributors for domestic rights only. So it is usually the case that a foreign sales company will get involved to sell the film territory by territory in the rest of the world. While there are some independent companies that have their own foreign-sales divisions, many filmmakers like to segregate the rights so that they have two revenue streams that are not cross-collateralized. Also, the type of product that may be valuable to the domestic division may be less valuable to the foreign division and vice versa.

All the independent companies have their particular strengths, and

I always encourage the filmmaking team to grill potential distributors about what they are planning to do with the film. While the deal offered is important, the filmmakers must be confident that the distributor they choose has a clear understanding of the target audience and be able to articulate this. The meetings that take place amount to a mutual interview where everyone makes sure they can collaborate well together. The interpersonal chemistry cannot be underestimated. After all, the filmmakers will be spending hours with the distribution executives, under stressful conditions, sometimes arguing, so it is natural for both sides to want to feel compatible and to feel a mutual respect. In the give-and-take it is the distributor's responsibility to exhaust every opportunity in the marketing of the picture; it is the filmmaker's responsibility to advise and be heard, since they have lived with the movie longer than anyone. I also encourage filmmakers to check references—to speak with other filmmakers whose pictures (both successful and unsuccessful) have been handled by the targeted distributors.

There are two basic types of distribution deals that are commonly used. One is a standard distribution-fee arrangement, similar to what the studios do, wherein the distributor pays an advance to the producer and covers all costs of prints and advertising, releases the film, takes a distribution fee (30%–35%) off the top from the theatrical receipts (fees are usually slightly different for the ancillary rights), recoups its prints and advertising outlay, and divides what is left with the producer. The actual percentage of the net that goes to the producers is negotiable depending on supply and demand, just like all the other aspects of any deal.

This is the deal used in most circumstances where the distributor is either paying a significant advance guarantee or a large amount of P&A (prints and advertising) money to support the film. It is formulated specifically to protect the distributor from huge losses on a film, because the assumption is that once this kind of deal is made, the distributor will be at greater risk on the film than the producer, although that is not always the case. Therefore, under this formula, it is entirely possible for the distributor to be in profit on a film (since it receives its fee plus P&A costs from first revenues) but for the movie to still be in a loss position from the point of view of the producer.

The second type of deal is the "costs off the top" deal, which is mostly used in situations where the distributor is committing very little money up front and therefore leaving the producer at greater risk. It is also used a lot for pickups of foreign-produced films, many of which can look at a U.S. deal as gravy. As the name implies, all distribution costs (prints and advertising) are deducted by the distributor, and the resulting proceeds are split (usually 50/50) between the producer and the distributor. Note that there is no distribution fee in this formula.

From a net-profits standpoint, the distributor is on an equal footing with the producer. Pictures released through this type of deal that went on to great success include Whit Stillman's *Metropolitan* and John Sayles's *The Brother From Another Planet.*

Negotiating the deal involves the amount of the advance, levels of P&A expenditures, fees and back-end net-profit percentages. Also discussed are the marketing strategy and other points that are important to the filmmakers. While no distribution company will give up control completely to anyone but the most important of filmmakers, generally speaking there will be more cooperation and collaboration with an independent company than with a major.

It is commonly the case that a film won't go to the highest bidder, but rather to the company with the best handle on the film or the best back-end participation. It depends on a filmmakers' faith in the eventual success of the film, and their cash needs, as well as how much demand there is for the film. Films that cost $2-million can get little or no advance, and films that cost $100,000 get advances that are twenty times their budget. It's all in the perceived value and has nothing to do with the budget.

Once a film is completed or acquired, the lead time necessary for the micro-marketing done by most independents is at least four months. During this time the filmmakers are constantly consulting with distribution executives on the advertising campaign, advertising expenditures, planning press interviews, the release pattern, numbers of theatres and other strategies. In all of these discussions logic must reign. Filmmakers regard their picture with the same devotion and emotional investment lavished on a baby. Distributors invest as much of themselves into the release of the movie, but after a while we learn to allow intellectual arguments to take over emotional ones. Ultimately the marketplace teaches us all the time, and we have to be listening to it. All of the selling tactics in the world will not overturn a negative response. Everyone in independent distribution and marketing has spent time banging his head against the wall trying to create an audience for a film he personally loved but that had not achieved any audience. Logic reigns.

Despite what is commonly thought, keeping the filmmakers happy is essential to a successful campaign, since we have to depend on them to go out on tour, doing interviews with regional press. Because independent movies are more director-driven and review-driven than studio pictures, much weight is placed on the filmmakers' shoulders to help us position the picture in the marketplace.

The true test of any picture, of course, is the gross. Profits and losses are tracked on a daily basis to determine whether the ongoing gross can withstand certain levels of advertising support. Each week's gross is compared to previous weeks' to get a gauge as to word-of-mouth and eventual profitability. Another advantage that independents can claim is the ability to make quicker adjustments based on performance. Not only is it easier for us to change gears than the majors, but our generally slower release patterns afford us more opportunity to learn from the marketplace and adjust.

One myth that can be shattered is the commonly held notion that the studios are better able to collect their share of box-office receipts from exhibitors than the independents. Exhibitors are notorious for paying slowly, and for holding off payment in order to force a renegotiation of terms on less successful pictures. The fact is that the larger, more stable independents have the same tool at their disposal to collect money as the majors—future product. In fact struggling studios, with dubious upcoming product lineups, will no doubt have more difficulty with collections than a secure independent. Further, because cash flow is so important to smaller companies, they tend to be more aggressive in collections and settlements with exhibitors than the studios. Indeed, smaller theatre chains are likely to pay smaller distributors more swiftly, simply because their mutual interests are at stake.

Accounting also fosters misconceptions. The case of *Buchwald vs. Paramount* has drawn attention to accepted motion picture accounting practices in that the odds of a net-profit participant receiving any money are remote at best. The common misconception is that accounting abuses create the lack of a net. In fact the real problem is that there is so much general mistrust in the Hollywood system that stars and star directors demand larger and larger gross participations. A film with these kinds of participations has what is called a floating breakeven point, which is responsible for the fact that there will never be a net profit on paper, even if the film is earning a lot of money. Again indies have the advantage. First of all, they are dealing with films that usually don't have personnel of the caliber that can command these gross participations. Also, the films are more modestly budgeted and efficiently marketed. Therefore if a film hits, there is a much greater chance that net points will be quite meaningful.

Fine Line Features was formed in 1991 as a division of one of the oldest and most stable of the independents—New Line Cinema. The idea was to expand the number of niches New Line is involved with, since, as discussed above, independent distribution is all about niche marketing. After the break-out success of the *Nightmare on Elm Street* and *Teenage Mutant Ninja Turtles* movies, which between them account for the top five independently distributed films of all time, the company, flush with notoriety and cash flow, addressed the question of growth.

Because distribution is a very labor- and overhead-intensive business, a huge success suddenly requires expansion in personnel to collect from theatres, to ship the prints and to perform other tasks. But with such expanded overhead comes a self-fulfilling pressure to release future pictures as if they will break out as well, and this is the catch-22 that has put many of New Line's competitors out of business. New Line's plan is to remain a niche marketer, but to add to the existing niches that got them to where they are by creating a distinct releasing unit devoted to upscale, adult-oriented product. Fine Line Features is that division.

The plan, which is modeled in many ways on the record business, is to use New Line's existing distribution apparatus, which is commonly held to be the best among independents. In this way the most overhead-intensive part of the operation is being utilized to its capacity, but the marketing and acquisition activities, which require a different sensibility and approach, are handled separately.

The difference in marketing and distributing adult-oriented product is that, since there is usually no selling hook of stars or genre (such as horror, sex or violence), each picture depends on quality to get its audience—anything short of excellence in the execution can be disastrous in the marketplace. To seek out product, Fine Line personnel cover all the film festivals, and rely on existing relationships within the business. Through an outreach program we also keep in touch with the output of film schools and the various filmmaking communities.

The movie business is a business where there are no rules, and the minute you think you've learned the rules, they change on you. That's why it's so important for an independent to be aware of the changing marketplace and be prepared to change course completely in response to it. Fine Line is a response to the current marketplace and will evolve as it changes.

I believe that the future of the movie business will further blur the distinctions between the majors and the independents. Because of demographic shifts in the nation's population the entire business will be catering to the growing, aging adult market. This is good news for independents, who excel at the kind of niche marketing necessary to reach this market.

I believe the focus for the entire business will be identifying and targeting submarkets of these aging adults and in making pictures that they will respond to. It's a fact that as they age, this swell in population, the so-called baby boomers, will possess more leisure time than they ever had since having children. If we can recapture that leisure time to moviegoing in the same way as when they were a generation of teenagers, we could be heading toward a significant increase in moviegoing, nothing less than the second golden age of the movies.

# HOME VIDEO

*by **RICHARD B. "REG" CHILDS,** president and chief operating officer of Nelson Entertainment and executive vice president of Nelson Holdings International Ltd. An alumnus of Stanford University and the graduate film division of UCLA, he began his career in 1968 as founder of Genesis Film Ltd., a subsidiary of Filmways, which distributed student films to colleges. Five years later he joined producer Bert Schneider to start RBC Films. Mr. Childs moved to Paramount when the studio purchased RBC, turning it into Paramount Non-Theatrical, which became Paramount Home Video in 1979. As that division's vice president and general manager, he helped guide Paramount's foothold into the home-entertainment arena. He left in 1983 to become vice president, worldwide ancillary sales, for the Samuel Goldwyn Company and the next year was named president of distribution at Embassy Home Entertainment. In 1985 he was promoted to president, production and programming. When Nelson Holdings acquired Embassy in 1986, Childs was named to his current position.*

*Domestic gross sales of all home videos are about twice that of box-office gross, but home video needs that theatrical launching of its titles in order to perform at that level. . . .*

With the advent of videotape after World War II, there were many experiments in formats that were the forerunners of today's videocassettes. By the 1970s inventions such as Cartrivision and EVR (Electronic Video Recording) had gained attention. Studios were invited to license their movies, or software, to these formats at a royalty rate, with no minimum guarantees, in order to help launch this new adventure in home entertainment, but they declined.

By 1978 Andre Blay of Magnetic Video was taking steps to license movies on videotape, notably an experimental deal with Fox for some fifty pictures with minimum guarantees of around $5,000 per title. One concern was that some retailers were renting movie titles on their own in the marketplace, and studios were not participating in that revenue stream. (That issue was to come to a head years later.) At the time, movie executives were evaluating this potential new source of revenue as videocassettes began being sold to customers over the counter in Beta (created by Sony) or VHS (created by JVC, the Victor Company of Japan) formats for $59.95 each.

This was a major historical step for motion pictures because up to that time movies were never sold; rather they were licensed to theatres or to television. It was traumatic to movie studios, since it was such a departure in their way of doing business, representing a "letting go" of their product through direct sale to consumers for the first time.

A year later Paramount made two important decisions. One was to license Fotomat to start selling videocassettes, first in southern California, then in eleven western states, then nationally. The other was to license chosen movies for purchase on Selectavision, RCA's competing disc technology. Initial jurisdiction for this new home-entertainment market was placed under nontheatrical, since the division had experience in shipping, handling and other logistics. By the summer of 1979, Michael Eisner, then head of Paramount, proposed tests for selling video product through direct marketing, mail order and two-step distribution, which is selling to a distributor who in turn sells to a retailer.

What was learned in this period was that retailers were buying cassettes from Fotomat stores around the country and then renting them at their own stores. This was allowed under the "first use" doctrine of the Copyright Act of 1976, which invoked copyright jurisdiction only upon the first sale of videocassettes; subsequent use such as rental would not generate income to the copyright holder. Rental revenue went free and clear to these entrepreneurial retailers.

By the end of 1979 Paramount had announced it was entering the video business directly through ten distributors. Fox, Warners and Columbia were also becoming active. Thirty-six titles were chosen for that Christmas, at a total of around 75,000 cassettes, and $9-million was billed in that first fiscal year. A new industry was being born.

Tracking the growth of videocassettes from 1980 is like following a model in a business textbook. Over the next years demand for this new product skyrocketed along with parallel sales of VCRs (videocassette recorders). There were experiments in pricing and marketing, and eventually the market matured and leveled off, finding its price and preferred format (VHS over Beta and laserdiscs). (For statistics, see "Industry Economic Review and Audience Profile, p. 377.)

But some studios saw unfair exploitation in the VCR, since it allowed home copying of copyrighted programs with no royalty payments to the studio owners. In 1977 MCA/Universal, joined by Disney, decided to sue the Sony Corporation because they were not receiving any revenue from recordings of their television programs made off the air. In January 1984 the Supreme Court held in a five-to-four vote in favor of Sony in *Universal vs. Sony.* Technically the decision permitted home videotaping of copyrighted programs with no royalty payments. In a larger sense it freed up the industry to proceed unencumbered. A fascinating and authoritative account of the history of the Sony decision can be found in *Fast Forward: Hollywood, the Japanese and the VCR Wars* by James Lardner.

Another significant event at the time was the doubling of video retail outlets in 1983–84, which has since leveled off at around 25–30,000, including mass retailers and record stores. As a result unit sales jumped as well.

Meanwhile pricing continued to escalate during that period, with MCA Universal reaching a high of $109.95. Mel Harris, in charge of video at Paramount, received permission from Michael Eisner to experiment to reduce prices on certain titles to $39.95 as a way of increasing unit sales. This was a daring strategy at the time, and it proved effective. All new Paramount titles were sold at the lower price, while the other studios maintained the $79.95 plateau. The lower price has since become known as "sell-through," calculated to encourage sales rather than rentals. The concept, pioneered by Paramount, was to stimulate the marketplace to sell more units. At the higher price there is a 2.5:1 ratio to reach the same net.

Pricing of laserdiscs has always been lower than videos as a way of subsidizing that format and stimulating hardware (playback machine) sales.

This new home-video industry took its rightful place in the revenue stream, right after theatrical. The sequential releasing pattern of any motion picture begins with theatrical release in New York, Los Angeles and Toronto. That's when the clock starts. Three or four months into theatrical is release to airlines and nonresidential pay-per-view (hotels). On the first day of the seventh month is the home-video release. Put another way, the video "window" starts on the seventh month after theatrical release. By this time, of course, the picture is usually played out theatrically. Within thirty to sixty days of home-video release is home pay-per-view, which, studies have shown, does not intrude upon home-video revenue. Next, free cable is released six to nine months after home video, followed by TV syndication or network release. Overseas, theatrical release may follow domestic by three to six months, benefiting from a picture's domestic success. Then the sequential marketing points occur in similar fashion, except that satellite distribution may precede cable in certain markets, depending on subscriptions.

The heart of any video division or company is the sales and marketing group, which determines the release schedule and advertising expenditures for a picture. Marketing comprises the creation of the in-store advertising campaigns, using the icons from theatrical release. Naturally video benefits from the $6–$15 million per picture spent on advertising theatrically. The home-video "key art" campaign may change slightly, but not appreciably, in order to benefit from the audience impressions established during theatrical exhibition.

In theatrical, the retailer is the theatre. In home video the retailer is the retail outlet. The marketing department creates one-sheets (posters), mobiles, counter cards, store displays and scripts for TV or radio ads.

The initial entry for a title into the home-video marketplace is usually priced in the $89.95 to $92.95 area. That level is considered a rental price, since purchases will be limited to retailers (to rent out) and collectors. The video company will share in revenue from that "first sale." But the real turnover will occur in renting that title, and that money will not be shared with the video company. A year later that price will be reduced to perhaps $19.95, and the title becomes "sell-through," which stimulates consumer purchase since the price is so low.

The key question is about price. Should the item be priced at the higher $89.95 to $92.95 range, positioning it for sales to retailers and collectors, encouraging rentals and after-rental sell-through, or will the initial run be at the sell-through price, encouraging consumer sales, in the lower $19.95 to $24.95 range? There is a tremendous revenue surge if a picture can be sold initially at the sell-through price (and a higher rate of returns), but it must be extremely popular, such as *The Little Mermaid* or *Teenage Mutant Ninja Turtles.* Most pictures are sold at the higher price. For example, *The Hunt for Red October* and *Ghost* were both priced at $99.95.

In sourcing product, a video company must have a clear idea of how many units to sell, which converts into dollars. Video is such an important revenue source that sometimes at the studio level and especially at the independent level the video arm is called upon to assess packages of script, director, cast and producer, before shooting, and to project numbers of units to be sold.

For each title the generally accepted form is to allocate 80% of acquisition costs in the first year. When acquiring rights, the acquisition cost is advanced as a guarantee against a royalty base of anywhere from 20%–40% payable to the producer. The advance to the producer is earned out of the producer's share before paying his royalty. Rule-of-thumb is that this figure averages around 40% of the picture's negative cost. A $20-million picture could expect an advance of $8-million for domestic home-video rights. Another 20% covering marketing and duplicating costs must be recouped, plus 3%–5% for television advertising buys. On that title for which $8-million has been advanced, if 200,000 units are sold, billing over $11.6-million, the video publisher will earn a distribution fee, recoup its marketing costs and its advance to the producer. Nelson has paid in excess of $8–$9-million as an advance to a producer for domestic video rights on a picture. Assuming that acquisition advance is recouped, if the royalty is 20% and the video company is billing $58 as the amount that comes back to it from the distributor on a cassette retailing at $92.95, they would be paying around $11.60 per cassette to the producer.

At a studio it's a matter of selling the studio's titles and allocating figures accordingly. The video division pays the parent company a royalty of perhaps 20%–25%, which appears on the producer's statement. If the studio's share of video sales is $10-million, that royalty can be $2–$2.5 million. Under a studio deal there is no video advance per se because the studio has financed the production, and revenue from all media are applied against the cost as well as marketing expenses. In a nonstudio deal an independent video distributor reads the script, assesses the budget, director, producer and cast, and then makes a judgment as to what the video performance of the package will be.

In earlier years an independent producer would make a domestic theatrical deal with a major (which might include television) and a separate domestic video deal with a video company. Both deals might be linked in that the advance for the video deal would often be earmarked to cover the prints and advertising expenditures theatrically. With this reduced risk, the theatrical distributor would lower their distribution fee. These deals are generally no longer being made because domestic rights are rarely being split.

An interesting marketing problem arises in the selling of home video. The day you ship your cassettes at the rental price of, say, $92.95 is probably the last day you sell a cassette at that price. The dilemma is, do you advertise before the video is on the shelves in order to stimulate sales into the rental market? Or wait until after those initial sales, and support the retailer with, say, TV advertising, encouraging the consumer to come in and rent, even though the video company does not participate in rental revenue and rentals do not increase sales? The answer is you should probably do both.

On a net basis a production company can expect to receive between 30% and 40% of a picture's negative cost from domestic home-video sales. Interestingly that 40% of negative cost is also the average cost of prints and advertising for theatrically releasing a picture. An equation that has developed has video in effect covering the P&A costs of a picture.

Today a major theatrical distributor will not make a deal without home-video rights. They want to control the product through sequential release, enhance their market share and cash flow, and keep their operations active. They also want to ensure reimbursement of their prints-and-advertising outlay through video revenue, invoking that equation.

There are four criteria the retailer uses to decide to purchase video titles: box-office, box-office, genre and cast. Box-office is repeated for emphasis because over the years, studies have shown a correlation in tracking video-sales revenue against box-office revenue. As to genre, action-adventure and comedy tend to sell better than others. With cast, certain actors may be featured in the video advertising who were not emphasized in the theatrical poster.

Companies know that if a picture's domestic box-office over the first two months of release is between $1-million and $3-million, they can expect to ship between 50,000 and 80,000 units. At that level the box-office is equal to the gross video dollars. From $4–$10-million, one can ship between 80,000 to 120,000 units; if the box-office is $10–$15-million, the range is between 100,000 and 150,000; at $20-million the range approaches 200,000 units. Of the four hundred pictures released annually, only about sixty pictures do this well. This is rarefied territory: a $20–$30-million box-office will ship between 200,000 and 230,000 units; between $30–$50-million, the figure can be 225,000 to 275,000; over $50-million can generate 300,000; between $55–$100-million, 400,000 units are a possibility. (Naturally, these figures can vary.) Over $100-million at the box office, the picture is a candidate for initial sell-through pricing, based on season, rating and whether it's collectible.

The high echelon of sell-through can range up to 9-million units, billing at, say, $14, generating $126-million to the video company. Even though costs of marketing and duplication would greatly increase, this is an idea of the high-end potential of this market. *Batman* reached this level; *E.T.* sold over 11-million units, billing over $150-million. Other examples of extremely popular titles include Disney animated pictures, *Total Recall* and *Pretty Woman.*

To review, at the low end of theatre performance video revenue will equal 100% of box-office; at the high end, video revenue equals perhaps 24% of box-office. With higher video units shipped, the video percentage of box-office declines. After all, there is a finite number of video stores at around 30,000 and a stable number of VCRs, at 70-million machines in over 70% of TV homes.

Sometimes a video title will outperform these box-office-based estimates. Steven Seagal pictures are examples, as well as sequels, generally.

Each month about eight to twelve pictures sell at the A-title level of 100,000 units, for a billing of $5.8-million, out of the thirty to thirty-five titles released. Roughly one-third are A-pictures, and the rest are revenue producers, but also-rans.

A title is announced three to four months before its release to the consumer video market, and advertising commitments are made. With an A-title a video company will spend between 12% and 15% of initial projected sales on advertising. 85% of ultimate sales will occur in the first two weeks. If sales are projected at 200,000 units, that translates into $11.6-million of sales revenue to the video company. Marketing costs would be between $1.2- and $1.5-million, including television and newspaper advertising.

There is a complete sales program that then takes place between the video company, or publisher, and the fifteen or so nonexclusive distributors. The term *publisher* is used because video distribution descended from book publishing and distribution, and record distribution, using the same middlemen, and adopting similar procedures and nomenclature. Commtron came from hardware distribution; Ingram and Baker & Taylor from book distribution.

Say a video title is set for July 1 release. Advertising is placed months in advance. A duplication order is made to cover half of the expected sales units; the other half is confirmed the day after orders from retailers are closed, around two weeks before the street date. The "street date" is the first date the product is available to be picked up or shipped to retailers. The manufacturer-publisher (video company) prints up the boxes (or sleeves) bearing the artwork and synopsis of the movie. The printed boxes are sent to the duplicator. The duplicator duplicates the cassettes, places them in the boxes/sleeves, shrink-wraps them, then ships the packaged cassettes to the wholesaler via truck in quantity designated by orders to the video company.

Each cassette is usually protected with some antipiracy technique, such as Macrovision, to prevent casual copying. Other systems might offer a small hologram identification logo (similar to the type that appear on credit cards), which is extremely difficult to reproduce.

The wholesalers ship the cassettes to or make them available for pickup at the 120 or so branch locations of distributors around the country. Some studios simplify the process. For example, MCA Universal does their primary video distribution through their five warehouses for MCA Records. Each title is packaged, shrink-wrapped and transported to the five locations, where regional distributors take over.

Both studios and smaller video companies sell to the same group of independent distributors, fifteen to twenty in all (with 120 or so branches), whose job it is to make sales to independent and chain retail stores. The top seven distributors are responsible for over 70% of video sales. Local retailers in turn either come in and pick up their order of units or have them shipped. *Rack-jobbers* are distributors who usually specialize in sell-through, nonfeature specialty and low-priced feature film product.

A video company has an accounts receivable billed to the distributor. (Certain large accounts, such as Music Plus or the Wherehouse, are sold directly. Blockbuster Video, the nation's largest retail chain, prefers to spread its business among three or four distributors.) Since the video company deals directly with distributors, it has no involvement with the paperwork of store sales from customers buying that hot title on July 1. The distributor is put on terms of thirty days for a cash discount, or sixty days, and must pay within sixty days. The video company watches the accounts for credit limits and collections. There is a return allowance, perhaps 10% of purchases for rental product and 20% for sell-through feature-film product. The distributor in turn has accounts receivables with the stores.

This process changes with an experiment called pay-per-transaction, wherein the video company sells a title for somewhat above cost and then actually participates in each rental fee. This concept has not been adopted by the industry on a full-time basis.

A suggested retail price is placed on the title, in the $89.95 to $92.95 range, which is sold to distributors at a 37% discount. This is arrived at through a "30 plus 10" formula, or 30% to the retailer and 10% to the wholesaler. Beginning with 100%, if 30% goes to the store and 10% of the balance (or 7%) goes to the wholesaler, the distributor gets a 37% discount off the suggested retail price.

If a title sells at $89.95, the video company gets 63% of that price, or $56.67; at $92.95 it's $58.56, the wholesale price, and at $99.95 its

$63. How a distributor divides its 30% between itself and a retailer varies greatly. Generally the distributor sells a $92.95 cassette to the store for around $65–$68, after taking its markup, making $6–$7 per cassette at the distributor level. The retailer then retains all the subsequent rental income.

These figures are valid for the domestic home-video market, covering the United States and Canada. The market breaks down into 92% from the U.S. and 8% from Canada, close to the population ratio. Videocassette recorder penetration is in over 70% of TV homes.

Of course, stores are in the rental business at the high retail-price level. A national average rental figure was recently around $2.54. Stores then calculate how many turns will be generated from a specific cassette. The so-called "heat period" for rentals is the first ninety days. If that cassette can be turned every other night at forty-five nights for, say, $2.50 per turn, that is $112.50 of revenue to the store, and a $50 profit on that cassette. Months later that used cassette will be sold for $20, happening well in advance of the video company's expected sell-through price on fresh cassettes, one year after initial release.

To trace the course of revenue to the video company from a unit selling at $92.95, begin with the 63% figure, or $58. The company then has to pay for the marketing costs, duplication, shipping, boxes, labels and transportation. Together those items are generally in the 20% of retail range, say $19. What remains is $39–$40 per cassette. That's what the video company has to work with to cover overhead and acquisition/production costs.

The retailer's challenge is to estimate how many units to buy. Usually that figure is lower than what the video company would like them to order. This is where the criteria of "box-office, box-office, genre and cast" come into play.

This is not really a seasonal business except that Christmas is a strong period for sell-through titles. Otherwise it depends on what product is in the stores, and that is determined by the six-month-after-theatrical time frame. Theatrical used to have three seasons, during school vacations: summer, Christmas and Easter. Today release schedules find certain pictures doing very strong business between these periods.

The nonfeature specialty market of home-video releases, including exercise, concert, documentary, TV programming and how-to tapes, has evolved parallel to feature videos. Jane Fonda's first exercise tape in 1982 proved this to be a very lucrative market. Specialty videos include titles financed and produced directly for home video and often sell at $19.95. A negative cost may be in the $250,000 range, and respectable sales at 40,000 units.

There is also a growing mail-order market in specialty videos licensed to such operations as Columbia House and Time-Life. A sample mail-order deal would pay the publisher manufacturing costs plus a 15%–20% royalty of the suggested retail price.

Overseas a distributor generally buys rights to a title in all media in a certain country. Overall those rights are generally worth from 35% to 50% of the negative cost. Leading markets are Japan, the United Kingdom, Germany, joining with Italy, France and Australia to represent around 50% of total overseas revenue. Action-adventure titles generally sell best in the international markets. But *When Harry Met Sally* sold very well overseas—with twice as many subtitles as usual—because the relationship was universal; however, *Bill and Ted's Excellent Adventure* didn't travel well in terms of theatrical and home-video sales.

A real problem overseas is video piracy, which is rampant in South America and the Far East. Domestic video piracy is also appalling, and the MPAA polices this with the FBI. Antipiracy or anticopying chip technology helps, but whenever a new one is devised, someone overcomes it.

Domestic home video is now a mature market. Its impact on the consumer and the movie business has been profound. The consumer embraced the idea of collecting and showing movies at home (as well as the magic of time-shifting, grazing and making video home movies), and VCRs have become a standard appliance in the home. After a slow start the movie business embraced the idea of selling titles, recognizing the huge financial potential, and the format has settled into place within the industry revenue stream. Domestic gross sales of all home videos are about twice that of box-office gross, but home video needs that theatrical launching of its titles in order to perform at that level.

On the horizon the pay-per-view industry and home video may well be at loggerheads, with each jockeying for a superior position in the sequential releasing pattern. Innovators are adopting new technologies, hoping to replace the half-inch video format, much like CDs have replaced vinyl records. Picture quality is improving with laser video discs, though there is not yet a recording capability. (See article by Martin Polon, "The Future of Technology for Motion Pictures," page 451.) Sony's 8mm video has arrived, and certain movie titles are available on the format, but that is a small market. Larger-screen and high-definition television (HDTV) will call for a compatible video format. The trend in video stores is for larger spaces in a supermarket environment, while mom-and-pop stores are on the decline. Specialty stores may arise much like specialty bookstores. The real challenge is to improve picture quality. While duplication (which occurs in real time, usually) is improving, we will see the introduction of an extension video format with a higher-quality picture, and ultimately we will be viewing movies at home on recordable technology the size of CDs.

IX
THE EXHIBITORS

# THE THEATRICAL EXHIBITOR

*by **A. ALAN FRIEDBERG,** a veteran exhibition executive who is chairman of Loews Theatres, a subsidiary of Sony Pictures Entertainment, Inc., comprising 866 screens in sixteen states and based in Secaucus, New Jersey. A summa cum laude, Phi Beta Kappa graduate of Columbia College who attended the Harvard Law School, Mr. Friedberg began his career with USA Cinemas in Boston as assistant manager in 1957 and rose through the ranks of management to become president of USA Cinemas in 1976 and owner in 1982 before moving to Loews in 1988. Mr. Friedberg is past president and chairman of the board of directors of the National Association of Theatre Owners; is on the Board of Overseers of the Boston Museum of Fine Arts; is chairman of the board, National Center for Jewish Film at Brandeis University; is a member of the board of the American Repertory Theatre at Harvard and has received many civic and cultural honors in his hometown of Boston, Massachusetts.*

***The distributor benefits greatly if the picture is a big success; the best the exhibitor can do is 10% above expenses. . . . If the box-office fails to generate more than the house expenses . . . the exhibitor is in a loss position. . . .***

The movie business was traditionally a vertically integrated industry wherein studios (the producer-distributors) also owned theatres. In fact the parent company of Metro-Goldwyn-Mayer was Loews, a theatre chain that acquired the studio to assure a regular flow of product to its screens. Fox owned Fox Theatres; Paramount owned ABC-Paramount Theatres; Warner Bros. owned Stanley Warner theatres.

Independent theatres were foreclosed from licensing films from studios that owned theatres in their markets, creating an anticompetitive atmosphere. As a result independents filed a lawsuit under the Sherman Anti-Trust Act that became *U.S. vs. Paramount, et. al.* and was finally resolved in 1952 through a number of Justice Department Consent Decrees, which the targeted studios entered into. These Consent Decrees called for the studios to divest themselves of their theatre holdings and enjoined them from engaging in certain anticompetitive practices, such as block booking, and restricted blind bidding (to one film a year for a period of three years), stating that movies are to be licensed on a theatre-by-theatre, picture-by-picture basis, without discrimination.

Under *block booking* a distributor would require an independent theatre to play all of its films, conditioning the licensing of one picture upon another. The Consent Decrees outlawed the practice. This is quite separate from *blind bidding,* wherein the exhibitor commits to a film without ever seeing it. When the restriction on blind bidding expired, studios reinstituted the practice, and by 1978 most studios were engaging in it, sometimes inviting bids as much as a year in advance of release. They would send out a bid-solicitation letter announcing the title, cast and release date and asking for bids in particular markets. Because so many theatre owners were being badly hurt financially with this practice, NATO, the National Association of Theatre Owners, spearheaded a drive in state legislatures to end blind bidding. As of this writing the practice has been outlawed in twenty-three states: Alabama, Arkansas, Georgia, Idaho, Indiana, Kansas,

Kentucky, Louisiana, Maine, Massachusetts, Missouri, Montana, New Mexico, North Carolina, Ohio, Oregon, Pennsylvania, South Carolina, Tennessee, Utah, Virginia, Washington, West Virginia, Prince George's County in Maryland and the Commonwealth of Puerto Rico.

One more term important to movie-theatre history is *product splitting,* an understanding among theatre owners in a given market to divide product. Exhibitor X may get first choice of the Christmas pictures by draw of a hat, and Exhibitor Y may have first choice of the Easter pictures. Another form of product splitting would find exhibitors assigning certain distributors to certain theatres, done with the tacit approval of the distributors. In 1980 a Justice Department memorandum pronounced splitting a per se violation of the antitrust laws.

Exhibitors become aware of new pictures directly from the sales staff of distributors or from tracking production schedules in the trade press such as weekly *Variety, Daily Variety, The Hollywood Reporter* or *Box Office* magazine. NATO's monthly paper carries a schedule of future releases as do subscriber services such as Exhibitor Relations.

Theatres contract for films in the form of licensing agreements as a result of direct negotiation or bidding. In the case of negotiations the terms may be reviewed after the conclusion of the run. In the case of bids they are firm, immutable. Either way a distributor will send a letter to an exhibitor announcing that a certain picture will be available on a certain day.

Today the majority of licenses are the result of negotiations, not bids.

The decision to negotiate directly with theatres or to solicit bids is a business judgment made by the distributor. The decision of the theatre owner as to what to offer in a negotiation or how much to bid is hardly an exercise in scientific precision, since no one can really predict the performance of a movie—even after seeing it. After all, conventional wisdom said that westerns don't perform, yet *Dances With Wolves* enjoyed strong box-office results. And no one predicted a $200-million gross for *Home Alone*. If one begins with the savvy premise that "you just don't know" how a picture will gross, it makes no sense to bid, since the exhibitor is held to that deal. In a nonbid situation there is generally an opportunity to renegotiate the terms after the fact in the form of a settlement if a picture performs poorly. The unpredictable elements of the movie business make it exciting.

Say there are only two exhibitors in a given city. It's in the film company's long-term best interest to have both of those operators coexist. If one of them bids, loses a lot of money and goes out of business, over the short term the film company has gained a few dollars, but in the long term the balance of power has shifted in favor of the surviving exhibitor. That's a good argument against bidding, from either point of view.

A different step a distributor might take in this market is to allocate its product, giving a Clint Eastwood picture to Exhibitor A and a Mel Gibson picture to Exhibitor B. If this is done to keep both theatres viable, and if both exhibitors agree, there's no problem. In fact, product allocation is frequently the case in the industry today.

Allocation also occurs when a specific theatre is considered desirable. For example, in Westwood Village in Los Angeles, the second-largest grossing area in the country, Mann's Village Theatre or Mann's Bruin Theatre, across the street from each other, are considered highly desirable, and a distributor will take great pains to play a hot picture in these houses. The same can be said for theatres in the number-one grossing area of the country, in Manhattan, where a number of "battleship" theatres have the capability of throwing off huge grosses at the box office.

Supply and demand also comes into play. In the highly competitive Christmas playing time, if there are sixteen pictures vying for eight theatres in a given city, distributors would jockey for position and screens in a buyer's market. If there are six pictures instead of sixteen, exhibitors would be hungrier and it would be a seller's market.

Ours is an industry built on relationships evolving from trust, integrity and loyalty. The exhibitor who remembers a distributor's help when there were fewer films is likely to return a favor to that same distributor when there are too many films. On the other hand, there are exhibitors and distributors with short memories and little sense of reciprocity.

It is no longer a seasonal business, although the highest-grossing potential comes when children are out of school. 60% of movie patronage is from the age group of fourteen to twenty-nine. As studios move their strongest releases ahead, Christmas starts in early November, and summer starts at Memorial Day. That's still the best playing time. But, as in any self-fulfilling prophecy, if a less-than-commercial film is released in a peak period, it won't do business. And *Home Alone* can open November 9 and become a blockbuster.

The bidding process, to some extent, is a public auction. In some cases, companies must open bids in public at a specific time, and theatre owner/bidders can be present to witness rival bids.

A "bid request" letter would be received by a theatre company for an important Christmas holiday picture two to six months in advance. It is usually sent by the distributor's branch manager to the film buyer in the theatre's home office which covers that market. A sample cover letter, which might feature a brief synopsis of the movie, along with the names of stars, director, producer and writer, would include the following excerpts:

> Studio A is pleased to offer the release of Picture B, starring Star C in an action-adventure thriller directed by Director D.

All bids must include

1. a schedule of admissions prices;
2. the number of shows for weekdays and weekends;
3. the screen number and the number of seats in the auditorium in which the picture will play.

Studio A reserves the right to reject any offer from an exhibitor indebted to us for film rental in excess of 45 days or whom we consider a bad paying account.

Be reminded that Studio A will not allow commercial screen advertisements (other than theatrical trailers, slides projected before a feature, and charitable and/or other public service announcements approved by us in advance) to be presented or performed at the theatre during the engagement of the picture.

As you know, public reaction to any motion picture is impossible to predict. Accordingly, we cannot and do not make any prediction or representation concerning this picture's grossing potential.

In the event that either no bid is accepted or bids are accepted for less than the number of runs sought, we may negotiate for the number of runs indicated.

Enclosed are suggested terms and other pertinent information relative to this bid offer as well as a list of exhibitors receiving same. Bids must be submitted in the enclosed envelope only. By submission of your bid, you agree that such an offer is noncancelable for ten business days following the day that the bid is opened. Bids will be opened for examination in our City E office on Wednesday, November 3, at 11:00 a.m. We look forward to receiving your bid.

On the attached page is the "bid request" which includes:

| | |
|---|---|
| Release: | Picture B |
| Availability: | December 10 |
| Area: | City E |
| Number of runs: | Ten |
| Run: | First |
| Bid Return Deadline: | 11 a.m., November 3 |
| Suggested Terms: | $50,000 guarantee, payable 14 days in advance of exhibition |
| Minimum playtime: | 8 weeks |
| Terms: | 90/10 over approved house allowance with minimums of<br>Three weeks at 70%<br>Two weeks at 60%<br>Two weeks at 50%<br>One week at 40%<br>Balance at 35% |

If the percentage terms for this bid are 90/10, the distributor receives 90% of the gross receipts after the exhibitor has deducted and retained his "approved *house allowance*," or nut. In the case of a multiplex, these figures are generally arrived at in that proportion that the number of seats in a specific auditorium bears to the total seating of all the houses. The house allowance is a figure negotiated between the distributor and exhibitor, which does not necessarily bear any relationship to actual expenses. It's usually more, which allows for some air in the theatre's profit margin. It differs from the "house expense," which is the actual cost of running the theatre including rent, payroll, maintenance, utilities, insurance, etc. (For an example of the Warner Bros. bid solicitation format and internal computer system, see illustrations in article by D. Barry Reardon, pp. 315–16.)

In running the numbers, the distributor will receive the "90/10 over allowance" computation or the stipulated minimum gross percentage for a given week, whichever is greater.

Assume house expenses of $5,000. If $10,000 comes in at the box office in the first week, the 90/10 formula would call for deducting $5,000, the house allowance, leaving $5,000 to be shared between distributor and exhibitor as follows: 90% to the distributor and 10% for the exhibitor. 90% of $5,000 is $4,500. However, the weekly minimums stipulated call for the first "three weeks at 70%" and 70% of $10,000 is $7,000, $2,500 higher than $4,500. The 70% figure would prevail, and $7,000 would go to the distributor as film rental. On that basis the exhibitor would be out-of-pocket $2,000.

Now assume a theatre takes in $50,000 that first week. Under 90/10, the exhibitor would retain the first $5,000. 90% of the remaining $45,000 is $40,500. Since 70% of the week's gross of $50,000 is $35,000, the 90/10 deal would be triggered, and the distributor would receive the higher figure, $40,500 in film rental. The exhibitor would wind up with $9,500, $4,500 more than its allowance for that week.

Let's take a picture that does not open as strongly, with a second week's gross of $5,000. In this case the 70% calculation would prevail, giving $3,500 to the distributor and $1,500 to the theatre. Assuming the operating expenses are still $5,000, the theatre would suffer a $3,500 loss.

The distributor benefits greatly if the picture is a big success; the best exhibitor can do is 10% above expenses for a hit picture. And that house expense figure is of course not guaranteed; if the box-office fails to generate more than the house expenses, as in the $5,000 weekly gross model, the exhibitor is in a loss position. Also, these figures don't include the theatre's share of local advertising, to be covered in a moment.

A *guarantee* is a nonrefundable amount of cash that an exhibitor must pay to a studio, often months before the release date, in order to secure a certain picture. If the guarantee is, say, $100,000 from a theatre, but the distributor's share of box-office receipts is ultimately only $75,000, that exhibitor takes an absolute loss of $25,000, since the entire $100,000 guarantee is retained by the studio. This differs from an *advance,* which is not so onerous in that it is refundable, to the extent that it is not earned in film rental by the distributor under the terms of the licensing agreement. Assume the theatre advances $100,000, and the distributor takes possession of that $100,000 before turning over the print. If the distributor's share of box-office receipts is again $75,000, $25,000 is not earned by the distributor under the terms of the agreement. That unearned $25,000 is returned to the exhibitor.

The issue of guarantees or advances recognizes that studios have spent millions of dollars in production and marketing costs and are paying interest on that outlay. Exhibitors are collecting cash at the box office in a cash business, and distributors are arguably entitled to start using the theatres' money, especially since settlement doesn't occur until perhaps months after a picture's run is over.

The guarantee figure solicited to be received two weeks before the picture opens is a way for distributors to begin offsetting the huge expenses of marketing and making the picture. A guarantee is nonrefundable; portions of an advance can be returned upon settlement. Naturally the distributor's share of gross receipts is applied against the guarantee. In this example the exhibitor is also asked to guarantee minimum playing time of eight weeks, hit or miss.

Another possible wrinkle some distributors have added is to specify a minimum *per capita* clause, which guarantees a minimum amount per admission of, say, $3.00 per adult and $1.50 per child. The distributor doesn't consider this price fixing, for the exhibitor is free to charge whatever he chooses. This is especially hard on discount theatre chains, which must avoid pictures with a per capita requirement.

After a deal is struck, a license agreement stating the terms is generated by the distributor and signed by both parties. If the deal was negotiated, this acts as a road map for possible subsequent renegotiation should the picture fail to perform. If it was the result of a bid, the deal is inviolate.

Here is another example of license terms, in more formal language from an exhibition license agreement:

> 90% of gross receipts for each week of the engagement over a house allowance of $3,000 per week, but in no event less than the percentage of gross receipts for each week of the engagement as follows:

| | |
|---|---|
| 1st through 2nd week | 70% |
| 3rd week | 60% |
| 4th week | 50% |
| If held, 5th week | 40% |
| 6th week (and all playing time thereafter) | 35% |

Holdover

If, in the 4th week of the engagement, the gross receipts equal or exceed $3,000 (the holdover figure), the engagement will continue for one week after the minimum run. Each week after the 5th week that the gross receipts equal or exceed the holdover figure, the engagement shall continue for an additional week.

Run/Clearance: first run

In this example assume a box-office gross of $4,500 in the third week. After deducting $3,000 for expenses, 90% of the remaining $1,500 is $1,350, lower than 60% of $4,500, or $2,700, which would go to the distributor as film rental, leaving the theatre with $1,800 for that third week. If that third week grossed a higher amount, say $10,000, 90/10 would control (90% of $7,000 is $6,300, higher than 60% of $10,000 or $6,000); but if it grossed a lower amount, say $2,500, the 60% would control (60% of $2,500 is $1,500) with $1,500 going to the distributor and $1,000 to the theatre, which would suffer a $2,000 shortfall in terms of its house allowance (notice "house allowance," not "house expense").

Notice that the fifth-week percentage is 40% "if held." This refers to the *holdover* clause, whereunder if the picture continues to perform above a minimum holdover figure (in this case $3,000), the theatre must continue to play it. This clause is rarely invoked. *Clearance* is not activated in this deal, but, as with a franchise, refers to competing theatres, which could not play this picture during the engagement. (In Los Angeles, for example, Westwood theatres demand clearance over Century City and vice versa.) Instead this deal specifies it is *first run*, or the first exhibition of the picture.

Although virtually every deal is a 90/10 deal, the weekly minimum percentages vary. A very tough deal for a projected blockbuster can command a minimum of sixteen weeks, with four weeks at 70%, four weeks at 60%, four weeks at 50%, four weeks at 40% and the balance at 40%. A softer picture might call for "one down sixty," suggesting the first week at 60%, second at 50%, third at 40% and fourth at 35%. That was the deal on *Star Wars,* because no one knew how valuable that picture would be, including its distributor, Twentieth Century Fox. Naturally the kind of percentage deal suggested by the studio sends a message to theatre owners as to its faith in a given picture.

Regarding advertising, distributors pay the lion's share and determine the content and placement of ads. This is fair, since movie advertising not only benefits theatrical but also radiates impressions in subsequent markets, notably home video. However, exhibitors do contribute to the local expenses of marketing a film, averaging about 20%, in the aggregate, of what the distributor spends locally. This does not include national advertising commitments, such as network television or radio buys.

That 20% is known as *cooperative advertising,* covering local newspapers, local television and local radio, and is divided among the theatres sharing a picture's run in a given market. To illustrate this, return to the example of a $10,000 gross, which paid 70% to the distributor or $7,000 in film rental. If each theatre's share in the ad campaign for the second week in that market was $1,000, our sample theatre would send the distributor one check for film rental at $7,000 for that week and another check for $1,000 for cooperative advertising expenses for that week. As another example, if a local ad campaign covering five theatres for a given week is $20,000, the distributor pays $16,000, and $4,000 is divided among the theatres for a per-theatre cost of $800 that week.

One of the areas of tension between distributor and exhibitor occurs when the exhibitor wants to remove a picture due to poor performance. Usually the theatre has choices of other pictures waiting to open and be slotted in, perhaps from another studio or from an independent distributor. It comes down to dollars and cents, sometimes amid threats from a distributor.

The real dance goes on once box-office figures are a matter of record. In the case of settlement renegotiation after a run (assuming the picture was not bid), reasons generally relating to expenses are offered on both sides—sometimes leading to acrimonious debate—as to why one party should ultimately receive a greater share than the original deal would allow. In the end, agreement is reached and payment is made.

The most important period in a picture's run is the opening weekend. If an opening is disappointing, you know this by Friday night at eleven P.M. Entertainment Data (EDI), a computer service, collects and distributes box-office figures. On Monday morning hard decisions are made based on comparative performance of theatres in prior years, as well as supply.

If a highly touted picture bombs and there is no film waiting behind it because everyone expected it to succeed, there is much jockeying to slot in some other picture. Sometimes the distributor insists on keeping the picture on screen, even though grosses are not generating enough for the theatre to pay its rent. This type of situation has become a major source of confrontation between exhibitor and distributor.

On the other hand, if a picture guaranteed for a twelve-week run continues to do well but there is a prior commitment for another picture to replace it after twelve weeks, that is sometimes a dilemma that arises between exhibitors and distributors.

Concession sales, which remain with the exhibitor and are never shared with the distributor, are the lifeblood of theatres. (See article by Stanley H. Durwood and Gregory S. Rutkowski, p. 352.) If one removed concession profits from exhibition and left only the amount retained of box-office revenues, the profit margin for movie theatres in the United States would be between −1% and 1%. If distributors ever wanted to share a piece of ancillary movie revenues—which are driven by theatrical exhibition—with theatre owners, perhaps theatre owners would consider sharing in concession sales.

Another business exhibition is involved in is real estate. Between 1986 and 1989, a period characterized by frenetic, overpriced theatre acquisitions and overbuilding, real estate was arguably as important as box-office grosses to some exhibitors, who subsequently faced financial crises. Since then there has been a return to rationality, with real estate abating in importance. Although the total number of indoor screens continues to rise in the United States, to around 24,000, that rate of growth has not been supported by an increase in attendance, which remains stable at around 1.2-billion admissions a year. Clearly this has put a squeeze on exhibitor expansion.

Movie-ticket pricing is always an issue. But according to the U.S. Department of Labor, movie-ticket prices represent a lower percentage increase than any other leisure item they track, something like 55% over ten years compared with a Consumer Price Index increase of 56% over the same period. There are experiments wherein admission is higher over the weekend than during the week, but this hasn't caught on.

Some cynics say a reason why distributors began to own theatres in the eighties is because theatres were dumping pictures from screens too soon, jeopardizing the huge production and marketing investments made by distributors. But no enlightened exhibitor wants to see any studio-distributor lose money on a picture, since any such financial reversal could spill over into the movie theatres. In the end, regardless of cost, the public makes the final decision as to a picture's success, and nothing except ego will keep a soft picture running in empty theatres. This is something exhibitors want to avoid. The rationale on the part of certain distributors to buy theatres in order to control playing time was fatally flawed. It is absurd to want to hold a picture doing no business. And it is interesting to open a newspaper and track which studio-owned theatres are playing whose movies. (Indeed, *Variety* has reported that some studios are seeking buyers for their theatres.) At the right price theatres can be a good investment for the studios or for other entities, since motion picture exhibition would appear to be a very viable business well into the future. But an investment in movie theatres at an inflated price for reasons of ego or control makes no sense.

In 1986 Loews Theatres was acquired by TriStar Pictures, a unit of Columbia Pictures Entertainment that, in turn, was bought by the Sony Corporation in 1989 and later renamed Sony Pictures Entertainment. As a result, though, there has been absolutely no change as to the philosophy or operation of the company. We are not told what to play, and if I ever got an order to favor Columbia or TriStar product, I would leave. As head of an exhibition company, my primary responsibility is to the bottom line, and the job is always to try to play the best available pictures at the lowest possible film-rental terms, regardless of ownership.

Exhibition is a very resilient business, having survived death predictions whenever a new delivery system entered the home: broadcast television, cable television, home video and satellites. Having survived all this, it is clear that movie theatres are here into the indefinite future. After all, people are social animals. There is a fundamental need to get out of the home, to interact with an audience and enjoy an entertaining movie on a big screen in a comfortable environment. And that is the unique service we provide.

# THE THEATRE CHAIN: AMERICAN MULTI-CINEMA

*by* **STANLEY H. DURWOOD,** *chairman and chief executive officer of AMC Entertainment Inc., a parent company of American Multi-Cinema, Inc. (AMC Theatres), based in Kansas City. A graduate of Harvard (B.A.), Mr. Durwood is an active leader in the civic and cultural affairs of his native Kansas City and is a director of the Motion Picture Association of America.*

*and* **GREGORY S. RUTKOWSKI,** *vice president, West Operations for AMC Theatres, based in Los Angeles. A 1971 graduate of the University of Missouri–Columbia (B.A.), he also holds a Master of Business Administration degree from the University of Missouri–Kansas City, 1974. Mr. Rutkowski joined AMC in 1975 and has spent virtually his entire career with the company.*

*... the rule of thumb [is] that roughly 70% of gross revenues is box-office; 30% is concession and "other"....*

The exhibition business can be condensed into two main issues: playing pictures (film booking) and finding and maintaining theatre sites (operations). A. Alan Friedberg's article focused on the film side (see p. 341); this one will capsulize the functions of the operations area at AMC Theatres.

American Multi-Cinema, Inc. (AMC Theatres) is a wholly owned subsidiary of AMC Entertainment Inc., based in Kansas City, Missouri, the largest American company exclusively devoted to motion picture exhibition (and second-largest operator of multiplexes), with 1,600 screens in 256 theatres in some twenty-four states. The company was originally known as Durwood Theatres, founded by Edward D. Durwood in 1920. In 1963 American Multi-Cinema opened the Parkway Theatres in the Ward Shopping Center in Kansas City, the first twin theatres in the world, with a common box office, lobby, concession counter and projection booth. In subsequent years, AMC efforts mirrored the recent history of advancements in exhibition: the first six-plex (1969); the first management training academy in exhibition (1979); the cup-holder armrest (1981); computerized box offices (1982); the first overseas multiplex, in England (1985); the first fourteenplex in the United States (1987); a national reward system for frequent moviegoers (1991).

For any theatre chain, expansion is a major decision. Site selection involves forecasting population trends and analyzing markets that can absorb more screens. For example, once an area of a city is targeted, we examine economic and demographic issues, such as population density, home ownership, income, age, education and occupations, in an effort to zero in on college-educated young people and families, the core moviegoing audience (and our future customers). When considering a site, we also review its proximity to competing theatres ("clearance") and access to complementary retail. Since distributors divide cities into exhibition "zones," or retail trade areas (our internal expression is "pockets"), another issue is whether a given proposed multiplex site will have easy access to prints.

After a site is selected, there are at least two ways to proceed. One approach might find AMC purchasing the land, or entering into a long-term ground lease and building the multiplex out of cash. Since our theatres can range from 40,000 to 60,000 square feet (over an acre of land), this is a major investment. The more typical approach might find AMC entering into a standard build-to-suit lease, wherein a developer would provide the financing and construction and either AMC would build, or the developer would build to our specifications. Costs including legal, architectural and preopening (commonly referred to as "soft costs") as well as furniture and fixtures (seats, screens, concession stands, wall coverings, booth equipment, carpeting, etc.) are usually borne by AMC in either case.

In the feasibility stage the theatre design is created, identifying the number of seats, screens and general configuration of the complex. This is usually done internally, and with the help of outside architects, always making sure it is compatible with adjacent retail. A set of plans is drawn up, and a contractor is chosen based on competitive bids. Construction typically takes five to six months, depending on the complexity of the building, with an extra month or two to "fixture" a multiplex. For maximum impact, theatre openings are generally targeted for the peak seasons of summer or Christmas.

Santa Monica, California, is a good example of expansion in an underserved zone, which benefited exhibitors and distributors alike. When the city decided to renovate and upgrade an old outdoor mall (the Santa Monica Mall) into what is today the Third Street Promenade, they invited movie theatres to build multiplexes to anchor the array of new restaurants, cafés and retail that they envisioned. The site was appealing to exhibitors because the demographics of Santa Monica and neighboring communities including West Los Angeles, Brentwood, Pacific Palisades and Malibu represented attractive target moviegoing audiences. The result was that, by 1989, three multiplexes had opened: AMC's Santa Monica 7, Mann Theatre's Criterion 6 and Cineplex Odeon's Broadway fourplex. Laemmle's fourplex was preexisting (see article by Robert Laemmle, p. 359), bringing the new screen count to twenty-one, with over 5,000 seats. The density and proximity of these new screens has created an exciting entertainment atmosphere at the Promenade, with much nighttime foot traffic, as pedestrians patronize restaurants and theatres.

Primarily, the costs of operating a theatre are comprised in three categories: direct operating expenses, film rental (the amount of box-office receipts belonging to distributors) and theatre rent.

Once a multiplex opens, attention turns to operating expenses. Direct operating expenses include payroll (usually around 10%–12%

of revenue), payroll taxes, concession (or food) costs, advertising (5%), repair/replacement costs, maintenance, utilities, film delivery (from the exchange to the theatre), supplies, xenon bulbs, postage and telephone, overnight mail, travel, union health and welfare payments and taxes-licenses-insurance costs.

The largest revenue source, of course, is the box-office receipts, followed by concessions; arcade income from electronic games or other machines; theatre rental to outside groups; and revenue from slide presentations, which may include some local advertising. Together, these represent the *total theatre gross*. For ease of illustration the preceding items are included under "concessions" in the following breakdown:

| | |
|---|---|
| Film Gross | 70% |
| Concessions | +30% |
| Total Theatre Gross | 100% |

For a sense of how costs relate to each other, here is a rough itemizing of estimated costs from a sample theatre. (These are for illustrative purposes; naturally, figures vary from theatre to theatre across the nation.) The "miscellaneous" item includes telephone, mail, postage, protective services, film delivery, checking services, arcade expenses and tickets. Remember, these figures are percentages of total theatre gross, *not* box-office gross:

| | |
|---|---|
| From Total Theatre Gross of | 100% |
| *Deduct*: | |
| Film Rental | 35% |
| Rent | 10% |
| Payroll | 10% |
| Concession Costs | 10% |
| Advertising | 5% |
| Taxes, Licenses, Insurance | 3% |
| Utilities | 2% |
| Repair, Maintenance | 1.5% |
| Miscellaneous | 1.5% |
| Payroll Taxes | 1% |
| Supplies | 1% |
| Theatre-Level Cash Flow | 20% |
| Less Overhead | −6% |
| Net Cash Flow (before depreciation, taxes or debt service) | 14% |

Running the division-wide numbers for a sample week, let us say that western-division screens (totaling some 400 plus) have an overall box-office gross of $3,150,000, while total revenue (adding concessions and other items) is $4,403,000. Box-office gross in this example is 71.5% of total theatre gross, confirming the rule of thumb that roughly 70% of gross revenues is box-office; 30% is concession and "other." To illustrate more statistics, the national ticket average is around $4.00, and the concession expenditure range per patron is between $1.50 and $1.70. Two key indicators used to track progress in exhibition are revenue per square foot and patronage per square foot. The division grosses in the area of $120 per square foot and plays to an average of twenty people per square foot.

Moviegoers sometimes wonder about how ticket prices are set. Pricing is solely the province of exhibition, which backs into it, based on normal costs of doing business compared with revenue projections. For example, with every rise in the minimum wage, payroll costs escalate. Also, supplies cost more over time. Added to that is the subtle but constant pressure from distributors to extract higher film rental for their product because of escalating production costs. The result is an ever-increasing pressure on exhibition's bottom line, which must be serviced by the price of admission. When the pressure becomes too great, theatres back into a price increase.

Concession sales are central to exhibition, since profits are solely ours and not shared with distributors. The concession supply business, in effect a trucking business, has changed dramatically in recent years, becoming highly centralized and automated. In earlier times regional concessionaires were the norm. Today many exhibitors turn to national companies, based on their ability to truck supplies from regional warehouses to theatres in a timely and cost-effective manner, thus offering a lower price to the exhibitor. Concession supply orders are generally made on a per-multiplex basis, via computer, each Monday for the following weekend.

Pricing of concession items is market-driven, but with an eye toward value-pricing; that is, the larger the size, the lower the per-ounce price the consumer pays. In an attempt to attract and satisfy the more upscale frequent moviegoer, special attention is paid to offering the finest moviegoing experience in terms of theatre amenities (wide screens, stereo sound and plush seats) coupled with a wider variety of quality foods and snacks. For example, AMC operates cafés in many of its multiplexes. These are a sort of eclectic concession stand, many with seating, which sell an assortment of desserts, bottled water, truffles, and other upscale, specialized goods.

In the future we will be seeing a wider variety of foods in the larger multiplexes, even possibly fast foods, in a more self-service, cafeteria-style environment. In an effort to meet the demands of our customers, this is intended to promote "ease of transaction" while minimizing costs.

Although the industry is extremely competitive and circuit (or brand) identification is important, we can unite for a mutual cause. As an example, recently a city-wide entertainment ticket tax was proposed in Los Angeles and was successfully defeated through the collective efforts of all area exhibitors.

Outside the exhibition industry it should be noted that distributors and theatre owners have never worked as closely together as they do now, with mutual respect, in a kind of basic compact: We provide the screens, they provide the product. We welcome a continuing supply of films not only from the major studios but from independent distributors as well. In the face of this, it is vitally important that we operate our theatres more efficiently. Our operating schedules must reflect the best utilization of our seating capacities so that revenue per seat and patrons per square foot are maximized.

This calls for very close cooperation between our two main corporate entities: Film, which handles negotiations and booking of pictures, and Operations. (Other functions within the company include marketing, training, finance, management information and accounting.) What kind of emergencies can arise? On the film side, a booking may be pulled at the last minute by a distributor; conversely a late booking will need immediate advertising support. On the operations side, daily crises could involve a delayed print delivery, which might lead to canceled performances, in which case angry customers must be placated. Other types of emergencies can occur; perhaps the janitorial crew arrives late or concession orders are incomplete, jeopardizing a primary revenue item.

At AMC we consider ourselves a value-added circuit, offering patrons a variety of services. The "TeleTicket" system is a service whereby a customer at selected locations can order tickets in advance via a computerized phone line and pay with a credit card, for a minimal service fee, then pick up the ticket at a will-call ATM at the theatre. We have also tested an ATM-style wall-mounted system where the customer, using a credit card, can buy a ticket (at no extra charge), as a sort of extension of the box office. Our box offices are totally automated, which streamlines ticket issuing for the patrons, as well as box-office reporting and closeout (over/short) procedures, which in turn provides immediate access to information for management. In the theatre auditoriums the HITS (High Impact Theatre) sound system and Torus screen (nonperforated and compound-curved) are AMC's proprietary innovations. "MovieWatcher" is a frequent-moviegoer program wherein the customer, after a qualification period of five visits, is issued a permanent mag-striped identification card that, when swiped on further visits, accumulates points toward significant discounts or bonuses. We track the behavior of our cardholders, including moviegoing patterns and frequency. "Silence Is Golden" is a response to a common customer complaint, noise in theatres. Our personnel are trained to remove disruptive patrons if there is no change in their behavior after receiving a warning. In addition, we offer a prefeature entertainment program, presently a combination of cartoons, updated newsreels or entertainment-related short subjects.

Our management training program is something that customers do not see, but enjoy the benefits of. There are five training academies around the country, one for each operating division (Northeast, Southeast, Midwest, Southwest and West). Each is staffed by a training director, responsible for the training of every management person in our employ. The program is six weeks in duration. The first three weeks involve operations and administration, and the second half covers the more philosophical, management side of our business. In a separate, continuing-education program, our multiplex managers, titled general managers, meet once a year in groups of ten at our academy in Los Angeles to review current management issues and exchange information.

Innovation in exhibition is ongoing and primarily due to the wishes of our customers. The future will find megaplexes with ample seats, screens, parking and concessions. As an industry, exhibition has found it easy to coexist with home entertainment, such as VCRs. For example, studies have shown that the avid moviegoer (representing 30% of our audience), who goes to the movies thirty-five to forty-five times a year, is also a person who rents as many cassettes. Those nay-sayers who predicted the demise of exhibition with the advent of home video were as wrong as those who predicted the end of theatres with the coming of television. The patient will not only live, but prosper.

# THE INDEPENDENT EXHIBITOR

*by **ROBERT LAEMMLE,** owner of the Laemmle Fine Arts Theatres, a Los Angeles–area theatre chain with sixteen screens that specializes in exhibiting first-run foreign language films, American independent films and quality Hollywood productions. He received a master of business administration from UCLA.*

*Reduced zones often result in less revenue per theatre (squeezing exhibitors), even though overall city grosses would be up (pleasing distributors). . . .*

Laemmle Theatres is made up of sixteen screens in eight theatres in Los Angeles, including one triplex, and two fourplexes. The growth of the company since 1964, when we owned one theatre, has occurred by adding screens every two to three years. The organization consists of me, my son, Gregory, and three office workers, who assist with flyer program layouts, publicity and public relations, and other miscellaneous office work.

Our theatres are broken down as follows: The three primary houses are the Royal in West Los Angeles, the Music Hall in Beverly Hills and the Fine Arts in Beverly Hills. We also have the Monica fourplex in Santa Monica; the Grande fourplex in downtown Los Angeles; the Town and Country triplex in Encino; and two theatres in Pasadena, the Esquire and the Colorado. A fiveplex in West Hollywood is under construction. We primarily show foreign films and fine-arts specialty films in all our theatres, except for the Grande, which usually shows first-run Hollywood product. If there's not a "good" Hollywood film to play there, we would show a foreign film. (Most distributors use the term *good* in relation to money-making; I use the term here in relation to quality.)

In Los Angeles there is hardly any bidding for motion pictures; most deals are the result of negotiation. Any large city is divided into zones, representing population centers of moviegoers. In negotiating to exhibit a film, a theatre in one zone may request clearance over those in other zones, for less competition. But with the expansion of theatre circuits and the natural shift in population densities, zone sizes are being reduced. For example, ten years ago on the west side of Los Angeles a film would have opened in Westwood only. Today, it might open in Westwood, Century City, Culver City, Marina Del Rey and Santa Monica, all at the same time. Reduced zones often result in less revenue per theatre (squeezing exhibitors), even though overall city grosses would be up (pleasing distributors).

To review our standards of business: we do not blind bid; we seldom bid on a film; we don't actively compete with the big circuits. Rather, it is to our advantage to let our competition book their screens; then we negotiate for that type of product that wants to play in our theatres. Also, we have a level or standard of product that we will show; and we don't believe in paying large guarantees. On this point we feel that if a film is good, it will find its proper film rental, and we will assist it in making the most money possible through our promotion work. We have a very good reputation for prompt payment of film rental. (One of the reasons for advances or guarantees is for distributors to be paid early, since many exhibitors delay payment from 60 to 120 days.) As to bidding, since we are a small outfit, we could lose more money on one bid including a nonrefundable guarantee than we could make the rest of the year in a particular theatre. However, we will pay a refundable advance for a picture, as long as we feel it's a reasonable figure. To determine this, we follow what a foreign film, for example, may be grossing in New York, to judge what size advance is reasonable for a Los Angeles opening.

When booking foreign films, distributors don't establish a release date in the same manner that is used for Hollywood films. A distributor will open a foreign film in New York, get the reviews, develop and test their advertising campaigns and then set opening dates in other big cities. Opening dates may depend on the availability of certain theatres, since only a limited number of theatres in each city specialize in showing foreign films. At the Royal in West Los Angeles, for example, we have two films waiting behind the current one. The opening dates on these films are not set; they follow in sequence, so the contract on the first film calls for it to follow the current film, as opposed to having a specific opening date. In many cases distributors negotiate with us specifically for the Royal or Music Hall Theatre because of the tradition and prestige of the house; they will wait for it to become available. Miramax is a company that is successfully broadening release patterns of foreign films, opening with over 100 prints at the same time, in a pattern that falls between the ten-print run of the traditional specialty distributor and the major studio release of over 1,000 prints. They will set an opening date in New York and then opening dates two weeks later in other key cities including Los Angeles, locking our theatre into a specific date, instead of waiting for the preceding picture to play itself out.

There are a lot of different independent distributors handling foreign films; some handle one film, some handle half a dozen. We get to know them all, and they know us. Sometimes we contact them as soon as we hear that they've acquired a certain film for domestic release; sometimes they call us and ask us to view a new film and consider playing it in one of our theatres. We also discover availability of new foreign films by reading the trades and other papers, attending several film festivals and going to private screenings, often prior to a picture's securing a distribution deal. In fact, we've been helpful in launching certain films that we've liked.

We generally avoid playing pictures requiring guarantees, but will give advances on occasion. A recent major foreign film involved a mid-five-figure advance plus minimum terms and a minimum playing time. It called for a split of 90/10 above our house allowance, with minimum floors for the first two weeks at 60%; for the third and fourth week, 50%; the fifth and sixth week at 40% and subsequent weeks at 35%. These floors may be slightly lower than for a Hollywood film, which starts at 70%.

Floors protect the distributor if the film's a disappointment; the 90/10 split also protects the distributor. For instance, let's say a theatre for one week has a $20,000 gross on a 90/10 deal with a 60% floor and a house nut of $5,000. Start with $20,000 as the gross, subtract $5,000, which is the house expense. Then split the remaining $15,000 90/10; that leaves $1,500 for the theatre and $13,500 for the distributor. How does that compare with the 60% floor figure? 60% of $20,000 is $12,000; 70% of $20,000 is $14,000. Since a net film rental of $13,500 (the distributor's share) is about 67% of $20,000, or about 7% higher than the floor, the 90/10 split will govern, with the theatre paying a higher percentage to the distributor than the straight 60% of gross.

The area of advertising is wide open for negotiation. Sometimes a theatre of ours might agree to contribute $500 to the first-run advertising campaign of a picture or $1,000 to the preopening and first week's campaign. Generally the distributor pays for the bulk of advertising because he is keeping the bulk of the generated film rental.

Another type of deal, possible for a small foreign film, would be on a basis of 25%–50%, using a "sliding scale," with advertising shared on a 50/50 basis with the distributor. If the film grosses $4,000 in a given week, the distributor would receive 25% of the $4,000, or $1,000. A sliding scale adjusts much like an income tax table; if the film grosses $5,000, the scale might peg at 30%, or $1,500 to the distributor. If the box-office gross is $6,000, it might be 40% on a scale, or $2,400 as film rental. The sliding scale is a preestablished and agreed-upon chart of figures that a particular theatre follows, based on its various expenses. Since the scale graduates upward as the gross increases, the film rental increases as well. The distributor would agree to such a sliding-scale chart in advance, and the figures naturally take into account the house nut.

The *nut,* or overhead, of one of our theatres includes payroll, rental on the theatre property, maintenance, insurance, utilities and standard advertising, which is a part of the usual, steady advertising each theatre bears.

Concessions are a much more significant profit center for Hollywood films than for foreign films, and at multiplexes, where one concession stand services multiple theatres and is constantly active. We regard concessions as a convenience to the customer, not the primary source of our income. With a Hollywood film it's often the other way around. Sometimes a theatre showing a Hollywood film would even be willing to give the 70% floor terms and perhaps 80% floors just to get a certain "popcorn movie"; then they make their money at the candy counter. After all, the markup on a bag of popcorn is probably around 75% (which does not, however, take into account the cost of running a concession operation).

Once we make a deal on a film, we put trailers on all our screens announcing the film in advance. In addition, there'll be posters at the theatres and fliers circulating with full reviews or excerpts; telephone calls will be made to certain civic groups, if they would respond to a particular film. For instance, if it's a German film based on Ibsen, we'll go after developing theatre groups and German-language groups. We've gained a good reputation for this type of special handling of foreign films.

As for prints, we try to receive a print at least a week before the run begins, for press screenings and to check it thoroughly for damage. There are usually no protection backup prints available for foreign films; usually, when a Hollywood film plays in Westwood, there's a backup print in case of tearing or other damage. Because a single 35mm print costs close to $2,000, the small, foreign distributor usually can't afford a backup.

In the day-to-day operation of the theatres, little emergencies arise, such as the breakdown of an air-conditioning system, plumbing problems, projection-equipment or sound problems, customer complaints, or employee problems. Otherwise the day is filled with the routine of planning advance work for future pictures.

There is one complaint that a theatre owner must voice, and that is with newspaper advertising policies. Newspapers charge higher rates for display advertising than for other forms of advertising. It's wrong that a department store can afford a half-dozen full-page ads in a city while a major film cannot because the same space just costs too much. This practice is discriminatory and should be abolished. It dates back to the days when amusement advertisers would leave town without paying their bills. Local theatres are here to stay, and they make a major contribution to the local economy; it's about time this advertising practice was corrected.

On the other hand, there is another practice that could be helpful. In some cities newspapers charge different rates depending on the number of theatres playing a given film; the ad rate is lower for a film playing on a single screen (usually an art house) and higher for a film playing on multiple screens. This is in effect a subsidy for art films, and should be encouraged in more cities.

Independent exhibition exists in a more difficult climate today due to the recent expansion of circuits. In some cases independents were acquired by larger chains. In other cases expansion through new construction has resulted in overscreening certain neighborhoods, reducing revenue from individual screens and threatening local independent theatres with extinction. Independent distributors must encourage the existence of independent exhibitors because their long-run livelihoods are interconnected. Circuits will only play specialized product when it suits them; independents will always want to play these films.

Finally let me voice concern that there are two elements of Hollywood domination overseas that are jeopardizing foreign-language films, the lifeblood of many art houses. First, in certain overseas markets (including France and Italy), attendance for American films is on the rise; for indigenous films it can be flat. This discourages local film production, and therefore there are fewer films available for import to the United States. Second, Hollywood attracts major filmmakers from all over the world who began in their native country and moved on to make Hollywood films. This is perhaps inevitable, since filmmakers are entitled to enjoy the success of their work, which usually means gaining access to a wider audience through American-financed pictures, but it can be unhappy for the local film industry that spawned them.

# THE HOME-VIDEO RETAILER

*by* ***VAN WALLACH,*** *senior editor and East Coast bureau chief for* Video Store *magazine, who is based in Westport, Connecticut. A graduate of Princeton University with a degree in economics, he started his career as a reporter-researcher for* Forbes, *has worked as a freelance writer on subjects ranging from finance to fur retailing, and has done celebrity interviews for* Whole Life Times, *including Howard Stern, Cesar Chavez and Holly Near. Joining* Video Store *in 1987, Mr. Wallach now directs the magazine's coverage of studios and sell-through video retailing and writes the "Video Stories" column.*

*Unlike the anonymity of the relation between moviegoers and exhibitors, video retailers need to know the exact identity of their customers. . . .*

The movie business is a cash business. Just as the box-office is the point of sale for theatrical exhibition, the video retailer is the point of sale (or rental) for home video.

It's been said that Hollywood is an industry built on relationships, a complex dance among producers, lawyers, agents, talent, financiers, directors and writers, to name a few. When a movie reaches home-video retailing, success or failure revolves around one relationship: a video retailer and the renting or buying customer. Multiply that give-and-take by 29,000 retail stores and millions of renters and buyers, and the essence of video retailing emerges.

Behind those millions of rentals and sales is a young, evolving industry. Video "specialists" are stores concentrating on video rentals, in contrast to the mass merchandisers (such as Wal-Mart and The Musicland Group), which sell videos but do not rent them. Video specialists range from the single-unit "mom and pop" stores to mighty Blockbuster Entertainment Corp., with over 1,600 stores in the United States. Whatever the size, specialists grapple with all the typical retail issues of investment, inventory, management, labor, marketing, merchandising and strategy. In this sense, when a movie becomes transformed into a video, it is in a business environment more akin to a bookstore or an auto-parts store than a movie theatre.

Unlike the anonymity of the relation between moviegoers and exhibitors, video retailers need to know the exact identity of their customers, for both security and marketing purposes. When somebody rents a movie, the videocassette, which can cost retailers over $70 (see article by Reg Childs, p. 328), physically leaves the premises. Before they can rent, retailers usually have prospective customers join a "club" to provide a modicum of security. Most retailers have dropped the membership fees common in the early years of home video, although some retain a one-time yearly charge, up to $50 per year.

A typical membership form requests basic information: name, address, home and work phone numbers, driver's license, often a credit card number. Many retailers ask for the names of family members who will be permitted to rent on the membership card. Renters can also place restrictions on the use of the card, such as limiting children to videos rated below R.

Such information serves a dual purpose. First, it provides some protection against theft or the slow return of cassettes. If a tape does not come back, a retailer can contact the customer to report that a tape is incurring late charges. In the case of refusal to return or the billing of late charges, membership data give the retailer a lead on where to send collection agencies. If a credit card number is in the system, the retailer can charge the value of the tape to the card.

Second, the membership data, combined with records of rental activity, provide a gold mine of marketing information. A common strategy is to identify customers who have not rented in a given time frame, say, a month, and send them a card good for a free or reduced-price rental. This reacquaints customers with the store and with the new titles stocked since their last visit. Retailers also use this as a way to learn if customers have complaints.

What are typical customers' rental habits? Generally several months or a year after buying a VCR, consumers cut back on rental volume, having satisfied the initial curiosity about the technology. Unless someone is either a film buff or a fan of a particular genre, the rental pattern is then limited to new releases, which are six-month-old theatrical titles. What has evolved is a video retailing environment marked by fewer new VCR owners (the heavy renters) and less activity by veteran VCR owners. Jack Landman, president of Take It Home, a chain based in Haltom City, Texas, outside Fort Worth, analyzed rentals recently and found a marked decline in weekly business from active accounts. The figure had fallen from renting an average of 3.5 units per week to 2.5.

How can video retailers increase this rental activity? They must recarve what has become a static pie by extracting more rentals from their existing customer base and by taking business from competitors.

To accomplish this, some retailers pay particular attention to steady accounts that have slackened in their rental activity. These differ from other accounts in that they had once produced regular income. As such, each is theoretically worth the same as several sporadic customers. By identifying them and attracting them back to the store, retailers can maintain their essential repeat clientele.

Another important measurement is by neighborhood. Analysis by and within zip codes can pinpoint areas of popularity. Retailers can then ask themselves why they don't draw from a particular neighborhood. This might be due to strong competition, a lack of visibility, even traffic patterns. If so, retailers can devise marketing strategies to raise awareness in those areas, through direct mail, billboards or advertising in local newspapers.

A third area for retailers to consider is age. Many driver's licenses indicate the year of birth. Tracking renters by age group can help retailers assess their inventory and marketing. For example, Terry Bettendorf, owner of the Movie Set chain based in St. Louis, noticed a move toward customers aged fifty to fifty-five in one of his stores. As a result he increased the selection of videos from the 1930s to the early 1950s. Similarly, Alan Daniels, owner of six Video Treats stores based in Poughkeepsie, New York, sorted his membership according to age groups and zip codes. Based on his findings, he cut back on radio advertising to concentrate on direct mail, television and daily newspapers.

Despite some moves toward diversification, rentals still contribute the bulk of retailers' revenues. According to a recent survey of retailers by *Video Store* magazine, rentals were 72.4% of gross revenues. Rentals of video games represented 10.2%, while video software (sell-through) were 5.1%. Hardware rentals came in at 3.1%, and alternative merchandise kicked in 2.0%.

Once money comes into stores from rentals and purchases, video retailers handle it as in any other retail business: They count it and take it to the bank. Mark Donovan, chief financial officer of the Total Video chain in Swartz Creek, Michigan, explained the details at his operation. Upon renting a tape a customer signs a receipt. He can either pay then or when returning the tape. Both the store and the customer have copies of the receipt. When the tape is returned, the clerk scans it back into the computer system, which notes whether a balance is due, and if so, the customer pays, concluding the transaction. Upon buying a tape, the customer takes a copy of the receipt. The store computer system reports that a certain amount of income has been received, and this is compared with an "over-and-short" tally of cash-register totals. The system also assists Total Video with tracking volume, inventory, customer traffic and special orders. The Total stores make two or three deposits at neighborhood banks each day. Then a "sweeping mechanism" moves the funds to Total's corporate bank.

How do retailers spend the accumulated funds? A recent *Video Store* survey found that the average store had revenues of $192,655. Another study showed stores spent $23,995 on video inventory, $4,000 on other inventory, $10,400 on rent and $18,000 on payroll and payroll taxes.

Mark Donovan provided a breakdown of Total Video's expenses. Software inventory purchases took the largest share, 25%–29%. Another 20%–25% went to employee wages, taxes and benefits. Rents, utilities, insurance and personal property taxes accounted for another

20%. The balance went to pretax general expenses, advertising and marketing.

Just as accounting is a source of controversy in Hollywood, it's an issue in video retailing, centering around depreciation of videocassettes. Since new cassettes garner most of their rentals in the first ninety days, they depreciate rapidly. The Total Video chain depreciates tapes over a three-year period, with most of the value written down in the first year, to accurately reflect the slackening demand for aging titles. In 1989 the Internal Revenue Service cleared up confusion over what was allowable in depreciating videocassettes in a ruling that approved straight-line and income-forecasting methods. Straight line calls for the retailer to depreciate the tape an equal amount per year over the life of the tape. With income forecasting, depreciation matches the revenue stream of the tape. On a tape with a wholesale cost of $60, straight line would yield $20 per year over three years. Using income forecasting, acknowledging the front-end nature of new release income, 75% might be taken the first year, or $45.

Selecting inventory is perhaps the most important in-store decision retailers make. Much as bettors handicap horses, retailers handicap titles, developing their buying strategy for a title by studying the performances of a genre and factors like cast, box-office, studio marketing and customer appeal, as well as whether a title is priced for rental or sell-through. Once purchased, they carefully monitor a tape's performance, tracking its turn rate, breakeven point and return on investment. As a rule, Jack Landman of Take It Home puts 25% of his gross sales back into inventory purchases. Customer input also helps him make choices. He posts lists of twenty upcoming titles at the checkout counters of his stores and asks customers to mark the ones they want to see.

Put all the factors together, and a title that does well at the box office often does well in home video. Paramount Home Video's *Ghost,* for example, was a theatrical blockbuster that shipped over 637,000 copies at $99.95. It had no suggested retail price; this is what the industry calls an "equivalence" price. Video distributors paid Paramount about $63 per unit, which returned an estimated $40.1-million to the studio.

Retailers must also identify that handful of performers whose number of domestic video units rented outperform their theatrical box-office tickets sold. These video stars, as reported by *Entertainment Weekly,* include Jim Belushi, Jean-Claude Van Damme, Patrick Dempsey, Brian Dennehy, Whoopi Goldberg, Gene Hackman, Bette Midler and Mickey Rourke.

But even with all this information, inventory selection remains a complicated business. Jack Landman says, "It's a hit-driven business and the hits cost more than ever. That's the problem." It's a problem because store activity focuses on the current hits, a narrow part of the inventory base. Retailers must buy enough units of a title to meet customer demand, but not so many that demand is quickly satisfied and the copies stop renting before they have at least earned their cost back.

Another issue of stocking has to do with used or "previously viewed" copies of titles, which retailers can "presell" to customers before the units actually arrive in stores. That is, customers can buy a copy of a movie, which they will receive after the store has rented it for, say, thirty to sixty days. By then demand will have cooled and the retailer will want to trim inventory. Studios have encouraged this activity with advertising campaigns letting consumers know retailers are willing to sell such movies. Examples include *Dick Tracy* and *Ghost*. In fact *Ghost* included an ad at the beginning of each tape directed to renters encouraging purchase.

Demographics also influence inventory decisions, especially on titles beyond new releases. Dramas and foreign language movies tend to have a following in upscale areas, while comedy and horror titles do well in blue-collar settings.

Pricing for rental is another way in which video retailing differs from theatrical. Theatres charge a flat rate for all movies, with some discounts for age and matinees. In video the newness of a title, length of rental period, time of week and genre are all factors in determining rental rates. Most stores have a tiered structure, rather than a single rate. New releases—hit titles that have been available for less than, say, three months—often cost more because of their appeal. "Catalog" titles, which are older, have lower rates. Children's titles cost even less to rent, while adult (X rated) titles are charged a premium. A store might rent out a movie for two or three days with a discount if it is returned early. Since the bulk of rental activity takes place on weekends, another strategy is to lower prices or offer two-for-one specials on weekdays. A modern rental scenario might price new releases at $3, catalog titles at $2, children's titles at $2 and adult tapes at $4. According to a recent *Video Store* survey, the average new-release rental price was $2.54; the average catalog rate was $1.98.

Maintaining customers, buying inventory and pricing the tapes are the three essential management issues in video retailing. At the beginning of the industry, in the early 1980s, a store could exist with a few hundred units in a site of under 1,000 square feet. By mid-decade there was a clear trend to larger stores, similar to how record stores had evolved. *Video Store* defines small stores as those that have less than 2,000 units; mid-size have 2,000–4,999; large have 5,000–7,499 and superstores have over 7,500 units.

Larger stores provide space for both additional titles and other entertainment lines used to attract an audience beyond video renters, including laserdiscs, audiocassettes, CDs, sell-through video, even children's play areas. But the larger size requires more labor, overhead and inventory costs.

As with all retailing, store location is crucial. Customers have to see the store and have easy access to it. Therefore video stores are either freestanding or part of strip centers. (Video stores within supermarkets are a variation on this.) While stores *selling* videocassettes are often found in malls, stores *renting* cassettes are more likely found on the street, since they need high visibility and quick access to attract customers on a steady basis.

Video retailing is an intensely localized business. A single store draws from about a three-mile radius. Given that, marketing occurs in a concentrated area. Common marketing tools are direct mail, either by the store alone or in a coupon pack, or community newspaper advertising. Only when a store is part of a local or regional chain do radio, television or daily newspaper advertising become cost-effective.

At the chain level, video stores can also make good use of cooperative advertising, funds provided by studios to the video distributors from whom retailers buy their tapes, usually in support of specific titles. While procedures vary, a retailer will generally run ads that meet the studio's specifications. Upon showing proof of performance to their video distributor, the retailer will be reimbursed in the form of a credit for purchase of more units. Studios and video distributors will also offer co-op funds for fresh or innovative advertising. For example, Take It Home devised a billboard campaign to which Disney's Buena Vista Home Video contributed to advertise *Pretty Woman*. The display lasted only a few days, but the chain sold more copies than usual of *Pretty Woman* during that period in stores closest to the billboard.

The complicated reporting procedures of co-op ads call for a store manager with specialized knowledge in advertising, which points up the ongoing professionalization of video retailing. While a jack-of-all-trades can manage one or two stores, sustained growth leads to a division of labor. The mom-and-pop stores give way to a corporate approach.

This is in evidence at the Endless Video Superstore Limited Partnership in Pembroke, Massachusetts, founded in 1989 as an independent chain and then converted to a Blockbuster franchise in 1991. Partners John Krainin and Demy Martin both worked at Xerox earlier in their careers, experience that guided their management approach at Endless Video. They knew they would be starting with a small staff, making all decisions themselves and then expanding with power delegated to line management. When the company reached twenty stores, Krainin and Martin had two operations managers to oversee ten stores each.

They also established a training program to nurture store managers, who are responsible for controllable costs at the store level, such as hiring, training and dismissing staffers, supply purchases and telephone costs. They also have input into corporate decisions on tape-inventory buys. Those who perform can move up. Endless Video challenges employees through "job enrichment exercises," which involve them in corporate issues. One day per week a promising manager will work on an area, such as real estate, in the main office or in the field. "We want to make the manager feel like he is part of the organization," Krainin says.

With music stores crossing over into home video, another question facing retailers is how to apply the best attractions of each type of entertainment software. One music chain, Tower Records, successfully adapted to the technological winds by starting a separate division in 1981, Tower Video, small in number of stores, but highly visible.

Super Club, based in Dallas, Texas (a subsidiary of Philips N.V., the Dutch electronics firm), assembled an entertainment conglomerate of both music and video chains starting in 1989. This included specialists Video Towne in Ohio; Alfalfa/Movietime in Louisiana; Movies at Home in Kansas; Best Video, a distributor, in Oklahoma; and music chains Turtle's, in Georgia, and Record Bar/Tracks in North Carolina. It also picked up several smaller video chains. Initially Super Club left them as fairly autonomous operations, and integrated the services, with Best Video supplying some sell-through product to the video locations and Turtle's supplying Super Club video stores with audio products and expertise. The inventory varied according to store location and demographics. Using the system developed by Turtle's, Best Video started an audio racking service for retailers outside the Super Club network, offering a model stock program with an opening inventory based on location demographics. Bar codes on products allow for sales tracking and stock replenishment. The program also addresses merchandising, fixturing and sales promotion.

A move to a home-entertainment center makes several demands on a video specialty store. Inventory dollars get spread over more product categories; merchandising becomes more complicated because shelf space must be found for audio products; and marketing has to reach music consumers.

Another issue among video retailers is sell-through. When these titles are released, often supported by strong studio marketing, it can be bittersweet for video specialists because mass merchandisers, such as Wal-Mart and Phar-Mor, often charge extremely low prices as "loss leaders." Customers come to buy a specific video and walk out with other items as well. How do video specialists counter this? They develop a wider sell-through business by stocking titles beyond those hits, including catalog titles that mass merchandisers do not price so aggressively. Larger retailers also reduce their risk by using rackjobbers, companies that place videocassettes in stores on consignment. Risk can be further cut by concentrating on "special orders," which have three advantages to the retailer. First, they eliminate inventory costs; second, they can charge the video at full list price; third, they develop a reputation for good customer service compared with the mass merchandiser, which may lack the staff to order tapes outside their current inventory.

A final concern of video specialists is the impact of new technology (see article by Martin Polon, p. 451). The concept of video on demand, through a pay-per-view or other sophisticated delivery system, sends shivers down specialists' spines. If consumers can order new releases at home whenever they want, that would eliminate the need to visit the video store. And since new releases are the dominant source of video stores' income, that would deliver a knockout punch. At present, pay-per-view is a minor vehicle for movie delivery because of the small number of "addressable" homes that can receive PPV signals and limitations on the number of titles available at any one time. It seems to work best for sporting events, such as wrestling and boxing and, in 1992, the summer Olympics. Other services indicate the continuing interest in alternative delivery systems. Examples include the "viewer-controlled cable television test" mounted in Denver by communication heavyweights Tele-Communications, Inc., AT&T and US WEST, testing the response of 450 customers to both video on demand and enhanced pay-per-view, and the direct-by-satellite plans of SkyPix Corp., based in Seattle. A serious challenge is years down the road, but it is there. For the present, video stores' efforts to develop catalog business and nonvideo sources of income make good sense as retailers continue to grapple for a piece of consumers' discretionary income in the ongoing dance for the entertainment dollar.

X
THE AUDIENCE

# INDUSTRY ECONOMIC REVIEW AND AUDIENCE PROFILE

## INDUSTRY ECONOMIC REVIEW

This chapter offers a statistical overview of American motion pictures, using 1990 as a model year, with comparisons to prior years. Data covering the United States are provided by the Motion Picture Association of America, Inc., Washington, D.C., through the courtesy of Jack Valenti, president. Included is a review of yearly domestic box-office grosses, admissions, negative costs, marketing costs, print costs, numbers of releases, theatre screens and employment. This is followed by a review of 1990 theatrical film rental for American pictures in the top ten international markets, courtesy of A. D. Murphy. Next is a survey of the American motion picture audience, conducted for the MPAA in the summer of 1990.

### United States Box-Office Gross

The year 1990 proved to be the second-highest-grossing year in history, reaching $5,021,800,000. The record-breaking year was one year earlier, 1989, with a box-office gross of $5,033,400,000, just $11.6-million higher than 1990, or 0.2%. Significantly both years broke through the $5-billion level. Over the decade the level of gross increased by 82.7%. The 1990 figure also reflected a continuing upward trend of moviegoers over forty, who make up around one-third of the audience, and a drop in audience share of moviegoers under thirty.

A quarterly comparison of the 1990 box-office performance with 1989 shows that 1990 was stronger in three out of four quarters. The first quarter showed a 0.3% gain; the second quarter was up 1.4% and the fourth quarter increased 7.5% over 1989. Only the third quarter lagged behind its 1989 counterpart, by 7.8%. Here is how the 1990 box-office totals compare with yearly numbers from 1980 to 1989:

## *United States Box-Office Gross*

| YEAR | BOX-OFFICE GROSS | YEARLY PERCENT CHANGE |
|---|---|---|
| 1990 | $5,021,800,000 | −0.2% |
| 1989 | $5,033,400,000 | +12.9% |
| 1988 | $4,458,400,000 | +4.8% |
| 1987 | $4,252,900,000 | +12.6% |
| 1986 | $3,778,000,000 | +0.8% |
| 1985 | $3,749,400,000 | −7.0% |
| 1984 | $4,030,600,000 | +7.0% |
| 1983 | $3,766,000,000 | +9.1% |
| 1982 | $3,452,700,000 | +16.4% |
| 1981 | $2,965,600,000 | +7.9% |
| 1980 | $2,748,500,000 | −2.6% |

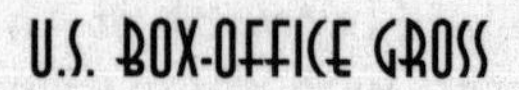


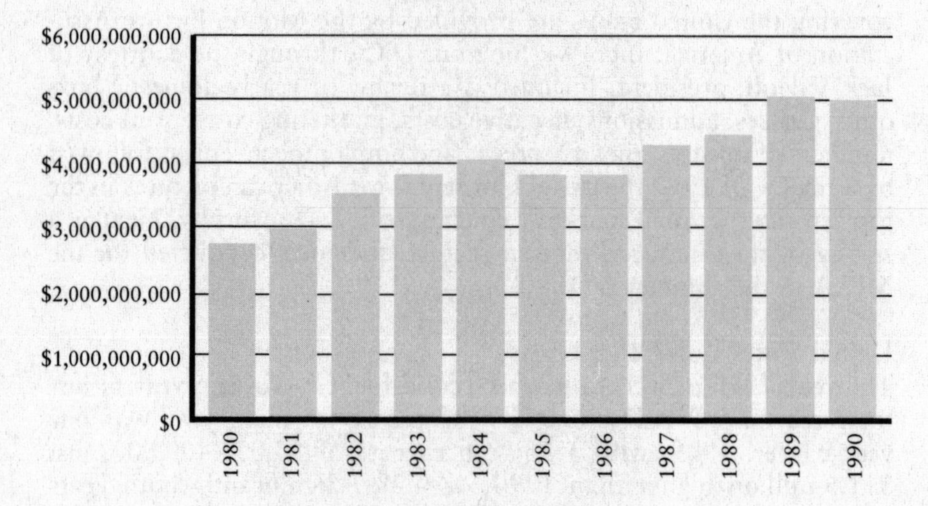

**Admissions**
For the fourteenth consecutive year theatre admissions topped the 1-billion mark.

| Year | Total Admissions | Yearly Percent Change |
|---|---|---|
| 1990 | 1,057,900,000 | −6.6% |
| 1989 | 1,132,500,000 | +4.4% |
| 1988 | 1,084,800,000 | −0.3% |
| 1987 | 1,088,500,000 | +7.0% |
| 1986 | 1,017,200,000 | −3.7% |
| 1985 | 1,056,100,000 | −11.9% |
| 1984 | 1,199,100,000 | +0.2% |
| 1983 | 1,196,900,000 | +1.8% |
| 1982 | 1,175,400,000 | +10.2% |
| 1981 | 1,060,000,000 | +4.5% |
| 1980 | 1,021,500,000 | −8.9% |

Over a decade numbers of movie admissions have fluctuated plus-or-minus 12%.

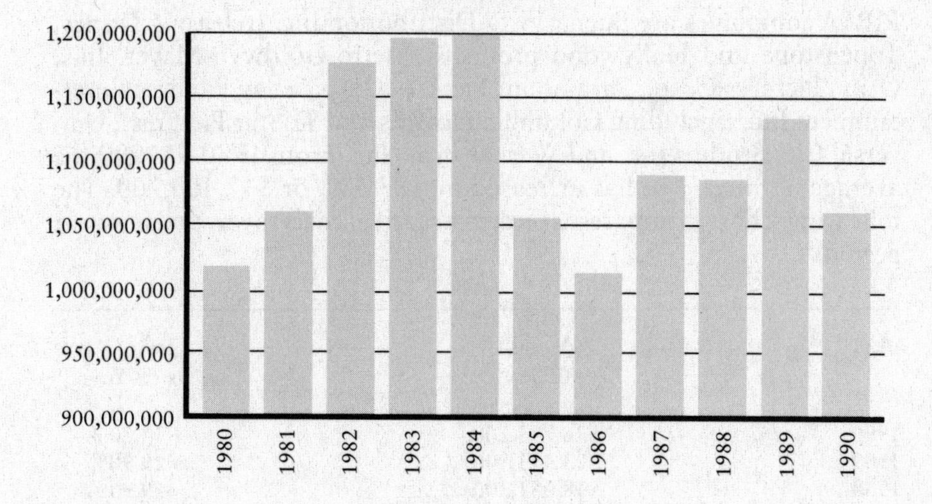


**Admission Prices**
Average admission prices rose by 6.8% to $4.747 for the year, compared with $4.445 per ticket in 1989. Over the decade admission prices rose 76.4%.

**High-Grossing Features**

In 1990 there were fifty-nine pictures that grossed over $10-million in the United States, compared with fifty-one in 1989. Thirty pictures grossed over $20-million.

| YEAR | *$10,000,000* OR MORE | *$20,000,000* OR MORE |
|---|---|---|
| 1990 | 59 | 30 |
| 1989 | 51 | 30 |
| 1988 | 45 | 18 |
| 1987 | 47 | 20 |
| 1986 | 41 | 17 |
| 1985 | 42 | 13 |
| 1984 | 35 | 14 |
| 1983 | 34 | 18 |
| 1982 | 36 | 14 |
| 1981 | 30 | 10 |
| 1980 | 36 | 17 |

**Negative Costs**

The average negative cost of producing a feature-length picture, financed in whole or in part by MPAA companies (including studio overhead and capitalized interest) rose to $26,783,200. This is an increase of 14.2% over the average 1989 cost of $23,453,500. The MPAA companies are Buena Vista Distribution Inc. (releasing Disney, Touchstone and Hollywood product), Metro-Goldwyn-Mayer, Inc., Orion Pictures Corp., Paramount Pictures Corp., Sony Pictures Entertainment Inc. (including Columbia Pictures and TriStar Pictures), Universal City Studios Inc. and Warner Bros. Inc. From 1980 to 1990 the average negative cost has increased by 185.5%, or $17,400,700. The following chart compares average negative costs over the ten-year period:

| YEAR | AVERAGE PRODUCTION COST PER FEATURE | % CHANGE OVER PRIOR YEAR |
|---|---|---|
| 1990 | $26,783,200 | +14.20% |
| 1989 | $23,453,500 | +29.90% |
| 1988 | $18,061,300 | −9.90% |
| 1987 | $20,050,500 | +14.87% |
| 1986 | $17,454,600 | +4.03% |
| 1985 | $16,779,200 | +16.42% |
| 1984 | $14,412,600 | +21.27% |
| 1983 | $11,884,800 | +0.30% |
| 1982 | $11,849,500 | +4.53% |
| 1981 | $11,335,600 | +20.82% |
| 1980 | $9,382,500 | +5.26% |

**Marketing Costs**

Per-picture marketing costs (prints and advertising) increased by 25.9% in 1990, to $11,641,000 per MPAA film from $9,248,000 in 1989. This was brought about by increases in the costs of both advertising and prints. Over the decade combined marketing costs have increased by 168.9%. The following charts offer details:

### *1. Combined Advertising and Print Costs*

| Year | Average Cost | Yearly Percent Change |
|---|---|---|
| 1990 | $11,641,000 | +25.9% |
| 1989 | $9,248,000 | +8.7% |
| 1988 | $8,509,000 | +3.1% |
| 1987 | $8,257,000 | +23.7% |
| 1986 | $6,673,000 | +3.4% |
| 1985 | $6,454,000 | −3.1% |
| 1984 | $6,651,000 | +27.8% |
| 1983 | $5,205,000 | +5.4% |
| 1982 | $4,936,000 | +12.0% |
| 1981 | $4,407,000 | +1.8% |
| 1980 | $4,329,000 | — |

### *2. Advertising Costs per MPAA Picture*

| Year | Average Cost | Yearly Percent Change |
|---|---|---|
| 1990 | $9,915,000 | +26.9% |
| 1989 | $7,812,000 | +9.7% |
| 1988 | $7,118,000 | +3.7% |
| 1987 | $6,865,000 | +26.3% |
| 1986 | $5,435,000 | +4.6% |
| 1985 | $5,241,000 | −2.3% |
| 1984 | $5,363,000 | +28.2% |
| 1983 | $4,182,000 | +3.0% |
| 1982 | $4,060,000 | +14.7% |
| 1981 | $3,541,000 | −0.1% |
| 1980 | $3,544,000 | — |

### *3. Total Print Costs per MPAA Picture*

| YEAR | AVERAGE COST | YEARLY PERCENT CHANGE |
|---|---|---|
| 1990 | $1,726,000 | +20.2% |
| 1989 | $1,436,000 | +3.2% |
| 1988 | $1,391,000 | −0.1% |
| 1987 | $1,392,000 | +12.4% |
| 1986 | $1,238,000 | +2.1% |
| 1985 | $1,213,000 | −5.8% |
| 1984 | $1,288,000 | +25.9% |
| 1983 | $1,023,000 | +16.8% |
| 1982 | $876,000 | +1.2% |
| 1981 | $866,000 | +10.3% |
| 1980 | $785,000 | — |

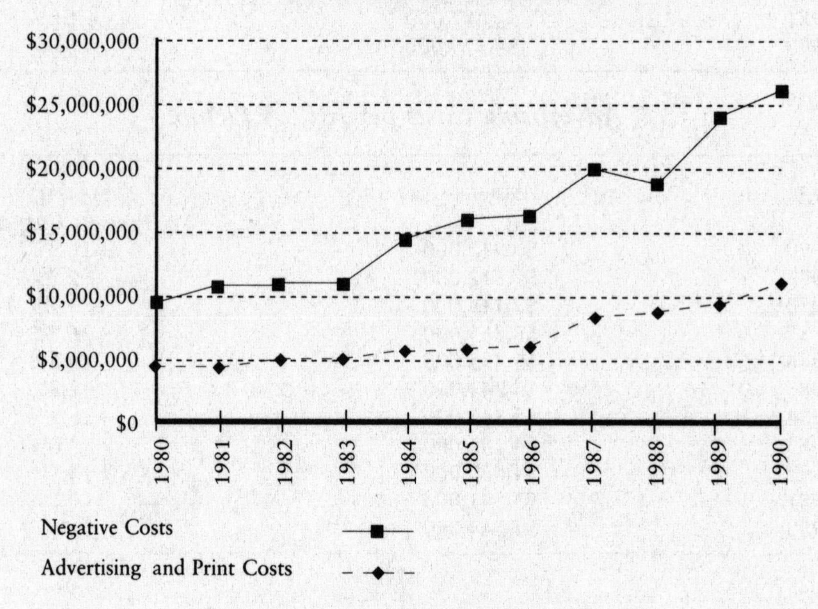

**Industry Advertising Expenditures**

Using 1989 as a model year, the total amount spent to advertise motion pictures in newspapers, on television, on radio, in magazines and via outdoor display in the United States was over $1.5-billion, an increase of 11% over 1988. The largest amount was spent on newspaper advertising, just over $1-billion, up 7.3%. The following chart illustrates the share of the 1989 U.S. movie-advertising dollar captured by each medium:

| *Medium* | *1989 % Share* | *1989 Amount* |
|---|---|---|
| Newspapers | 67.3% | $1,066,400,000 |
| Network TV | 13.5% | $214,400,000 |
| Local TV | 12.9% | $204,400,000 |
| Syndicated TV | 2.1% | $32,900,000 |
| (All TV) | (28.5%) | ($451,700,000) |
| Radio | 2.0% | $31,000,000 |
| Magazines | 0.4% | $6,700,000 |
| Cable | 1.5% | $23,500,000 |
| Outdoor | 0.3% | $4,400,000 |
| Total | 100.0% | $1,583,700,000 |

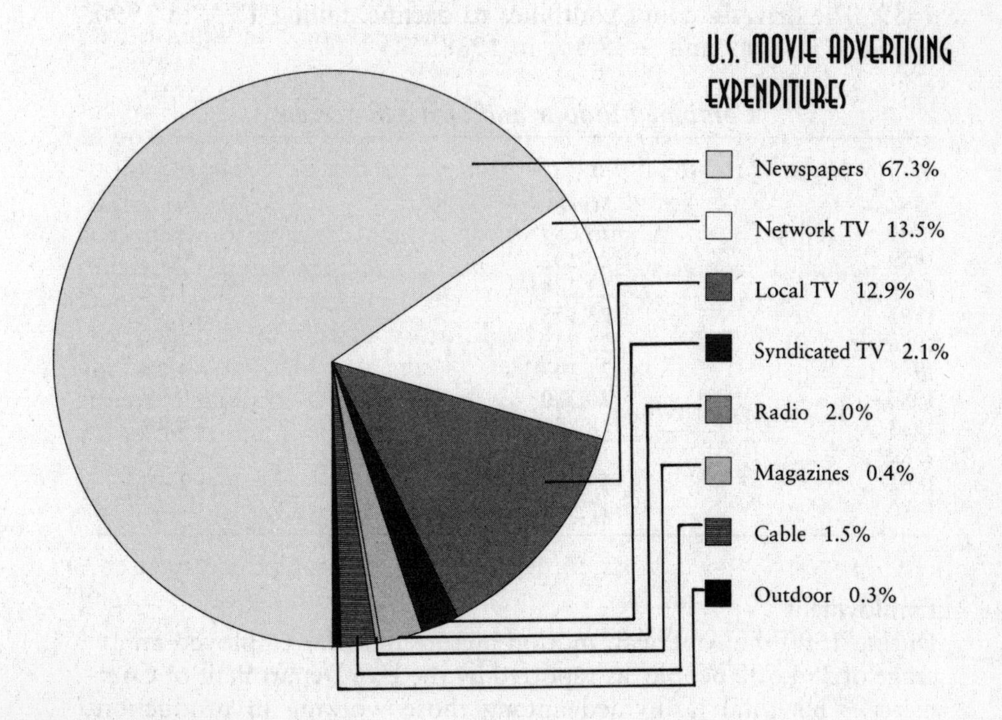

**Releases**

In 1990 the MPAA computerized tracking system reported 387 new theatrical releases from all identifiable distributors, including MPAA companies, the lowest figure in five years. This is a decrease of 14% from 1989, when 449 new films were released.

| Year | New | Reissues | Total |
|---|---|---|---|
| 1990 | 387 | 23 | 410 |
| 1989 | 449 | 43 | 492 |
| 1988 | 472 | 19 | 491 |
| 1987 | 487 | 20 | 507 |
| 1986 | 419 | 32 | 451 |
| 1985 | 389 | 81 | 470 |
| 1984 | 408 | 128 | 536 |
| 1983 | 396 | 99 | 495 |
| 1982 | 361 | 67 | 428 |

**Theatre Screens**

As of the end of 1990 there were 23,689 total screens in the United States, an increase of 2.4%, or 557 screens, compared with 1989. Since 1971 theatre screen count has increased by 68.5%, or 9,634 screens. As has been the pattern for many years, this is the result of the rising number of indoor screens, which were up 745, or 3.4%, from 1989. The drive-in count continues to decline, falling 17% in 1990, −28.6% in 1989 and −38.4% in 1988.

*Combined Indoor and Drive-In Screens*

| Year | Total Screens | % Change Over Prior Year |
|---|---|---|
| 1990 | 23,689 | +2.4% |
| 1989 | 23,132 | −0.4% |
| 1988 | 23,234 | −1.4% |
| 1987 | 23,555 | +3.5% |
| 1986 | 22,765 | +7.7% |
| 1985 | 21,147 | +4.7% |
| 1984 | 20,200 | +7.0% |
| 1983 | 18,884 | +4.8% |
| 1982 | 18,020 | −0.1% |
| 1981 | 18,040 | +2.6% |
| 1980 | 17,590 | — |

**Employment**

During 1990 the American motion picture industry employed an average of 394,500 people, as reported by the U.S. Department of Commerce. This total is divided among those working in production/

services areas (154,800), distribution/videotape (131,800) and exhibition (107,900). Over the past several years total employment has not varied significantly, with decreases in movie theatre employment offset by increases in the production/services and distribution/videotape sectors.

**Export Markets**

The following charts illustrate the performance of American movies in the top-ten export markets over a five-year period from the journalism of A. D. Murphy (*Daily Variety,* June 10, 1991). Notice that figures cover theatrical *rentals* only (not box-office grosses), representing the theatrical distributors' share of box-office, and are converted from local currency into U.S. dollars. Each yearly listing includes a summary combining export market totals with U.S. theatrical rentals, for a total global rental figure. (All figures are rounded off to the nearest $100,000.)

*1990*

| Rank | Country | Theatrical Rentals |
|---|---|---|
| 1. | Japan | $236,700,000 |
| 2. | Germany (West only) | $175,200,000 |
| 3. | France | $164,200,000 |
| 4. | Canada | $148,300,000 |
| 5. | United Kingdom/Ireland | $144,400,000 |
| 6. | Italy | $117,000,000 |
| 7. | Spain | $110,400,000 |
| 8. | Australia | $70,400,000 |
| 9. | Brazil | $48,400,000 |
| 10. | Sweden | $39,800,000 |

| *1990 Summary* | Theatrical Rentals |
|---|---|
| Total Export Markets | $1,649,500,000 |
| U.S. Market Only | $1,829,000,000 |
| Total Global Rentals | $3,478,400,000 |

| *1989 Rank* | *Country* | THEATRICAL RENTALS |
|---|---|---|
| 1. | Japan | $201,600,000 |
| 2. | Canada | $152,500,000 |
| 3. | France | $127,600,000 |
| 4. | Germany (West only) | $117,500,000 |
| 5. | United Kingdom/Ireland | $115,300,000 |
| 6. | Spain | $94,500,000 |
| 7. | Italy | $84,700,000 |
| 8. | Australia | $73,900,000 |
| 9. | Sweden | $39,400,000 |
| 10. | Brazil | $34,000,000 |
| | *1989 Summary* | THEATRICAL RENTALS |
| | Total Export Markets | $1,346,900,000 |
| | U.S. Market Only | $1,780,100,000 |
| | Total Global Rentals | $3,126,900,000 |

| *1988 Rank* | *Country* | THEATRICAL RENTALS |
|---|---|---|
| 1. | Japan | $141,900,000 |
| 2. | Canada | $125,200,000 |
| 3. | Germany (West only) | $100,800,000 |
| 4. | France | $98,800,000 |
| 5. | United Kingdom/Ireland | $90,300,000 |
| 6. | Italy | $73,400,000 |
| 7. | Spain | $67,500,000 |
| 8. | Australia | $45,300,000 |
| 9. | Sweden | $27,400,000 |
| 10. | Brazil | $18,700,000 |
| | *1988 Summary* | THEATRICAL RENTALS |
| | Total Export Markets | $1,020,300,000 |
| | U.S. Market Only | $1,413,600,000 |
| | Total Global Rentals | $2,433,900,000 |

| *1987 Rank* | *Country* | THEATRICAL RENTALS |
|---|---|---|
| 1. | Japan | $138,100,000 |
| 2. | France | $101,000,000 |
| 3. | Germany (West only) | $98,000,000 |
| 4. | Canada | $96,700,000 |
| 5. | United Kingdom/Ireland | $73,300,000 |
| 6. | Italy | $70,400,000 |
| 7. | Spain | $49,400,000 |
| 8. | Australia | $31,900,000 |
| 9. | Brazil | $24,100,000 |
| 10. | Sweden | $24,000,000 |
| | *1987 Summary* | THEATRICAL RENTALS |
| | Total Export Markets | $935,100,000 |
| | U.S. Market Only | $1,244,500,000 |
| | Total Global Rentals | $2,179,600,000 |

| *1986 Rank* | *Country* | Theatrical Rentals |
|---|---|---|
| 1. | Japan | $102,600,000 |
| 2. | France | $98,500,000 |
| 3. | Canada | $86,800,000 |
| 4. | Germany (West only) | $64,700,000 |
| 5. | Italy | $64,600,000 |
| 6. | United Kingdom/Ireland | $49,100,000 |
| 7. | Spain | $48,200,000 |
| 8. | Australia | $27,400,000 |
| 9. | Brazil | $24,500,000 |
| 10. | Sweden | $17,100,000 |
| | *1986 Summary* | Theatrical Rentals |
| | Total Export Markets | $798,300,000 |
| | U.S. Market Only | $1,165,100,000 |
| | Total Global Rentals | $1,963,400,000 |

## U.S. VS. EXPORT MARKETS: 5 YEAR SUMMARY OF TOTAL THEATRICAL RENTALS

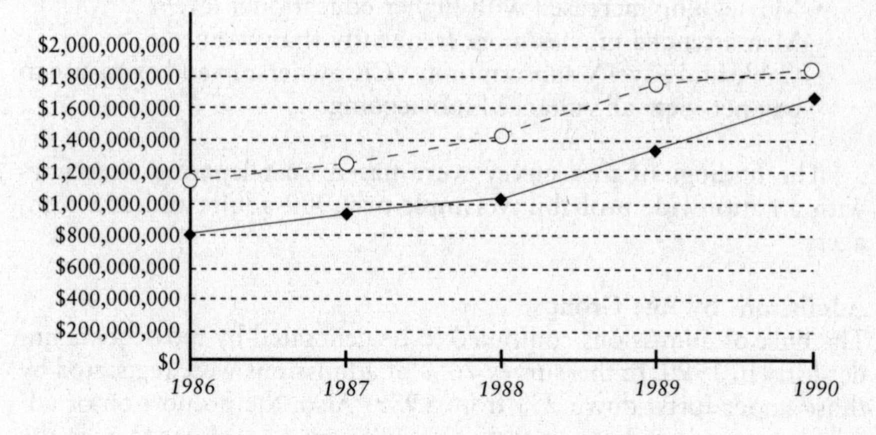


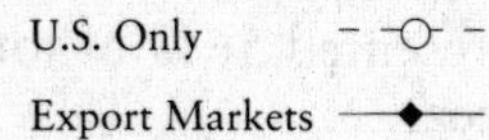

## AUDIENCE PROFILE

A survey of the American motion picture audience conducted during the summer of 1990 by the Opinion Research Corporation (ORC) of Princeton, New Jersey, for the Motion Picture Association of America, reported the following findings:

- Moviegoing is enjoyed by two-thirds of the population over the age of twelve, the second-highest level over the prior decade. (A moviegoer is defined in this study as someone who has seen at least one film at a theatre within the last twelve months.)
- The number of moviegoers has remained essentially stable, at 136,300,000, up 0.2% over the prior year.
- Admissions slowed by 5.9% to just under 1-billion.
- The slowing was caused by fewer average admissions per moviegoer.
- Those over forty continue to increase, accounting for over one-third of all moviegoers.
- Moviegoers under thirty show a decline in their share of admissions.
- As in the past, single persons were more frequent moviegoers than married persons.
- Moviegoing increased with higher educational levels.
- Men attended movies more frequently than women.
- Cable and pay-TV subscription, VCR ownership and exposure to pay-per-view all enhanced moviegoing.

The findings of this survey were based on telephone interviews with a nationwide probability sample of 2,009 adults and 252 teenagers.

### Admissions by Age Groups

The bulk of admissions continued to be generated by moviegoers under forty in 1990. In this survey 76% of admissions were registered by those under forty, down 2% from 1989. Also, the proportion of admissions among those over forty was 24%, up 1%. Moviegoers in the twelve-to-twenty-nine age group represented 56% of total yearly admissions, down 4% from the prior year.

The following table portrays the pattern of admissions defined by age groups over three years:

### *Admissions by Age Groups*

| Age | Percent of Total Yearly Admissions | | | Percent of Resident Civilian Population as of January 1990 |
|---|---|---|---|---|
| | *1990* | *1989* | *1988* | |
| 12–15 years | 11% | 11% | 12% | 6% |
| 16–20 years | 20% | 19% | 20% | 9% |
| 21–24 years | 11% | 14% | 12% | 7% |
| 25–29 years | 14% | 16% | 13% | 11% |
| 30–39 years | 20% | 18% | 20% | 20% |
| 40–49 years | 12% | 12% | 11% | 15% |
| 50–59 years | 5% | 4% | 5% | 11% |
| 60 years & over | 7% | 7% | 7% | 21% |
| 12–17 years | 18% | 19% | 19% | 10% |
| 18 years & over | 82% | 81% | 81% | 90% |

**Frequent Moviegoers**

The proportion of frequent movie attenders among both the total public and the adult public was essentially the same in 1990 as in 1989. Frequent attendance among teenagers decreased by 1% compared with the prior year, according to the ORC. Based on responses regarding yearly movie attendance, less than half (45%) of American teenagers attended at least once a month, a proportion over twice that for the general public aged eighteen and over.

The survey defines *frequent* moviegoers as those who attend at least once a month; *occasional* moviegoers as those who attend once in two to six months; and *infrequent* moviegoers as those who go less than once in six months.

The following table covers frequency of attendance according to age groups over three years:

### *Frequency of Attendance*

| | Total Public Aged 12 and Over | | | Adult Public Aged 18 and Over | | | Teenagers Aged 12–17 | | |
|---|---|---|---|---|---|---|---|---|---|
| | *1990* | *1989* | *1988* | *1990* | *1989* | *1988* | *1990* | *1989* | *1988* |
| Frequent | 23% | 24% | 20% | 22% | 22% | 18% | 45% | 46% | 38% |
| Occasional | 31% | 32% | 30% | 30% | 30% | 28% | 38% | 42% | 43% |
| Infrequent | 12% | 10% | 11% | 12% | 10% | 11% | 8% | 8% | 12% |
| Never | 33% | 32% | 38% | 36% | 35% | 42% | 5% | 2% | 3% |
| Not Reported | 1% | 2% | 1% | — | 2% | 1% | 3% | 2% | 4% |

In terms of marital status, single persons (including widowed and divorced) continued to be more frequent moviegoers than married persons, as in previous years. But the proportion of nonattendees among married persons (36%) is only slightly greater than among single persons (35%).

**_Marital Status—Adults_**

| | Married 1990 | Married 1989 | Married 1988 | Single 1990 | Single 1989 | Single 1988 |
|---|---|---|---|---|---|---|
| Frequent | 16% | 16% | 14% | 26% | 29% | 23% |
| Occasional | 32% | 32% | 29% | 29% | 30% | 28% |
| Infrequent | 15% | 12% | 13% | 9% | 9% | 8% |
| Never | 36% | 38% | 43% | 35% | 30% | 40% |

Families with children (ages eighteen and under) attended more than families without children, a pattern unchanged in recent years.

**_Adults With Children_**

| | 1990 | 1989 | 1988 |
|---|---|---|---|
| Frequent | 22% | 25% | 18% |
| Occasional | 39% | 39% | 35% |
| Infrequent | 15% | 10% | 15% |
| Never | 23% | 25% | 31% |

**_Adults Without Children_**

| | 1990 | 1989 | 1988 |
|---|---|---|---|
| Frequent | 20% | 21% | 18% |
| Occasional | 26% | 26% | 24% |
| Infrequent | 11% | 10% | 9% |
| Never | 43% | 40% | 48% |

Moviegoing frequency continued to increase with higher educational levels among adults.

**_Education—Adults (over 18)_**

| | Higher Education (At Least Some College) | | | High School Completed | | | Less Than High School Completed | | |
|---|---|---|---|---|---|---|---|---|---|
| | 1990 | 1989 | 1988 | 1990 | 1989 | 1988 | 1990 | 1989 | 1988 |
| Frequent | 27% | 29% | 24% | 17% | 18% | 14% | 11% | 9% | 7% |
| Occasional | 34% | 35% | 33% | 30% | 31% | 28% | 19% | 18% | 15% |
| Infrequent | 13% | 9% | 13% | 13% | 11% | 11% | 9% | 11% | 7% |
| Never | 26% | 26% | 29% | 40% | 37% | 46% | 60% | 56% | 70% |

Higher-income families (with annual incomes of $25,000 or more) in the United States attended movies more frequently than did others.

As in previous years, males more than females tended to be frequent moviegoers.

**Sex**

| | Ages 12 and Over | | | | | |
|---|---|---|---|---|---|---|
| | *Male* | | | *Female* | | |
| | *1990* | *1989* | *1988* | *1990* | *1989* | *1988* |
| Frequent | 25% | 28% | 23% | 21% | 20% | 19% |
| Occasional | 32% | 33% | 31% | 31% | 30% | 27% |
| Infrequent | 12% | 9% | 11% | 12% | 11% | 11% |
| Never | 30% | 28% | 34% | 35% | 36% | 42% |
| | *Ages 18 and Over* | | | | | |
| | *Male* | | | *Female* | | |
| | 1990 | 1989 | 1988 | 1990 | 1989 | 1988 |
| Frequent | 23% | 26% | 20% | 17% | 18% | 16% |
| Occasional | 32% | 32% | 30% | 30% | 30% | 26% |
| Infrequent | 12% | 9% | 11% | 13% | 12% | 11% |
| Never | 33% | 30% | 38% | 39% | 39% | 46% |

On some of the above tables, those persons not reporting their movie attendance have been omitted, which accounts for columnar tallies of under 100%.

**VCR Ownership and Cable (Basic, Pay, Addressable) Subscription**

Over half of the households in the United States with TVs subscribe to a cable-television service. Penetration is greater in households with an annual income of $25,000 or more than in those below this income level, and more likely to be found in households with children than without. The level of videocassette-recorder ownership continues to climb, exceeding 70% of TV households.

The following four charts demonstrate the leveling-off of the number of VCR households, mirroring the maturation of VCR penetration in the United States, followed by the extent of attendant sales of prerecorded and blank videocassettes. The numbers trace the virtual birth of VCR popularity in 1980 to its position in over 70% of all TV homes.

### 1. VCR Households

| Year | VCR Households | % Change Over Prior Year |
|---|---|---|
| 1990 | 65,356,200 | +5.0% |
| 1989 | 62,259,600 | +10.8% |
| 1988 | 56,200,000 | +22.7% |
| 1987 | 45,800,000 | +40.9% |
| 1986 | 32,500,000 | +38.3% |
| 1985 | 23,500,000 | +56.7% |
| 1984 | 15,000,000 | +80.0% |
| 1983 | 8,300,000 | +72.9% |
| 1982 | 4,800,000 | +92.0% |
| 1981 | 2,500,000 | +35.1% |
| 1980 | 1,850,000 | — |

### 2. VCR Penetration Rates in U.S. TV Households

| Year | VCR Households | TV Households | Percent Penetration |
|---|---|---|---|
| 1990 | 65,356,200 | 93,100,000 | 70.2% |
| 1989 | 62,259,600 | 92,100,000 | 67.6% |
| 1988 | 56,200,000 | 90,400,000 | 62.2% |
| 1987 | 45,800,000 | 88,600,000 | 51.7% |
| 1986 | 32,500,000 | 87,400,000 | 37.2% |
| 1985 | 23,500,000 | 86,100,000 | 27.3% |
| 1984 | 15,000,000 | 85,300,000 | 17.6% |
| 1983 | 8,300,000 | 84,200,000 | 9.9% |
| 1982 | 4,800,000 | 83,700,000 | 5.7% |
| 1981 | 2,500,000 | 81,900,000 | 3.1% |
| 1980 | 1,850,000 | 78,000,000 | 2.4% |

### 3. Sales of Prerecorded Videocassettes to U.S. Dealers

| Year | Prerecorded Cassettes | % Change Over Prior Year |
|---|---|---|
| 1990 | 220,000,000 | +10.0% |
| 1989 | 200,000,000 | +48.1% |
| 1988 | 135,000,000 | +22.7% |
| 1987 | 110,000,000 | +31.0% |
| 1986 | 84,000,000 | +61.5% |
| 1985 | 52,000,000 | +136.4% |
| 1984 | 22,000,000 | +131.6% |
| 1983 | 9,500,000 | +58.3% |
| 1982 | 6,000,000 | +9.1% |
| 1981 | 5,500,000 | +83.3% |
| 1980 | 3,000,000 | — |

### 4. Sales of Blank Videocassettes to U.S. Consumer Market

| Year | Blank Cassettes | % Change Over Prior Year |
|---|---|---|
| 1990 | 325,000,000 | +16.1% |
| 1989 | 280,000,000 | −6.7% |
| 1988 | 300,000,000 | −11.8% |
| 1987 | 340,000,000 | +21.4% |
| 1986 | 280,000,000 | +53.8% |
| 1985 | 182,000,000 | +67.0% |
| 1984 | 109,000,000 | +91.2% |
| 1983 | 57,000,000 | +130.8% |
| 1982 | 24,700,000 | +9.8% |
| 1981 | 22,500,000 | +50.0% |
| 1980 | 15,000,000 | — |

After the boom years of the 1980s the number of basic-cable households, pay-cable subscriptions, individual pay-cable subscribers and addressable-cable households in the U.S. appears to have reached saturation, as indicated by the next four charts.

### 5. Basic Cable Households

| Year | Basic Cable Households | % Change Over Prior Year |
|---|---|---|
| 1990 | 54,929,000 | +4.5% |
| 1989 | 52,564,470 | +8.1% |
| 1988 | 48,636,520 | +4.3% |
| 1987 | 46,627,100 | +5.9% |
| 1986 | 44,030,000 | +10.4% |
| 1985 | 39,900,000 | +8.1% |
| 1984 | 36,900,000 | +14.9% |
| 1983 | 32,100,000 | +18.0% |
| 1982 | 27,200,000 | +16.7% |
| 1981 | 23,300,000 | +18.8% |
| 1980 | 19,600,000 | — |

### 6. Pay-Cable Subscriptions

| Year | Pay-Cable Subscriptions | % Change Over Prior Year |
|---|---|---|
| 1990 | 41,900,000 | +1.2% |
| 1989 | 41,400,000 | +11.3% |
| 1988 | 37,200,000 | +10.1% |
| 1987 | 33,800,000 | +4.0% |
| 1986 | 32,500,000 | +2.5% |
| 1985 | 31,700,000 | +5.3% |
| 1984 | 30,100,000 | +7.5% |
| 1983 | 28,000,000 | +35.9% |
| 1982 | 20,600,000 | +36.4% |
| 1981 | 15,100,000 | +69.6% |
| 1980 | 8,900,000 | — |

### 7. Pay-Cable Subscribers

| Year | Pay-Cable Subscribers | % Change Over Prior Year |
|---|---|---|
| 1990 | 26,626,600 | −1.7% |
| 1989 | 27,077,400 | +3.0% |
| 1988 | 26,300,000 | +6.9% |
| 1987 | 24,600,000 | +3.8% |
| 1986 | 23,700,000 | −2.1% |
| 1985 | 24,200,000 | +19.8% |
| 1984 | 20,200,000 | +12.2% |
| 1983 | 18,000,000 | +34.3% |
| 1982 | 13,400,000 | — |

### 8. Addressable Cable Households

| Year | Addressable Cable Households | % Change Over Prior Year |
|---|---|---|
| 1990 | 22,000,000 | +4.8% |
| 1989 | 21,000,000 | +9.9% |
| 1988 | 19,100,000 | +25.8% |
| 1987 | 15,100,000 | +29.7% |
| 1986 | 11,640,000 | +29.0% |
| 1985 | 9,020,000 | +40.9% |
| 1984 | 6,400,000 | +77.8% |
| 1983 | 3,600,000 | +140.0% |
| 1982 | 1,500,000 | — |

## 10 YEAR GROWTH OF TV HOUSEHOLDS VS. VCR, BASIC AND PAY CABLE

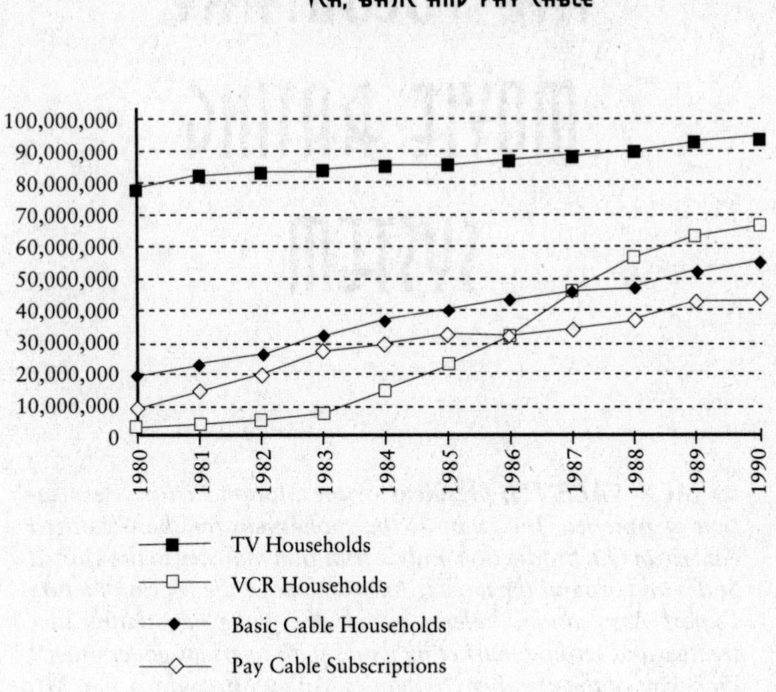

# THE VOLUNTARY MOVIE RATING SYSTEM

*by **JACK VALENTI**, president of the Motion Picture Association of America, Inc., who is the spokesman for the organized American film production and distribution industry in the United States and around the world. As president of the Motion Picture Export Association, Valenti travels the globe negotiating film treaties and settling marketing issues with overseas governments on behalf of the member companies. An author and teacher, Mr. Valenti was named president of the MPAA in 1966, after serving as special assistant to President Lyndon B. Johnson. MPAA member companies are Buena Vista Distribution Inc. (releasing Disney product), Metro-Goldwyn-Mayer, Inc., Orion Pictures Corp., Paramount Pictures Corp., Sony Pictures Entertainment Inc. (Columbia and TriStar), Twentieth Century Fox Film Corp., Universal City Studios Inc. and Warner Bros. Inc.*

***The voluntary rating system is not a surrogate parent, nor should it be. . . . Its purpose is to give prescreening advance informational warnings, so that parents can form their own judgments. . . .***

## HOW IT ALL BEGAN

When I became president of the Motion Picture Association of America (MPAA) in May 1966, the slippage of Hollywood studio authority over the content of films collided with an avalanching revision of American mores and customs.

By summer of 1966 the national scene was marked by insurrection on the campus, riots in the streets, rise in women's liberation, protest by the young, doubts about the institution of marriage, abandonment of old guiding slogans and the crumbling of social traditions. It would have been foolish to believe that movies, that most creative of art forms, could have remained unaffected by the change and torment in our society.

**A New Kind of American Movie**

The result of all this was the emergence of a "new kind" of American movie—frank and open, and made by filmmakers subject to very few self-imposed restraints.

Almost within weeks in my new duties I was confronted with controversy, neither amiable nor fixable. The first issue was the film *Who's Afraid of Virginia Woolf?* in which, for the first time on the screen, the world *screw* and the phrase *hump the hostess* were heard. In company with the MPAA's general counsel, Louis Nizer, I met with Jack Warner, the legendary chieftain of Warner Bros. and his top aide, Ben Kalmenson. We talked for three hours, and the result was deletion of *screw* and retention of *hump the hostess,* but I was uneasy over the meeting.

It seemed wrong that grown men should be sitting around discussing such matters. Moreover, I was uncomfortable with the thought that this was just the beginning of an unsettling new era in film, in which we would lurch from crisis to crisis, without any suitable solution in sight.

The second issue surfaced only a few months later. This time it was Metro-Goldwyn-Mayer, and the Michelangelo Antonioni film *Blow Up*. I met with MGM's chief executive officer because this movie also represented a first: the first time a major distributor was marketing a film with nudity in it. The Production Code Administration in California had denied the seal. I backed the decision, whereupon MGM distributed the film through a subsidiary company, thereby flouting the voluntary agreement of MPAA member companies that none would distribute a film without a Code seal.

Finally, in April 1968, the U.S. Supreme Court upheld the constitutional power of states and cities to prevent the exposure of children to books and films that could not be denied to adults.

It was plain that the old system of self-regulation, begun with the formation of the MPAA in 1922, had broken down. What few threads there were holding together the structure created by Will Hays, one of my two predecessors, had now snapped. From the very first day of my own succession to the MPAA president's office, I had sniffed the Production Code constructed by the Hays Office. There was about this stern, forbidding catalog of dos and don'ts the odious smell of censorship. I determined to junk it at the first opportunity.

I knew that the mix of new social currents, the irresistible force of creators determined to make "their" films (full of wild candor, groused some social critics) and the possible intrusion of government into the movie arena demanded my immediate action.

Within weeks discussions of my plan for a movie rating system began with the president of the National Association of Theatre Owners (NATO), and with the governing committee of the International Film Importers & Distributors of America (IFIDA), an assembly of independent producers and distributors.

Over the next five months I held more than one hundred hours of meetings with these two organizations, as well as with guilds of actors, writers, directors and producers, with craft unions, with critics, with religious organizations and with the heads of the MPAA member companies.

## THE BIRTH OF THE RATINGS

By early fall I was ready. My colleagues in NATO joined with me in affirming our objective of creating a new and, at the time, revolutionary approach to how we would fulfill our obligation to the parents of America.

My first move was to abolish the old and decaying Hays Production Code. Then, on November 1, 1968, we announced the birth of the new voluntary film rating system of the motion picture industry, with three organizations, NATO, MPAA and IFIDA, as its monitoring and guiding groups.

The initial design called for four rating categories: G for general audiences, all ages admitted; M for mature audiences—parental guidance suggested, but all ages admitted; R for restricted, children under sixteen (later raised to under seventeen years of age, and varying in some jurisdictions) would not be admitted without an accompanying parent or adult guardian; X for no one under seventeen admitted.

The rating system trademarked all the category symbols, except the X. Under the plan, anyone not submitting his or her film for rating could self-apply the X or any other symbol or description, except those trademarked by the rating program. (The X rating's name has been changed to NC-17, a trademarked symbol: No Children Under 17 Admitted. This is explained more fully later.)

Our original plan had been to use only three rating categories, ending with R. It was my view that parents ought to have the right to accompany their children to any movie the parents chose, without the movie industry or the government or self-appointed groups interfering with their rights. But the National Association of Theatre Owners urged the creation of an adults-only category, fearful of possible legal redress under state or local law. I acquiesced in NATO's reasoning, and the four category system, including the X rating, was installed.

The emergence of the voluntary rating system filled the vacuum left by my dismantling of the Hays Production Code. The movie industry would no longer "approve or disapprove" the content of a film, but we would now see our primary task as giving advance cautionary warnings to parents so that they could make the decision about the moviegoing of their young children. That decision is solely the responsibility of parents.

## CHANGES IN THE RATING SYSTEM

We found early on that the M category (M for "Mature") was regarded by most parents as a sterner rating than the R category. To remedy this misconception, we changed the name from M to GP (meaning General audiences, Parental guidance suggested). A year later we revised the name to its current label, "PG: Parental Guidance Suggested."

On July 1, 1984, we made another adjustment. We split the PG category into two groupings, PG and PG-13. PG-13 meant a higher level of intensity than was to be found in a film rated PG. Over the years parents have approved of this amplifying revision in the system.

On September 27, 1990, we announced two more revisions.

First, we introduced five-to-eight word explanations of why a particular film received its R rating. Since, in the opinion of the Ratings Board, R-rated films contain adult material, we believed it would be useful for parents to know a little more about that film's content before they allowed their children to accompany them. These explanations would be available to parents by calling the local theatre playing the picture or by inquiring at the box office.

Second, we changed the name of the X category to NC-17: No Children Under 17 Admitted. The X rating over the years appeared to have taken on a surly meaning in the minds of many people, a meaning that was never intended when we created the system. Therefore we chose to go back to the original intent of the design we installed in 1968, in which this "adults only" category explicitly describes a movie that most parents would want to have barred to viewing by their children. That was and is our goal; nothing more, nothing less.

We have now trademarked "NC-17: No Children Under 17 Admitted" so that this rating symbol and the legend can be used only by those who submit their films for rating. Other rating symbols are already federally registered trademarks. Those who do not choose to participate in the rating system can take their film to market using any letters or descriptions they desire, except the trademarked symbols and legends of the rating system.

## THE PURPOSE OF THE RATING SYSTEM

The basic mission of the rating system is simple: to offer to parents some advance information about movies so that parents can decide what movies they want their children to see or not to see. The entire rostrum of the rating program rests on the assumption of responsibility by parents. If parents don't care, or if they are languid in guiding their children's moviegoing, the rating system becomes useless. Indeed, if you are seventeen or over, or if you have no children, the rating system has no meaning for you. Ratings are meant for parents, no one else.

The Ratings Board does not rate movies on their quality or lack of quality. That is a role left to film critics and audiences. Had we attempted to insert ourselves into judging whether a film is "good, bad or indifferent," we would have collapsed the system before it began.

The criteria that go into the mix that becomes a Ratings Board judgment are theme, violence, language, nudity, sensuality, drug abuse and other elements. Part of the rating flows from how each of these elements is treated on screen by the filmmaker. In making their evaluation, the Ratings Board does not look at snippets of film in isolation, but considers the film in its entirety. The Ratings Board can make its decisions only by what they see on the screen, not what is imagined or thought.

There is no special emphasis on any one of these elements. All are considered. All are examined before a rating is applied. Contrary to popular notion, violence is not treated more leniently than any of the other material. Indeed, many films rated X in the past, and NC-17 now, have at least tentatively been given the "adults only" rating because of depictions of violence. However, most of the directors/producers/distributors involved have chosen, by their decision, to edit heavier violent scenes in order to receive an R rating.

## HOW THE RATINGS ARE DECIDED

The ratings are decided by a Ratings Board located in Los Angeles. They work for the Classification and Rating Administration, whose funding comes from fees charged producers/distributors for the rating of their films. The MPAA president chooses the chairman of the Ratings Board, thereby insulating the board from industry or other group pressure. No one in the movie industry has the authority or the power to push the board in any direction. As the MPAA president, I take no part in rating decisions and do not overrule or dissuade the board from any decisions it makes.

One of the highest accolades to be conferred on the rating system is that, from its birth in 1968 to this hour, there has never been even the slightest jot of evidence that the rating system has ever deliberately fudged a decision or bowed to pressure. The Ratings Board has always conducted itself at the highest level of integrity. That is a large, honorable and valuable asset.

There are no special qualifications for board membership, except that members must have a shared parenthood experience, must be possessed of an intelligent maturity and, most of all, must have the capacity to put themselves in the role of the average American parent so that they can view a film and apply a rating that most parents would find suitable and helpful in aiding their decisions about their children's moviegoing.

No one is forced to submit a film to the board for rating, but the vast majority of producers/distributors do in fact submit their films for ratings. Any producer/distributor who wants no part of any rating system is free to go to the market without any rating at all or with any description or symbol they choose, as long as it is not confusingly similar to the G, PG, PG-13, R and NC-17 rating symbols, which may not be self-applied.

### The Board Votes on Ratings

The board views each film. Each member present estimates what most parents would consider to be that film's appropriate rating. After group discussion the board votes on the rating. Each member completes a rating form spelling out his or her reason for the rating. Each rating is decided by majority vote.

The producer/distributor of a film has the right under the rules to inquire as to the reason that particular rating was applied. The producer/distributor also has the right, based on the reasons for the rating, to edit the film and come back to the board to try for a less severe rating. The reedited film is brought back to the board and the process goes forward again.

**The Rating Appeals Board**

A producer who for any reason is displeased with a rating can appeal the decision to the Rating Appeals Board, which sits as the final arbiter of ratings.

The Appeals Board comprises twenty-one members, men and women from the industry organizations that govern the rating system. They gather to view the film and hear the appeal. After the screening the producer whose film is being appealed explains why he or she believes the rating was wrongly decided. The chairman of the Ratings Board states the reasons for the film's rating. Both the producer and the Ratings Board representative have an opportunity for rebuttal. In addition the producer may submit a written presentation to the board prior to the oral hearing. After Appeals Board members question the two opposing representatives, they are excused from the room. The board discusses the appeal and then takes a secret ballot. It requires a two-thirds vote of those present to overturn a Ratings Board decision. By this method of appeal controversial decisions of the Ratings Board can be examined and any rating deemed a mistake set right.

The decision of the Appeals Board is final and cannot be appealed, although the Appeals Board has the authority to grant a rehearing on the request of the producer.

## WHAT THE RATINGS MEAN

**G: "General Audiences—All ages admitted."**

This is a film that contains nothing in terms of theme, language, nudity and sex, violence, and so forth that would, in the view of the Ratings Board, be offensive to parents whose younger children watch the film. The G rating is *not* a "certificate of approval," nor does it signify a children's film.

Some snippets may go beyond the language of polite conversation, but they are common, everyday expressions. No stronger words are present in G-rated films. The violence is at a minimum. Nudity and sex scenes are not present; nor is there any drug-use content.

**PG: "Parental Guidance Suggested; some material may not be suitable for children."**

This is a film that clearly needs to be examined or inquired about by parents before they let their children attend. The label PG plainly states that parents may consider some material unsuitable for their children, but that the parent must make the decision. Parents are warned against sending their children, unseen without inquiry, to PG-rated movies.

There may be some profanity in these films. There may be violence, but it is not deemed so strong that everyone under seventeen need be restricted unless accompanied by a parent. Nor is there cumulative horror or violence that may take a film all the way into the R category. There is no drug-use content. There is no explicit sex in a PG-rated film, although there may be some indication of sensuality. Brief nudity may appear in an unrestricted film, but anything beyond that puts the film into R.

The PG rating, suggesting parental guidance, is thus an alert for special examination of a film by parents before deciding on its viewing by their children.

Obviously the line is difficult to draw. In our pluralistic society it is not easy to make subjective judgments without incurring some disagreement. So long as parents know they must exercise parental responsibility, the rating serves as a meaningful guide and as a warning.

**PG-13: "Parents strongly cautioned. Some material may be inappropriate for children under 13."**

PG-13 is thus a sterner warning to parents to determine for themselves the attendance in particular of their younger children, since they might consider some material not suited for them. Parents, by the rating, are alerted to be very careful about the attendance of their under-teenage children.

A PG-13 film is one that, in the view of the Ratings Board, leaps beyond the boundaries of the PG rating, but does not quite fit within the restricted R category. Any drug-use content will initially require at least a PG-13 rating. In effect the PG-13 cautions parents with more stringency than usual to give special attention to this film before they allow their twelve-year-olds and younger to attend.

If nudity is sexually oriented, the film will generally not be found in the PG-13 category. If violence is rough or persistent, the film goes into the R (restricted) rating. A film's single use of one of the harsher sexually derived words, though only as an expletive, shall require the Ratings Board to issue that film at least a PG-13 rating. More than one such expletive must lead the Ratings Board to issue a film an R rating, as must even one of these words used in a sexual context. These films can be rated less severely, however, if by a special vote the Ratings Board feels that a lesser rating would more responsibly reflect the opinion of American parents.

PG-13 places greater responsibilities on parents for their children's moviegoing. The voluntary rating system is not a surrogate parent, nor should it be. It cannot, and should not, insert itself into family decisions that only parents can, and should, make. Its purpose is to give prescreening advance informational warnings, so that parents can form their own judgments. PG-13 is designed to make such parental decisions easier for those films between PG and R.

**R: "Restricted, under 17 requires accompanying parent or adult guardian."**
In the opinion of the Ratings Board this film definitely contains adult material. Parents are strongly urged to find out more about this film before they allow their children to accompany them.

An R-rated film (the "under 17" age varies in some jurisdictions) has adult content that may include hard language, or tough violence, or nudity within sensual scenes, or drug abuse or other elements, or a combination of some of the above. That is why parents are counseled, in advance, to take this advisory very seriously.

**NC-17: "No children under 17 admitted."**
This rating declares that the Ratings Board believes that this is patently an adult film. No children will be admitted. NC-17 (the age varies in some jurisdictions) does not necessarily mean "obscene or pornographic" in the oft-accepted meaning of those words. The board does not and cannot mark films with those words. These are legal terms, for the courts to decide. The reasons for the application of an NC-17 rating can be strong violence or sex or aberrational behavior or drug abuse or any other element that most parents would want to be off-limits for viewing by their children.

**Appraisal**
In any appraisal what is "too much" becomes very controversial. How much is "too much" violence? Are classic war films too violent with scenes of marines storming a beach and slaying hundreds, wounding thousands? Is it the graphic cop killing, the gangster shoot-out, or the slap across the face of a woman that determines "too much"? How much emphasis is spilled blood to be given? Where is the line to be drawn between "this is all right" and "this is not all right"?

The same vexing doubts occur in sex scenes or those where language rises on the Richter scale, or where behavior not considered "normal" is revealed on the screen. What follows is disagreement, inevitable, inexorable and often strident. That is what the rating system has to endure and confront. We understand that. We try to do our level best so that most parents would find our ratings mostly accurate and mostly useful.

But, importantly, we urge and implore parents to care about what their children see and watch, to focus their attention on movies so that they can know more about a film before they consent to their children watching it.

To oversee the Ratings Board, the film industry has set up a Policy

Review Committee consisting of officials of MPAA and NATO. These men and women set guidelines for the Ratings Board to follow and make certain that the board carries them out reasonably and appropriately.

Because the rating program is a self-regulatory apparatus of the film industry, it is important that no single element of the industry take on the authority of a "czar" beyond any discipline or self-restraint.

**Advertising and Trailer Policy**

Film advertising is part of the film industry's self-regulatory mechanism. All advertising for rated motion pictures must be submitted to the Advertising Administration for approval prior to its release to the public. This includes, but is not limited to, print ads, radio and TV spots, pressbooks, videocassette packaging and theatrical and home-video trailers.

Trailers are an important aspect of the program. They are approved either for "all audiences"—which means they may be shown with all feature films—or for "restricted audiences"—which limits their use to feature films rated R or NC-17. In "all audiences" trailers there will be no scenes that caused the feature to be rated PG, PG-13, R or NC-17.

Each trailer carries at the front a tag that tells two things: (1) the audience for which the trailer has been approved, and (2) the rating of the picture being advertised. The tag for "all audience" trailers will have a green background; the tag for "restricted" trailers will have a red background. The color is to alert the projectionist against mismatching trailers with the film being shown on the theatre screen.

## HOW THE RATING SYSTEM IS USED BY THEATRE OWNERS AND VIDEO RETAILERS

Motion picture theatre owners, who cofounded the rating system in 1968, were the first group in the entertainment industry to voluntarily enforce its guidelines. NATO estimates that about 85% of the theatre owners in the nation subscribe to the rating system and will not admit a child under seventeen to an R-rated motion picture without an accompanying parent or adult guardian, or admit a child under seventeen to an NC-17-rated motion picture under any circumstances.

In the mid-1980s, as watching movies on videocassettes at home soared in popularity, video retailers joined theatre owners in embracing the voluntary guidelines of the rating system. Parents who relied on the rating system to determine which films their children viewed in theatres found the information provided by the rating classifications equally helpful in home video. To facilitate its use, ratings are displayed on both the videocassette package and the cassette itself. The Video Software Dealers Association (VSDA), which is the major trade association for video retailers in the United States, has strongly endorsed the observance of the voluntary movie rating system by video retailers.

## THE PUBLIC REACTION

We count it crucial to make regular soundings to find out how the public perceives the rating program and to measure the approval and disapproval of what we are doing.

Nationwide scientific polls, conducted each year by the Opinion Research Corporation of Princeton, New Jersey, have consistently given the rating program high marks by parents throughout the land. The latest poll results show that 75% of parents with children under eighteen found the ratings to be "very useful" to "fairly useful" in helping them make decisions for the moviegoing of their children. This is an all-time high for the rating system. On the evidence of the polls, the rating system would not have survived if it were not providing a useful service to parents.

The rating system isn't perfect, but in an imperfect world it seems each year to match the expectations of those whom it is designed to serve—parents of America.

# MERCHANDISING

*By **STANFORD BLUM,** who is president of the Imagination Factory, a licensing and merchandising company based in Los Angeles. A graduate of the University of Baltimore Law School, Mr. Blum drew attention to the revenue potential in the infant industry of exploiting merchandising rights of motion pictures and sports events through his company the Image Factory, which licensed properties such as* Rocky, Star Wars *and the 1980 and 1984 Olympics. He served as executive vice president of features, television and marketing for Cannon Films from 1986 to 1988 before being named executive vice president of television, marketing and licensing for Nelvana Entertainment, where he produced such network animated series as "Little Rosey" and "Beetlejuice," and licensed properties including Babar the Elephant. Mr. Blum is a consultant to Marvel Comics and New World Television and, at the Imagination Factory, is developing projects including* Fungle the Gnole, Paparazzi Samurai *and* Cyberpunk.

*The* Teenage Mutant Ninja Turtles *toys, which predated the movies, grossed over $1-billion for all licensing in 1990. . . .*

Merchandising as an industry has generated $87-billion in sales since 1977. But the merchandising of a motion picture, one of the most valuable of ancillary rights, was largely ignored (except for Disney) until the huge success of *Star Wars* in 1977, licensed by Twentieth Century Fox, which awakened the movie business to the great wealth to be derived from merchandising the picture logo or characters in toys, posters, T-shirts, clothing, jewelry, watches and other products. As a result of *Star Wars* and *Rocky,* studios set up ancillary divisions to properly exploit this newly lucrative area. To date, *Star Wars* products have grossed at the retail level over $2-billion from merchandising rights worldwide.

On an average picture, posters and T-shirts are the most popular items. But if a film has characters that can be made into toys, merchandising can be a bonanza, as in *Teenage Mutant Ninja Turtles.* This was an American underground comic book that was made into a syndicated television series and then a live-action feature. The *Teenage Mutant Ninja Turtles* toys, which predated the movies, grossed over $1-billion for all licensing in 1990. In another permutation, the Care Bears began as characters in a line of American Greeting cards, which then led to movies, at which point licensing took off and led to the television series. And while *The Little Mermaid* was a successful Disney movie, it was only after the home video was released that merchandising revenue exploded and became a phenomenon. In an example of how high toy deals can go, Mattel advanced $4-million in 1991 for the rights to make toys of the characters from *Hook*, directed by Steven Spielberg.

The essential elements of merchandising a picture are the rights to the title and logo (or artwork) of the picture. These rights must be licensed from the owner (usually the financier-distributor) in order to exploit them on a toy, poster, T-shirt, video game or in promotional tie-ins with a major fast-food restaurant or a major on-package promotion. Certainly copyright and trademark protection must be obtained by the owner not only to protect the exhibition of the movie itself but also to cover the use of the artwork title, created for advertising, in its adaptation to various items of merchandise. Further, if a performer's likeness as a character in the film is to appear on such merchandise, as in *Dick Tracy,* the rights to such likeness must be obtained from the performer, usually in the form of a deal that includes a percentage share to that performer of everything that is licensed or merchandised bearing his likeness.

If a picture is developed and fully financed by a financier-distributor (studio), that company will retain all merchandising rights and the rights to all characters and actors' likenesses as those characters. The studio will have its own merchandising licensing division make the appropriate deals. If the producer realizes there is merchandising potential in his movie, he should try to negotiate for a percentage (usually 25%–50%) of that revenue as part of his deal with the studio. However, the studio will probably cross-collateralize expenses against what the merchandising revenue would be.

The first step in making a license work for a picture that is geared for kids is to arrange for a deal with a major toy manufacturer. This would ideally be in the form of a master licensing agreement covering a line of toys designed to be released prior to the picture's opening. Subsequent licensing would cover T-shirts, bags, back-to-school items, lunchboxes, jewelry, clothing, linens and towels.

A typical master licensing agreement would find a major toy company such as Hasbro, Kenner, Mattel or Playmates putting up between $250,000 and $1-million for the rights to make a line of toys from a given movie. A typical toy line includes PVC figurines, action figures, windups and play sets. The deal is structured with a guarantee and a royalty of 5%–6%. In the case of a studio, that 5%–6% royalty would flow down the revenue stream, and the producer's share may be between 25% and 50% of that. Timing is important, as is promotion. Obviously studio promotion is a huge machine that can support retailers while marketing the picture, as in Warners' attention to *Batman,* with toys from Kenner and many other licenses.

For a sense of the timing involved, a deal with a toy company made in June would allow for enough time for tooling and manufacturing the prototypes (including the changes made on both sides to get it right) to show mock-ups to the major retailers in September, then introduce the finished products at the major trade show, Toy Fair in New York, in February of the following year, when orders are made for a line that would hit the stores by May for a summer picture, or a September TV series. (When Kenner Toys was licensed by Twentieth Century Fox for *Star Wars,* neither party knew it would be the landmark picture for merchandising. Because of the lead time needed at the time in creating toys, Kenner didn't have the merchandise ready for stores by the Christmas after the picture's summer release, 1977. Rather, they sold a nonrefundable certificate for customers to return the following April to pick up the merchandise, and the response was enormous.)

A subsequent licensing agreement may cover T-shirts, which also includes a line of sweats. A sample deal for such a transaction would find a four-color silkscreen T-shirt manufacturer (licensee) putting up a $100–$150,000 guarantee, payable $25–$50,000 up front, with a royalty of 8% of the wholesale price, which would accumulate to meet the $100–$150,000 guarantee. On a typical T-shirt transaction the manufacturer would sell it to the retailer for $5. The store then marks it up and sells it for perhaps $20. The 8% royalty in this case would return 40¢ to the licensor/rights holder per transaction. This formula applies to all types of items. If a cup bearing your logo sells for $5 and the manufacturer/wholesaler's price was $2, your royalty of 8% of wholesale would be 16¢ per cup.

There are two tiers in selling to stores, mass merchandising and upscale department stores. The country's main mass-merchandising stories are J. C. Penney, K Mart, Wal-Mart and Target. Their strength is in volume, and they are serviced by certain licensees. J. C. Penney's runs 2,500 stores nationwide; K Mart has around 2,400 stores, Wal-Mart has 1,600 and Target has 350. One chain's opening order could total 100,000 pieces. Using our T-shirt example, that returns $40,000 to the rights holder. Other licensees cover upscale stores such as Bloomingdale's, Neiman-Marcus, Saks Fifth Avenue and F.A.O. Schwarz.

The arithmetic for a poster is similar, involving one transaction with a printer. Assume that the Imagination Factory is the entity that owns the poster rights to a motion picture. We arrange for the printing and then sell the posters, at 45¢ each, to distributors, who mark them up and sell them to department stores, which charge the customer $2.50 to $5.00. Of that 45¢ per poster, about 15¢–20¢ covers printing costs. The rights holders would receive 10% of our price to the distributors, or 4.5¢, and the balance represents our overhead and profit.

A typical deal with a licensee-manufacturer requires three key elements: front money, a royalty and a guarantee. Front money is good-faith money that can range anywhere from $2,500 to $500,000, depending on the deal. It is really an advance against the royalty. The royalty covers everything sold at the wholesale price and is based on 10% of net sales. *Net sales* refers to gross sales less quantity discounts and returns. If the company's also going to agree to a guarantee beyond the front money, they will sell aggressively because a guarantee is nonrefundable. For example, a deal with a clothing manufacturing company to make a line of clothes may call for $100,000 up front, an 8%–10% royalty, and a $250,000 guarantee. The guarantee is due one year after the contract is signed. However, there's no standard deal. Most toy manufacturers won't step up to a 10% royalty rate; they would settle for a royalty between 5% and 6%. In that case, all planning and designing is done by the toy company, with approval of samples vesting in the licensor and producer.

There are two further elements in a manufacturing deal: approvals and exclusivity. If an item contains an actor's likeness, a sample contractually must be shown to the actor for his approval. Many companies that license don't give exclusivities; that is, they give out licenses to more than one company for the same product, such as a tote bag. It could be argued that this increases competition, but I believe it only creates unnecessary tension in the marketplace if more than one national manufacturer is licensed for a certain item. Every licensing deal I make is exclusive.

Now let's turn to an independent producer, who may be gathering financing for a picture partially from a distributor and partially through other sources. Assuming he retains merchandising rights, this producer can try to sell licenses to his picture on his own, but he would be at a competitive disadvantage in an unknown market. My advice would be for him to make an overall deal with one of the major licensing companies experienced in licensing to handle all merchandising exploitation of his picture.

The producer and the licensing agent would enter into a representation agreement in which the agent would agree to solicit and enter into agreements with manufacturers for merchandising, subject to the approval of the producer. There is usually no difference between basic terms of a studio-negotiated license agreement and one negotiated by an independent licensing agent. For its services the licensing agent would receive a fee ranging from 30% to 50% of all monies due from the manufacturing companies under all license agreements negotiated by the agent. Licensing agreements usually provide for a royalty payment to the producer's company of between 5% and 10% of all net sales by the licensee, which is generally accompanied by a cash advance against royalties, as mentioned.

When a producer approaches my company to handle merchandising for a picture, I first determine that he and his project are valid and ongoing. Then the essential questions in determining whether the picture qualifies for a merchandising effort are: How does the screenplay read? Who are the lead actors? Who will be advertising and distributing the picture? If the screenplay content is violent, downbeat or heavily sexual, it is probably not a picture for merchandising because the vital chain stores would not carry items relating to such subject matter. If the picture is heroic, family-oriented or youth-oriented, it has a better chance at successfully licensing products. Suggestions can be made in the screenplay stage that would enhance the merchandising potential of the movie.

Next who's starring in the picture? The mass-merchandising stores want to know if the lead actors have wide appeal or are up-and-coming stars because that will influence how they sell the products in their stores. They easily advertised *Batman* into a merchandising bonanza that included posters, iron-ons and clothing. This fed the merchandising success in areas such as dolls and other toys. The producer's obligation is to tie up all merchandising rights for the actors in the very first stage of negotiating their employment agreements.

The next important issue is who is distributing the picture, because the essential advertising that will be used in merchandising flows from the distributor. As mentioned, the royalty payment to the distributor on every merchandised item covers the use of the recognized title and artwork (logo) of the picture, which has been created by the distributor for advertising. Planning a product campaign is so sophisticated now that the major retailers want to know where and when a picture is breaking so that their stores can be prepared and stocked. When a picture opens with over 1,000 prints, a release schedule detailing theatres and opening dates will help them sell posters in each city.

A producer should not assume that a potential licensee will come to him. Rather, buyers from the key mass-merchandising and retail chains should be invited to an early screening of the picture, along with manufacturers who want to obtain merchandising rights. Product ideas and merchandising angles should be solicited from them. If the retailers are excited over the market potential, the manufacturers will be eager to make deals. The producer should review those manufacturers whose products enjoy screen coverage in the movie. With stills of such coverage, steps can be taken that might lead to a promotional or premium tie-in with companies such as Coca-Cola, McDonald's or Burger King; there are substantial royalties in this area. Again, a separate screening of the movie for such company representatives can lead to generating tie-in ideas. Publicity kits on the picture mailed to manufacturers can also solicit merchandising interest.

Where does the producer make the profits in merchandising? If a financier-distributor is doing the licensing, the producer's deal may call for 25%–50% of all revenue from merchandising. If the producer is free to deal with an independent licensor, the producer would receive 50%–60% of all front monies, guarantees or royalties negotiated by the licensor, who would receive the balance, 40%–50%.

This discussion has centered upon retail merchandising, which is one market. A separate market is the product tie-in or premium and promotion market. An example of a promotion has found a series of four Babar collectible PVC figures available exclusively with the purchase of certain sandwiches at Arby's. Children would come in and collect the full set of figures. In this type of deal the licensor/rights holder benefits from the impression made by the campaign advertising the promotion, and receives 10% of the 25¢ wholesale price of the premium items, or 2.5¢ per piece. A highly successful promotion can turn over some 18-million pieces, generating $450,000 to the rights holder. Arby's gains the benefit of purchases enhanced by those new patrons who want to collect the figures and are introduced to their menu.

Another type of merchandising deal is the sweepstakes deal. A sweepstakes deal would be a flat deal involving perhaps a $100,000 payment from Anheuser-Busch to the licensor for rights to use the name and artwork of the picture in their contest. Then they would make a deal with a company that would organize the contest.

Merchandising is now branching out into other areas. For *Fungle the Gnole,* created by Alan Aldrich, being developed as a combined live-action and animatronic movie, there will be a line of nutritious food for kids that can be heated upon opening the packaging.

Finally there is the deal wherein manufacturers may simply want to use a picture logo in their advertising. This would also be a flat deal. If, for example, Procter & Gamble wanted to use a logo in their advertising for a product, the cost for that license would depend on the media involved. If such advertising were to include radio, television and print, the fee could range from $25,000 to $500,000.

If an unauthorized company uses a picture logo in advertising or on retail goods, it's a federal offense, a violation of copyright law. One response is to send "cease and desist" letters, which have little practical power. A stronger strategy is to sue the company and report them to the FBI. The offending company would have to pay a royalty on everything sold, and all remaining items would be confiscated.

Motion pictures are harder to merchandise than television shows. A TV show can reach 120-million people in one night, whereas a motion picture could take six months to reach the same size audience. Also a motion picture is widely available for only six months, while a hit TV series stays on a network for perhaps five years and lives on in syndication.

The merchandising industry is a mature industry, but people are learning more about it every year. It's important for a producer to understand the value of this potential market in order to protect this ancillary right.

XI
INTERNATIONAL

# THE GLOBAL MARKET

*by **PETER J. DEKOM.** For a biographical note on Mr. Dekom, see page 123. Special thanks are expressed to Mark Damon for his contributions to this article.*

*Even though Japan represents the number-one export market for American films, the name of the game in that country is home video. . . .*

With four out of the seven major studios in foreign hands, the big story for motion pictures as we look to the turn of the century is the cultural and economic impact of overseas markets. Twentieth Century Fox is owned by Rupert Murdoch's Australia-based News Corporation; the Sony Corporation of Japan owns Columbia Pictures and therefore TriStar; MGM, which includes United Artists, is in transition after a shaky acquisition by Italian Giancarlo Paretti; and Matsushita Electric Industrial Company of Osaka, Japan, owns MCA-Universal. This leaves Disney, Orion (reputed to be up for sale), Paramount and Time Warner as the remaining American-owned entities.

Several years ago a dollar traded for 250 yen, but today it is worth roughly half that, and there is nothing on the horizon that suggests a material increase in the strength of U.S. currency. In fact, as high-definition-television systems become available, the potentially devastating multibillion-dollar impact on American balance of payments from the likely Asian electronics imports, when coupled with expected U.S. budget deficits, will do nothing to strengthen the dollar against Western European and Japanese currencies.

This points to the single most important economic modification in the way the movie business works today. Where once we looked to the American marketplace to determine which project should be made and which actors and directors should be engaged, increasingly the question must incorporate international tastes.

Since the dollar is worth relatively less, the revenue streams overseas, because of currency fluctuation alone, have increased by between 50% and 100% in value. And this is simply the tip of the iceberg. Suddenly the rest of the world has responded to the impact of in-home viewing of motion picture product. Home video has become a routinized part of many international households, and free television, especially in those countries where the airways are not government-controlled, has become another means for people to enjoy movies at home. In fact in most nations theatrical attendance has been in significant decline. For example, in the late 1980s attendance at French motion picture theatres dropped by almost two-thirds, with Italy, Spain, Holland and Belgium not far behind. By the early 1990s this had bottomed out, and there are recent modest gains in each market. Theatrical attendance in Japan has also suffered, though there are some bright spots. In Germany, for instance, theatrical attendance has improved, although the ancillary markets are not as significant there as in other countries.

Aside from currency fluctuations, events such as the growth of pay-television overseas, the privatization of television and the increased competition with government channels as well as the increase of home-video viewership in many areas have further increased the value of "foreign rights" to American motion picture product.

For American producers the news is still good, even though the buying spree in overseas free television has slowed. In Europe alone American movies account for approximately 50% of all motion picture product, based on revenues. Movies of sufficient quality are generating between 50% and 100% of their negative costs from overseas prelicensing agreements, with significant potential for overages. No wonder companies such as Carolco, Morgan Creek, Hemdale, Largo Entertainment and Vision International have placed the overseas marketing of their product at the cornerstone of their financial business plans. Note that each of these companies has utilized major studio distribution, in whole or in part, for theatrical release of their product in the United States. Thus the fact that they may be fractionalizing their foreign distribution deals does not deny them access to major studio release in the U.S.

Another trend has emerged that sounds a serious word of caution: The worldwide marketplace, domestic included, in virtually every single medium, has become an A-title business. Whether selling pay-television, home video or syndicated product in the United States, or trying to secure presale financing in foreign territories, the revenues for low-level, B or C pictures are rapidly disappearing. And while the definition of A-product may be somewhat elusive, it is clear that movies distributed in a wide release pattern by majors in the United States, and/or movies with significant recognizable stars and established directors, have a better chance of achieving that A status than any other form of motion picture.

Whether a picture is capable of transcending cultural barriers is another important consideration in foreign licensing, but there are no longer entire "categories" of product that are eliminated completely. Comedy, for example, especially physical comedy, but also relationship comedies, are traveling better than ever before, especially if they have major stars and world-class directors. Naturally the rapport between the overseas sales agent and the overseas distributors is crucial, especially when local distributors attempt to back out of deals after viewing the finished product.

To put the overseas market in perspective, taking a sample year as 1990, revenue from all movie companies and sources (theatrical, home video, cable and TV) equaled $4.5-billion, which translated in trade parlance into a $3.5-billion balance of payments. In descending order of revenue potential (and remember, this list varies year by year), the markets are Japan, the United Kingdom and Ireland, Germany, France, Italy, Spain, Australia, the Far East (not counting Japan), Scandinavia, Latin America, Benelux, India, Portugal, South Africa, Greece, Israel, Turkey and Eastern Europe.

For overseas dealings the American majors are affiliated through their trade association, MPEAA, the international arm of the MPAA. American independents, along with the most active overseas importers of English-language films, are organized as members of the American Film Marketing Association. The AFMA was established in 1980 to enhance the sales of such pictures overseas and address issues of trade barriers and piracy, as well as the spiraling costs of attending the traditional film markets at the Cannes Film Festival and MIFED in Milan. This led to the creation of the American Film Market in Beverly Hills one year later, which has become a principal marketplace where some 200 companies gather to buy and sell English-language pictures. In addition to American companies, AFMA includes companies from Great Britain, Canada, Australia, France, Italy, Germany and Hong Kong, among others. Taking our sample year of 1990, the AFMA reported total independent overseas sales in all media as $1.13-billion, which is an indication of how the independents carve out a sizable portion of the $4.5-billion overall overseas revenue.

Positioning one's product can be a daunting challenge for international producers. Those who are not aligned with a major studio via output deals must either piece together territorial distributors or hire a sales agent to knit the tapestry. Others find security in numbers. An example is the Summit Export Group, the overseas sales organization comprising producers Arnon Milchan's Regency International, Andy Vajna's Cinergi, Bernd Eichinger's Constantin Film and Beacon Communications.

A note of caution: When traveling along the highly changeable and evolving road of product values and emerging formats, remember that dollar amounts are subject to ebbs and flows. They have been included in this article as signposts (using the same year 1990) and for illustrative and comparison purposes. By the time you read this, things will have changed. For current figures, consult the international trade press or executive practitioners along the highways and byways of international distribution. With the foregoing in mind, let us explore the overseas marketplace in greater detail, emphasizing the principal arenas of Japan, Western Europe and Australasia.

## JAPAN

Even though Japan represents the number-one export market for American films, the name of the game in that country is home video. Theatrical film rentals have been slowly declining over the last few years (in fact per-capita attendance is extremely low, although there has been some recent recovery), while home video has been on the rise. It's not unusual for annual home-video sales to be double or triple theatrical rentals. Nevertheless many techniques are being used to lure young people back into theatres, including building very plush environments with gaming and restaurant facilities and expanding viewing time (normally, because of train schedules, Tokyo theatres have only one showing per evening). In fact Warner Bros. has entered into a joint venture with supermarket chain Nichii to build thirty multiplexes around the country, and they are planning to keep the theatres open longer hours. But the handwriting is on the wall. The major revenue streams from Japan for the near future appear to be those other than from the 1,800 movie theatres. Yet, with ticket prices averaging $11, and given the benefits to ancillary values of advertising a picture in theatrical release, it is clear that the theatrical market cannot be ignored, even though it is subject to the vagaries of public whim.

Because it is such an important overseas market, all of the American majors have a theatrical-distribution presence in Japan. There is UIP (which represents Paramount and Universal), Warner Bros. (which handles Disney and Orion product), Columbia (including TriStar) and Twentieth Century Fox. In addition, local companies like Toho-Towa, Nippon Herald and Shochiku-Fuji handle independent American product and are prominent in this market. There are a number of indigenous production companies involved in distribution, but the above-mentioned companies are the primary ones handling American films.

The theatrical marketplace in Japan is an extremely advertising-and-promotion-intensive market, often costing millions of dollars to release a big American picture. However, a small release is possible for a vastly lower price. Further, the huge theatrical rental dollars generated in Japan are concentrated on no more than five or ten American films per year, with most U.S. product receiving no or an extremely modest theatrical release.

As noted, the big story in Japan is home video. With wholesale ranging between $90 and $100 and retail generally between $135 and $150, the revenues from this market alone are obviously significant. A healthy share of the video market for any particular title would be upwards of 10,000 units. As noted, the sky's the limit for video in Japan, a country where *Top Gun* is reported to have sold at least 130,000 units, *Die Hard* close to 150,000 and *E.T.* is selling in excess of 200,000 units.

With CIC Victor Video (including Universal and Paramount titles) entering a phase of direct sell-through of videos at about $28 per unit, it will be very interesting to see whether this reduced pricing will become a trend, increasing the number of video units sold and, if so, whether this will significantly increase the aggregate dollar base generated from the video market. Early reports indicate that sell-through units are exceeding the number of units sold for rental purposes. Laserdiscs have a much higher penetration in Japan than in the U.S. and account for 15% of the market (even excluding the sing-along product).

There are some changes taking place in Japan involving television that should probably disturb American producers. Recently American motion pictures have not fared well on television against locally produced product in local ratings wars and are likely to be sold for fewer dollars in the future.

Generally the six networks in Japan do not purchase motion picture product until they have tracked theatrical performance. These networks, NHK (public broadcasting in Japan), NTV (Nippon TV), TBS (Tokyo Broadcasting), Fuji-TV, Asahi Broadcasting and TV Tokyo 12 usually license motion pictures in a package, often driven by a "locomotive," or hit movie, a practice similar to U.S. TV syndication. Some stations will argue for separate prices on separate movies, but this appears to be the exception, not the rule. When a motion picture performs very well in the Japanese theatrical marketplace, a percentage of film rentals (generally around 10%) can be negotiated, but this is also the exception. Most films are simply sold at a flat price as part of a total package.

Cable is just starting out, regulated by the Ministry of Post and Telecommunication. There are approximately 7.5-million cable homes, just 10% of all TV homes in Japan. A new NHK DBS (Direct Broadcasting Satellite) system recently debuted, with approximately 1.6-million initial subscribers, on a voluntary collection system (i.e., they haven't enforced collections yet), and a much smaller service, Japan Satellite Broadcasting, was recently introduced. It is expected that these satellite transmissions, including motion pictures, will form the backbone of a new delivery system to Japan in the near future.

Japanese Satellite Broadcasting, capitalized at some $300-million from 250 shareholder companies—leading players in Japan's broadcast, banking, newspaper, electronic and insurance industries—is an interesting case in point. The Japanese consumer is now afforded a choice among two NHK pay channels and JSB, but this does not come cheap. A JSB customer (subscribing through an appliance store, not directly with the company) makes an initial investment of some $550. This covers a tuner and antenna costing around $350, plus a $200 onetime membership charge, plus a monthly fee of about $15 for the service. With some 220,000 initial subscribers, *Variety* reported that JSB projects its customer base as reaching 5-million by 1996. However, certain technical glitches and errant satellite launchings may abridge that figure somewhat. As a measure of JSB's far-flung reach for programming, it was an investor in Broadway's Tony Award–winning *The Will Rogers Follies,* which it plans to air to subscribers. For American movies, customers tune into the Wowow Home Theater Channel. Its acquisition entity, Wowow Programming Inc., is half owned by New York–based Media Enterprises, through which movie buys are generated.

As in the United States, pay-TV in Japan has altered the order in which movies proceed through the various windows of availability, taking its position following home video and prior to free television. It goes without saying that, through its satellite delivery, JSB is in a position to beam high-definition television to its subscribers, a bonus currently being offered experimentally only to customers of NHK. Some observers optimistically predict a combined pay viewership of 20-million subscribers for both NHK and JSB services by 1997, according to *Variety*. Such numbers could cut into or call for rethinking the overall strategies for broadcast television and home video in Japan.

## THE EUROPEAN ECONOMIC COMMUNITY

### England

England appears to be driven by home video and television. The video market is third in the world (after the U.S. and Japan), serviced by some six thousand retail stores. Major video players are CBS/Fox, RCA/Columbia, Warner and CIC. Generally it is very difficult to separate home-video rights from theatrical in England (where VCRs are in 65% of the over 21-million TV households), and frequently television is sold as part and parcel of the entire rights package. With wholesale video at somewhere around £50 per unit (retail around £80 per unit), and with the pound trading at approximately $1.65, it is easy to see how lucrative this market can be with top motion pictures achieving sales of between 50,000 and 75,000 units and average films performing in the 10–20,000 range. Nevertheless, as in the United States, the field is entirely dominated by A-product.

The theatrical market in the U.K. has shown signs of restored health recently, with box-office admissions increasing slowly, although production continues on the decline. 1990 represented a ten-year high with 91-million tickets sold, increasing 3.4% from the 1989 level of 88-million and continuing an upward trend since 1984. *Variety* estimated that 80% of the £240-million 1990 gross went to American product. The major distributors in England are Cannon Releasing, Columbia/TriStar Film Distributors, Rank Film Distributors, UIP (Paramount and Universal), Twentieth Century Fox, Palace, Guild, Entertainment Film Distributors, First Independent (formerly Vestron) and Warner Bros. Distribution in a joint venture with Disney. Cannon, Rank and Brent Walker also control exhibition chains. Although American Multi-Cinema built eighty multiplex screens in England recently, it has left the market, transferring ownership of the screens to CIC/UA. The total number of U.K. screens is approximately 1,800, with an average ticket price of £2.40. It costs between $200,000 and $400,000 to release an average film wide in England, but the expected revenues cannot be averaged, since this varies greatly depending upon performance.

The interesting story in the United Kingdom is television, both free and pay. While pay-television is just getting under way as a result of new satellite launches, free television has gone into a veritable feeding frenzy for programming because of the privatization of British television and the anticipated government sanctioning of some fifteen regional franchises. Don't be surprised to find such American stalwarts as NBC, Disney and Time Warner's Home Box Office represented as investors along the way. As of this writing, there are three TV buyers in the U.K., the BBC (British Broadcasting Corporation), the ITV (Independent Television) companies and Channel 4. Channel 4 and ITV tend to use the same buying services, except for specialized product. The top prices for licensing movies for television are in the $1-million range. According to one buyer for the BBC, 30% of the U.K. theatrical film rentals is a good rule of thumb for setting price.

The recent combining of BSB (British Satellite Broadcasting) and Rupert Murdoch's Sky Channel into BSkyB (British Sky Broadcasting) solved an ongoing, intense rivalry that was threatening both entities. Each company had made astronomical deals to acquire pay-TV rights to studio libraries, frequently as high as $1-million for recent library titles. Percentage of theatrical rental deals had also been negotiated, with guarantees in the $1–$2-million range for top films against a percentage of film rental, again on the order of magnitude of 30%. Since the merger the company has made its intention clear to go back to the studios and renegotiate these deals downward. Because the combined entity is a monopoly, the sellers will have little alternative. Further, some experts believe it will take years longer than originally expected to generate the threshold subscriber base needed to make this system viable.

**France**

In addition to the U.S. major studio distribution arms, France has five major theatrical distributors: Gaumont, MK2, AAA, AMLF, and UGC. All of the major distributors control television and home-video distribution companies or handle these formats directly. Video distributors include Disney, Warner, CIC, CBS/Fox, TF-1 Video and RCA/Columbia. Normally theatrical and home-video rights are tied together. The video market in France is not nearly as lively as the markets in Japan, Australia or England. Here a sale of 7,500 units is excellent, although the pricing structure is not that much different from that in the United Kingdom. Home-video penetration hovers just under 50% and has the potential of going higher. As for the high-end possibilities of video purchases, *Variety* reported that *E.T.* has sold over 400,000 units in France.

There are approximately 4,500 cinemas in France, most of which are older screens in need of refurbishing (which is slowly getting under way). Although the population is about the same as in the U.K., there are far more screens. The major French exhibitors are Gaumont, UGC and Pathé. Attendance in movie theatres has been abominable, with sharp declines of 30% occurring in the late eighties. However, there has been recent recovery in the theatrical business.

Free television is thriving in France via five principal buyers: LaCinq, M-6, TF-1, FR3 and Antenne-2 (both state-owned). The prices here also vary, depending on product. Free-television prices are decreasing in France, and the devastation in the theatrical marketplace has certainly had an impact on the value of product in that arena.

The big story is pay-television, where Canal Plus, the most successful pay-TV company outside the United States, boasts over 3-million subscribers and is looking to expand beyond France to Germany and Switzerland. In fact Canal Plus has been reported as paying approximately $1-million for the rights to blockbuster American movie titles. In addition, they have been actively investing around the world to ensure product flow. In 1991 they joined producer Arnon Milchan and Germany's Scriba & Deyhle to infuse Warner Bros. with some $600-million in production financing. Separately they own 10% of Carolco Pictures, are involved in pay-TV operations in Spain, Belgium and Germany, and have a cofinancing arrangement with Mel Brooks's Brooksfilms through its production entity, Studio Canal Plus.

In another move of global positioning, French industrialist Francis Bouygues, the largest shareholder in TF-1, has created CIBY 2000, based in the United States, for the production and financing of international motion pictures (for example, a $20-million David Lynch film).

When combining free and pay-television rights, it is still possible to garner between 4% and 8% or even more of the budget from the

French marketplace, depending on theme. For an average big-budget A-title picture, one can expect from $500,000 to $1-million or more in a prelicensing agreement.

One difficulty in selling television product to France is their national quota system. 50% of their television programming must be French in origin, and 60% of European origin, which considerably reduces the ability of American producers to sell to the market.

Releasing a movie in France frequently costs between $50,000 and $600,000, considerably less than the averages spent in the United States. Should French television be opened up to advertising for movies in the future, these numbers could more than double. This is principally because French distributors tend to advertise heavily only in urban areas, allowing the picture to filter out into suburban and provincial areas subsequently. This practice can be seen in Germany as well. Generally, when licensing to France, one is also licensing the rights to the surrounding French-speaking countries, including French-speaking Switzerland and French-speaking Belgium.

**Germany**

German distributors generally cover Austria and German Switzerland as well. These companies include Warner Bros., Scotia Cannon, UIP, Columbia/TriStar, Fox, Tobis, Neue Constantin, Jugend Film, Delta Film, Connexion and Senator Film. There are approximately 4,000 screens in the German territory and, like France, advertising dollars are not very high. But unlike France, the theatrical marketplace is clearly the most lucrative in this territory.

Since unification, East German film and TV companies have been positioning themselves to compete on a broader basis, but it's an uphill climb. One such company is Progress Film, which had been an East German distribution monopoly subsidized by the government and releasing pictures by the DEFA Studio in Babelsberg, outside Berlin.

Video-distribution companies in Germany include Neue Constantin, Warner Bros., RCA/Columbia, CBS/Fox, CIC and VCL. Video sales are not particularly high, with good sales in the 5–10,000 unit range and maximum sales at 15,000 units. Pricing is structured around $125. Cable can be found in over 17-million homes throughout the country, broken down between approximately 16-million in the west (more than half of TV homes) and 1.3-million in the east.

The least lucrative area in Germany is television, which is still mainly controlled by the government. Prices for free television vary from $30,000 to $150,000 depending on the nature of the motion picture and the buyer in question. Expected privatization in the future would change this, however. Two nascent systems, satellite-sourced channels SAT 1 and RTL-Plus, were the first forays into this market. Germany's single pay-TV channel, Premiere, was launched in 1991.

Partnered in this venture are the Kirch Group, the ubiquitous Canal Plus and Bertelsmann's UFA motion picture and TV entity. Clearly the future pattern of EC delivery will include much more pay-television.

A good German advance for an American motion picture can be in the low-end seven-figure range.

**Italy**

For the most part the numbers in Italy are not particularly high except for television rights. The video market and theatrical market are smaller than their French and German counterparts, primarily because Italy was the first European country to enjoy TV privatization. Viewers in Italy can receive over twenty channels of TV programming. This factor has hurt theatrical extensively and made home video a very late bloomer.

After years of decline in the number of theatrical admissions, 1989 and 1990 enjoyed slight percentage increases, but these numbers are one-eighth of the 800-million tickets sold in the record period of the mid-1950s. Hollywood product dominates Italian releases, with *Pretty Woman* one of the strongest recent titles. Movie theatres are down to 3,600 throughout the country (with ticket prices at $9), half of the 7,500 in 1980, and many are in need of modernizing. Italian producers complain that theatres are too eager to remove pictures that don't open strongly, echoing a complaint often heard in the U.S.

Pay-television in Italy is nascent, with TelePiu formed by Silvio Berlusconi and Vittorio Cecchi Gori in 1991. (In another example of global reach, these two gentlemen have also established an American production company, PentAmerica, as a source of product.) As far as broadcast television is concerned, there are seven nation-wide networks. Silvio Berlusconi has also launched three private free-television networks, increasing competition to the three existing government-owned (RAI) stations and the seventh, Tele-MonteCarlo. The national Telcom Ministry plans to raise the number of TV networks to fifteen, and increase and organize local stations as well. In movie sales to Italian network television, the prices have been quite significant, with some guarantees going over the $2-million mark, a disproportionate sum for this territory. Future guarantees, however, seem destined to come down.

## AUSTRALASIA

The principal territories are Australia and New Zealand, with other islands such as New Guinea and Fiji included for good measure. Australia is clearly the dominant market in the area and often controls distribution into lesser territories. However, the market can be characterized as depressed, and a producer with an A-title that will travel internationally will find that Australasia might supply less than 4% of the total budget for a motion picture in the form of presales and prelicensing. The two main independent theatrical buyers have been Village Roadshow and Hoyts. When Hoyts recently found themselves in financial difficulty, Village Roadshow found themselves in a near monopoly, and dropped the license fees they were willing to pay.

In a country that boasts some 850 theatrical screens with average ticket prices hovering around $5 U.S., it is obvious that the cost of releasing a movie in Australia is much less than in the United States. However, the upside of top movies there can still be significant. The strongest performer, not surprisingly, was the first *Crocodile Dundee* picture, which generated just under $20-million Australian. A title like *Three Men and a Baby,* another strong all-time performer, generated more than $7-million Australian, was released with 104 prints, with about Aus.$200,000 spent to open the picture and around Aus.-$500,000 in total releasing costs over its entire theatrical life.

The principal exhibitors in Australia are Village Roadshow/GUO (Greater Union Organization), Hoyts/CIC, and Wallace Theatres. Now that Village Roadshow and Greater Union have merged, they are the undisputed king of theatrical exhibition, with around 170 screens between them.

The home-video market in Australia is significant and dominated by these distribution companies: Village Roadshow (handling Disney, including Touchstone and Hollywood), CBS/Fox, RCA/Columbia/Hoyts (including Orion and Columbia/TriStar), CEL (including MGM product), Warner Home Video, CIC Taft (including Paramount and Universal) and Palace Video. Over 70% of the approximately 5-million television households in Australia have VCRs. Video titles are higher priced than in the U.S., with wholesale prices (about 65% of retail) varying between Aus.$80 and Aus.$110 for major product. While blockbuster titles will sell more than 20,000 units and the all-time champ *Crocodile Dundee* sold around 31,000 units, a more typical sales figure for major studio titles hovers around 10,000.

Since pay-television is still years away in Australia (although sporting events are available via satellite in certain clubs), the bulk of television revenues come from the highly competitive privately owned stations, Channels 7, 9 and 10, as well as two government-owned stations, SBS and ABC. Because each of the privately owned TV networks has changed hands over the last several years, competition has been great, and the debt service from these acquisitions has caused severe cash-flow problems. These private networks also bought a number of film packages from American majors in the late 1980s, at a time when American features and television programs were doing exceedingly well in this TV market, with probably 60% of any advance received from Australia attributable to the expected television revenues generated there. (The balance of any such advance would probably be allocated to home video, less any losses that would be expected from the theatrical release of the respective film.) But a severe downturn followed in the television market, causing a ripple effect throughout the entertainment industry in the country, a less-than-positive reality for American producers.

## OTHER COUNTRIES

Scandinavia, Spain and South Korea are the other significant territories, where the presale dollars can still add a few percentage points to the budget of a typical feature film.

TV is the big story in Scandinavia. Up until the mid-1980s, Sweden, Norway, Finland and Denmark were each served by one state-owned TV network. Today there are at least nineteen channels competing for viewers, delivered via broadcast, satellite and cable. In Sweden the chief TV services are TV4 Nordisk Television, SVT 1 and 2, Nordic Channel, Scansat TV3, SF Succe and Film Net. Finland's outlets include Helsinki Television and YLE 1 and 2; Denmark has TV2 Danmark, Kanal-2 and DR/TV; Norway viewers have TV Norge and NRK/TV. As for theatrical, 80% of box-office revenue in Scandinavia goes to American movies, while locally produced product is nearly entirely subsidized by the respective governments.

In Spain television is also in the headlines. Three new private networks now compete with the two channels of RTVE, Spanish Television, run by the government. The new networks are Antera 3-TV, Silvio Berlusconi's Telecinco and the ubiquitous pay channel, Canal Plus. As of yet, Spain has no cable TV, and home video seems on the downturn, a victim of video piracy. In the theatrical marketplace Spain is ranked approximately sixth among countries importing American films.

In South Korea, for A-pictures it is possible to make deals in the $300–$500,000 range, and a blockbuster title can garner $1-million for all rights. The principal sources are theatrical and home video, since television is not yet a major factor. All in all, these numbers appear to be on the ascent. Another market that bears watching is the Benelux countries, with prelicensing about half of those for South Korea.

For all practical purposes (and as chauvinistic as this may appear), the rest of the world's markets are not nearly as significant.

Though the Latin American populations are large, and demand for product strong, one still faces the fact that these are Third World countries where currencies are frequently restricted and the revenues generated disproportionately small, given the populations in question. Africa and the Near and Far East (except Japan, the Philippines, Taiwan and Hong Kong) are equally insignificant in the determination of which product should be made. Compounding the question is the issue of video piracy, which is rampant in the Third World. Obviously all of these markets purchase rights to American films, and presales are possible, but this is not where the significant revenue stream is derived.

# OVERSEAS TAX INCENTIVES AND GOVERNMENT SUBSIDIES

*by* **NIGEL SINCLAIR,** *a founding partner at Sinclair, Tenenbaum & Co., a law firm in Beverly Hills specializing in entertainment, entertainment finance and international business. A graduate of Cambridge University, Mr. Sinclair began his law practice in the London office of Denton Hall Burgin & Warrens, moving to California in 1981 to open their Los Angeles office. Eight years later he helped found Sinclair, Tenenbaum. Mr. Sinclair serves on the board of directors of a number of entertainment companies and was executive producer of* Daddy's Dyin', Kill Me Again, *and* The Blue Iguana.

*and* **STEVEN GERSE,** *senior counsel at Walt Disney Pictures in Burbank, where he handles business and legal affairs for motion pictures under the Disney and Touchstone labels. A 1983 graduate and Harno Scholar from the University of Illinois College of Law, Mr. Gerse was in private entertainment practice as a partner at Sinclair, Tenenbaum & Co., has written movie business articles and was coinstructor (with Nigel Sinclair) of a course on the international film business in the producer's program at UCLA.*

*. . . the particular financing program may impose certain distribution restrictions, such as requiring the producer to give away distribution rights for the country concerned. . . .*

Besides the traditional sources of financing available to an American motion picture producer, such as studio financing and presales, the producer may be able to take advantage of various financing opportunities overseas. Several countries have instituted tax incentives or subsidy programs that directly or indirectly provide funds for motion picture production in an effort to stimulate the local film industry and promote the country's culture. While such programs are not designed with American producers or American movies in mind, the U.S. producer may be able to obtain the benefits of such programs by assembling the necessary creative elements or by entering into foreign coproduction arrangements.

Along with the obvious advantage of providing precious financing, such overseas opportunities offer other attractive lures. A producer who has raised substantial financing overseas, for example, may be able to obtain more favorable distribution terms because he may not have had to presell key rights or markets in order to raise production financing. In the case of a tax-related coproduction, for example, the owner of the picture need only be concerned with covering his net exposure after the tax benefit rather than covering his entire investment.

Although there are a variety of foreign financing opportunities that are not created by government action, such as below-the-line facilities or currency deals, this chapter focuses on tax-incentive and direct-subsidy programs legislated by the governments of certain countries. Tax incentives for motion picture investment vary greatly from country to country, but each has the central feature of providing tax relief on film investments that would not otherwise be deductible or on profits that would otherwise be taxable. Direct-subsidy programs involve a cash investment by a government agency to finance the production of qualified films. In some countries investment from private industry has been much more significant than contribution from government in recent years, and this chapter will also note the significant developments in this area.

Such tax incentive and subsidy programs seem to undergo a three-stage life cycle, and the appeal of a particular program may depend upon which state of the cycle the program is in. In the first stage the foreign government initiates the incentive or subsidy program in order to encourage a national film industry and stimulate local employment. The tax incentives or subsidies are initially available on a rather unrestricted basis, and investors and producers from around the world are quick to capitalize, not always promoting the same goals as the country concerned. Then, after the program has been abused, typically with much of the available funds having gone to U.S. producers to make Hollywood films, or after a large number of films have been produced solely for their benefit as tax shelters and without regard to their cultural or artistic merit, the government reacts by imposing a number of restrictions on the program. In this second stage, for example, the government may reduce the amount of the tax write-off or subsidy, or impose strict local content requirements based on the subject matter of the picture or the nationality of the artists, crew and locations, or in terms of distribution restrictions. Then, after filmmakers have found it difficult to produce pictures with true international appeal, a new round of political lobbying leads to the third stage, in which government policy seeks to become more practical, and more benevolent to "international" pictures. The result is usually some form of formal coproduction program, often consisting of specific coproduction treaties with individual countries, or modified tax write-offs with a watered-down "local content" test.

There are several threshold facts that a U.S. producer must realize before considering the myriad foreign financing opportunities. First, it is unlikely that the producer will be able to rely solely on the tax incentives or subsidy to finance the film. The net value to the production of such benefits is typically between 10% to 33⅓%. The project generally must also have some compatibility with the foreign country in which the cofinancing benefit arises, in terms of either subject matter, location or nationality of the director or star. Further, the terms of the cofinancing may restrict the extent to which the American producer can fulfill the producer function and may require a foreign coproducer in order to obtain the benefits of that country's film program. Finally the particular financing program may impose certain distribution restrictions, such as requiring the producer to give away distribution rights for the country concerned.

The following country-by-country survey of the principal overseas tax-incentive and subsidy programs is by no means exhaustive. Although the information is current as of this writing, the benefits and requirements of the programs in this area are constantly evolving and are subject to sudden change, so the reader is encouraged to seek counsel regarding the latest details of the particular country concerned.

## AUSTRALIA

Although Australia remains an important source of film financing, its tax-relief program has been greatly reduced over the past several years. Introduced in 1980, Division 10BA of the Australian Income Tax Assessment once offered significant tax concessions to investors and contributed greatly to the growth of Australia's film industry. The 10BA program originally provided the Australian taxpayer with a 150% deduction on his capital expenditure in a qualifying "Australian" picture. If the investor received profits from the film, the investor would also receive a tax exemption on such profits of up to 50% of the original investment. To be a qualifying Australian picture, the film had to generally be based on an Australian script with significant Australian "content" and use predominantly Australian actors and locations. Most of the acclaimed Australian films of recent years, such as *Road Warrior* and *The Man From Snowy River,* were financed from 10BA-inspired funds. The 10BA provisions soon resulted in a huge tax loss to the Australian government, however, and the 150% deduction/50% exemption was reduced first to 133%/33%, then to 120%/20%, and finally to a 100% deduction in 1988. Because Australia's top marginal rates of tax were significantly reduced from 60% to 49%, 10BA is no longer an attractive proposition for investors. Recent attempts to raise 10BA funds from investors have fallen short of expectations. Not surprisingly, Australian film production has dropped significantly in recent years.

In the meantime the Australian government has pledged to replace the original 10BA benefits through the use of the Australian Film Finance Corporation (FFC) investment group, which is now Australia's primary source of motion picture financing. Funded in part by the Australian government, the FFC supports certain films with equity investment, production or print and ad loans, guarantees or a combination. In return the FFC will have an equity participation in which to recoup its investment and may also require Australian distribution rights. The film must be a "Qualifying Australian Film," which can apply to a coproduction pursuant to Australia's extensive network of coproduction treaties. Significant Australian content or creative participation will be required. In addition to numerous Australian TV programs and local feature films, the FFC has recently funded a number of major international movies including *Green Card, Until the End of the World* and *Map of the Human Heart*. Government funding for the FFC, however, has declined from a high of Aus.$70-million in 1989 to Aus.$35-million for the 1992–93 financial year. Over recent years the FFC has also engaged in some small public fund-raisings, with the FFC providing part of the subscription on higher risk terms, resulting in further public investment in films and TV in Australia.

The Australian Film Commission has established a formal program for coproductions between Australian producers and overseas producers, although such productions have become less attractive to investors after the demise of 10BA. The project must have at least 40% Australian personnel and financial equity, and the number of Australian personnel must be at least in proportion to the Australian financial equity. The film must also qualify for 10BA certification and thereby satisfy certain threshold requirements regarding Australian content and creative participation. The coproduction program operates strictly on a government-to-government basis, so the program is restricted to countries with a national film commission such as Canada. The United States does not qualify. An American producer may be able to utilize such a coproduction structure, however, and access 10BA, 51(1) (see below), or FFC funds through third territories such as the U.K. or Canada. The attractiveness of the Australians (as well as the British and the Canadians) as coproduction partners is that the project can be shot in English. Australia recently entered into significant coproduction treaties with the United Kingdom and Canada, and negotiations are currently under way with Germany, the former Soviet Union, Italy and Israel.

Despite the virtual disappearance of 10BA in its original form, however, Section 51(1) of the Income Tax Assessment Act still allows a 100% write-off to the Australian investor over two years, for an investment in a motion picture. The particular appeal of 51(1) to the American producer is that the film is not required to be an Australian-content picture or even be shot in Australia. In some instances American producers have entered into limited partnerships with Australian investors, who receive a tax deduction under 51(1), to raise production financing for American projects.

## CANADA

Canada's system of tax preferences for motion picture investment has remained relatively stable over the years, although a recent change has modified the tax benefit somewhat. A Canadian who invests in a certified Canadian feature could previously write off 100% of his investment over two years against all of his income. Recently, however, the capital cost allowance applicable for certified productions has been reduced from 100% to 30% against nonfilm income, but an additional allowance of up to the remaining cost of the films may be claimed against the taxpayer's film income. In another recent change the Canadian government declared that prelicense guarantees (such as presales) no longer reduce the investor's tax deduction as long as such amounts are payable within a limited period. Therefore a Canadian investor can receive the full write-off, even though not all of the investment is at risk.

To qualify as a certified Canadian film, the project must satisfy criteria based on the number of Canadians who hold the key creative positions and the percentage of the costs incurred in Canada. A detailed point system is used by the Canadian Film and Videotape Certification Office to measure the extent of the Canadian elements involved in the project, and the production must amass a minimum of six points out of a possible ten points in order to be certified as a Canadian film. A Canadian director is worth two points, as is a Canadian writer. One point each is awarded if the leading performer, second leading performer, production designer, director of photography, music composer or editor is Canadian. As a threshold requirement the producer and all individuals fulfilling producer-related functions must be Canadian, although in limited circumstances a non-Canadian may be able to receive a courtesy credit for a producer-related function, such as "executive producer" or "executive in charge of production." Either the director or the writer *must* be Canadian, and at least one of the highest paid or second highest paid performers must be Canadian. However, the Canadian Film and Videotape Certification Office (an office of the Canadian Department of Communications) on application may recognize a production as a Canadian film if either the director and writer or both leading performers are non-Canadian, as long as all other key creative functions are filled by Canadians.

In addition to satisfying the points test, at least 75% of the total remuneration paid to individuals, excluding the producer and the key creative personnel listed above, as well as payments for postproduction services, must be paid to Canadians or in respect of services provided by Canadians. Also, a minimum of 75% of the processing and final preparation costs in connection with the picture must be paid for services provided in Canada.

The province of Quebec allows a deduction, which, like the federal system, was recently reduced to 30% in the first year, for investment in a qualified Quebec production. To earn recognition as a Quebec film, the project must be produced and distributed within Quebec, by a Quebec company, and satisfy similar points and expenditure tests as used for purposes of Canadian certification, as described above, except that the individuals must be domiciled, and the expenditures must be made, in Quebec. Although Quebec has been the most active of the provincial governments in providing tax and other incentives for local film productions, other provinces have various programs. It should be kept in mind that these provincial programs can be used in addition to, and not instead of, the federal program.

Canada also features a subsidy program for certain projects. Through Telefilm Canada, the Canadian government will provide up to one-third of the budget of a certified Canadian film, and up to 49% of the budget of a Canadian film that scores 10 out of 10 points on the Canadian certification system. The investment is made directly into the film, pari passu with other investors. At the beginning of April 1991, Telefilm started a new fund of Can.$5-million for broad-range commercial films that need not meet all of the strict content requirements of the other funds.

Canadian producers may also take advantage of co-production treaties which Canada has with England, France, Germany and Israel, which allow preferential tax treatment of production expenses.

## ENGLAND

In terms of governmental financial assistance to motion pictures, the United Kingdom is more significant as a historical reference point than as a present source of funds. Once one of the most significant government subsidy programs, England's Eady Plan provided subsidies to certain qualifying English pictures. In order to qualify, the producer had to be a British resident or the production company had to be registered in and controlled in the United Kingdom. The subsidy money was collected by the government through a levy on box-office receipts in all British theatres. Unfortunately for the British film industry, the government discontinued the Eady Plan in the early 1980s.

More significantly, England's film-leasing arrangements have also disappeared. Under such arrangements an investor who purchased an unused new film negative of a qualified British motion picture could claim a capital allowance (tax deduction) of 100% of the investment in the first year. The purchaser would then lease the film to a subsidiary distribution company. To qualify as a British picture, the project had to be made by a British company, satisfy a labor-content test and limit its studio work to the United Kingdom. The film-leasing arrangements raised financing for a large number of British films, including *Gandhi* and *Never Say Never Again*. Eventually, however, the government reduced the 100% write-off to 75%, then reduced it to 50% and beyond, so that its benefits soon were not worth the trouble.

The effect of England's loss of the film-leasing arrangements and Eady levy is demonstrated by the fact that U.K. feature-film production has fallen by 40% since 1983. The only significant government assistance presently available in England is through British Screen Finance, which receives a direct grant of £1.5-million per year from the British government and has provided about £3-million of unsecured financing to between fifteen and twenty films per year. *A World Apart* is a recent example of a film partly financed by British Screen. The Business Expansion Scheme, which offers tax write-offs to a U.K.

taxpayer who invests in share subscriptions in companies carrying on a qualifying trade, was meant to encourage outside investment in certain industries (including the film industry). However, the benefits of the scheme have been recently reduced, so that no company can raise more than £500,000 in a year and an individual can only deduct £40,000 per year. This reduction, along with the lowering of the U.K.'s personal and corporate tax rates, has left the numbers too small to provide much support to the industry. The Business Expansion Scheme also cannot be used by a single-picture production company, as the company must continue to be a qualifying company for three years and the individual must hold onto the company shares for five years.

There are a few regional subsidies available in England, such as the Northern Media Fund, which requires that 30% of the total production budget must be spent in the Northeast region.

## FRANCE

The French government provides tax incentives for individuals who invest in a qualified SOFICA, which is a company formed for the financing of audiovisual products. Individuals who have their main tax residence in France can deduct 100% of their cash contribution to a SOFICA, up to a maximum of 25% of their income. If the individual transfers his shares before five years have passed, however, he will lose the tax benefit. A company that invests in a SOFICA can subject 50% of their cash contributions to depreciation and amortization. If the company holds more than 25% of the share capital of SOFICA, however, it will lose the tax deduction. Also, in the event of a decrease in capital or dissolution of the company, the French Ministry of Finance can order the reinstatement of the amounts previously deducted. In 1987 SOFICA companies provided over $30.2-million to fifty-three French films.

The SOFICA program has strict French cultural content requirements, since a SOFICA can invest only in a qualified "French or European (EEC country) Production" or "International Production." For a "French or European Production," the company producing the movie must be a French company or a subsidiary of a foreign company incorporated in France and with a French manager; the creative, artistic, and technical crew must be French or from an EEC country; and at least 50% of the production expenses must be incurred in France. For an "International Production," the French term for a coproduction, the French contribution must be at least 20% of the production costs, the number of French (or EEC) artists must be directly proportional to the amount of the French (or EEC) contribution; and at least 20% of the production expenses must be incurred in France. It is theoretically possible for an American producer to team up with a

French producer who can negotiate with the French Ministry of Culture for an "International Production," particularly if the production will use one of the French studios such as La Victorine in Nice. However, the administration of the SOFICA system is highly political, and there is strong resistance to its use by disguised English-language/Hollywood-type films.

France also provides a subsidy administered from receipts collected from a levy on cinema tickets. The funds raised are distributed to French producers to be used on films meeting the requirements set forth above and can be quite significant. *Green Card* and *Until the End of the World* are recent examples of French coproductions intended for the world market.

To a large extent, private industry has been a more active film investor than the government. Recently, for example, French media conglomerate Canal Plus entered into a $600-million agreement with Warner Bros. and producer Arnon Milchan to fund twenty films.

## GERMANY

A German producer may be eligible to receive not only a national subsidy but also one of the three "state" subsidies from Berlin, Munich or Hamburg. Although the national subsidy does not have formal German-content requirements and the production need not be filmed in Germany, it is usually necessary that the production have at least a German director, a German featured performer and a German crew. A German producer or production company must be involved, although an American producer could enter into a coproduction arrangement with a German producer, who would then apply for the subsidy. If a film is eligible for the subsidy, a German television company will often contribute additional amounts for the German television rights. *Paris, Texas* is an example of a film that was financed in part by the German national subsidy.

As an example of the significance of the three state subsidies, in 1987 the city of Berlin made $12-million in subsidies available to various productions, and the city of Munich provided about $9-million. For feature films the subsidy comes in the form of a line of credit of up to 30% of the production costs. However, the credit line cannot exceed $1.2-million per picture, and at least one and one-half times the subsidized loan amount must be spent in Berlin. The subsidy "loan" plus interest must be repaid from 50% of all of the producer's receipts from the film, although the producer may hold back up to 20% of the receipts to recoup his investment as well as any outside investment containing unconditional repayment obligations, and to cover distribution guarantees and advances. All obligations to repay the loan cease five years after the commercial release of the picture, and the obligation to pay interest terminates twelve months after the film's commercial release or eighteen months after delivery of the answer print, whichever is earlier.

The state subsidy programs are also available for coproductions with foreign producers, particularly if the film is to be shot in that particular city or in one of the local studios. Even if the film is not shot in the applicable city, it is expected that the production will use at least part of the local production and postproduction facilities. For an international coproduction the subsidy is often in proportion to the German participation in the project. In Bavaria, for example, the subsidy is limited to one-third of the total German investment.

## NEW ZEALAND

Under the New Zealand Income Tax Act an investor can deduct 100% of his investment in the year the film is completed if that film is certified by the New Zealand Film Commission as a New Zealand film. In order to obtain such certification, the film must have "significant New Zealand content," for which the NZFC considers the subject of the film, the locations, and the nationalities and residence of the talent and crew, the owners of the production entity and the eventual copyright owners of the film. Other factors include the sources of the financing and the ownership of the equipment and technical facilities to be used. Although the final certification (which must be obtained in order to take the 100% deduction) will not be issued until the fine cut of the film has been completed, a provisional certification may be obtained at any stage before completion of the film in order to provide some measure of security to potential investors. A coproduction with a country for which New Zealand has a formal coproduction arrangement, such as Australia, Canada and France, can also be eligible for certification as a New Zealand film. A non–New Zealand qualifying film may be written off over two years.

As the New Zealand tax rate is presently being lowered to 28%, the tax incentives have become less enticing to investors. Also, the pool of available investment is limited by the size of the country. In some instances the New Zealand Film Commission will provide direct-loan or equity finance for the development and production of a distinctly New Zealand project.

## JAPAN

The Japanese are relative newcomers to the film-financing arena, and not surprisingly the private sector has been the primary catalyst. The most significant recent Japanese activity, of course, has been the purchase by Japanese companies of two of America's major motion picture studios. In 1989 Sony Corp. purchased Columbia (including TriStar), and Matsushita purchased MCA/Universal. Japanese industry has invested heavily in American independent film companies as well, such as the huge cash infusion by JVC/Victor into Largo Entertainment, and from an offering sold by Nomura Babock & Brown into Morgan Creek.

Taking advantage of Japanese taxation laws, which allow the owner of a film to depreciate the film over two years, Japanese investors may form a partnership to invest in a motion picture and then deduct 90% of their investment over a two-year period. The applicable individual tax rate is approximately 68%, making the subsidy extremely attractive. Such partnerships also generally provide a return of the investor's investment through fixed license fees from the distributor over a certain period (which payments are often guaranteed by a bank), as well as a royalty stream. On one such proposed arrangement, for example, the partnership was to receive 15% of the film's gross receipts up to an agreed maximum, and 1.5% thereafter.

Japanese investors (through Nomura Babcock & Brown) have entered into a joint venture with Walt Disney Pictures and Interscope Communications to produce, finance and distribute pictures, and Disney also recently concluded a $600-million joint venture utilizing equity investment from a Japanese limited partnership through Yamaichi Securities. In addition, Japan's Ascii Corporation has made a major investment in the Edward R. Pressman Film Corporation.

# AN INDEPENDENT WITH A GLOBAL APPROACH: SOVEREIGN PICTURES

*by **BARBARA BOYLE**, president and cofounder of Sovereign. A graduate of UCLA Law School, Ms. Boyle began her career as a corporate counsel for the independent production and distribution company American International Pictures (AIP). In 1967 she went into private practice as a partner in the entertainment law firm Cohen & Boyle, representing talent including independent producer-director Roger Corman. In the mid-1970s Ms. Boyle joined Corman's New World Pictures as executive vice president and general counsel, later chief executive officer. She moved to Orion Pictures in 1982 as senior vice president of worldwide production. Four years later she became executive vice president of production for RKO Pictures before cofounding Sovereign in 1988. Based in Los Angeles, Ms. Boyle is active in Women in Film, the UCLA Law School Entertainment Advisory Council, the Independent Feature Project/West and the Hollywood Women's Political Committee.*

***. . . Sovereign's business plan . . . is . . . to finance generally half the negative cost of a picture in exchange for all international distribution rights in perpetuity. . . .***

Sovereign Pictures was formed by Ernst Goldschmidt to offer independent filmmakers a new approach to financing and distributing their pictures, a logical extension of what he had been perfecting as head of international distribution at United Artists (and later Orion), which in turn can be traced to his involvement with Francis Ford Coppola and lawyer Barry Hirsch in connection with the international distribution of *Apocalypse Now.*

By 1988 major distributors in all media outside the United States and Canada were concentrating on A-pictures—high-quality, high-visibility films. These buyers determined that there was an opportunity for a new international distribution company to emerge. At the same time filmmakers had become keenly aware of escalating revenues from overseas markets and wanted their pictures handled in a coordinated fashion in these territories by one guiding entity. After the major studios there was no company positioned to do this; Sovereign offered that service. (Presale agents generally sell all rights to individual territories, usually to raise production funds, but by and large are uninvolved with the distribution-exhibition process.)

In summary the growth of the international market, the desire of filmmakers to retain control over distribution in that market, the plan to distribute A-pictures as a continuing supplier (not on a project-by-project presale basis) and the advantage of developing a library of such commercial pictures joined to define the business plan of Sovereign Pictures.

The agents and the lawyers of important filmmakers saw two advantages to Sovereign's approach: first, the control that comes through distribution rather than presale; second, the economic advantage of bifurcating rights, that is, separating U.S. and overseas distribution.

Under a worldwide studio distribution deal profits from overseas are used to recoup fees and costs from domestic, and vice versa, because all revenues are cross-collateralized. Under the Sovereign model, with the territories divided, this cannot happen. *The Name of The Rose, The Never-Ending Story* and *The Last Emperor* are just a few examples of European pictures that predictably performed far better overseas than domestically. And because U.S. rights were sold separately from overseas rights, those global profits flowed back to the filmmakers. But consider *Innerspace,* a completely American movie, which doubled its theatrical revenue outside the U.S. Because it was distributed worldwide by a single entity, all revenues from stronger territories flowed back to offset losses in weaker ones. Such examples were not lost on filmmakers Alan Parker, Jim Sheridan, Noel Pearson, Mel Gibson, Franco Zeffirelli, Oliver Stone and Ed Pressman (to name just a few), who began seeing advantages in the dual aspects of improved economics and control offered by the Sovereign formula.

The company has two offices, one in Los Angeles, the other in London; production, finance and corporate are based in Los Angeles; marketing and distribution is in London.

Sovereign works through a network of distributors throughout the world, to whom it provides marketing and advertising materials. A comprehensive marketing plan is agreed upon between Sovereign and each of its distributors to maximize the potential theatrical revenue.

Among Sovereign's theatrical distributors are Rank Film Distributors, Ltd., in the U.K., Warner Bros. in Japan and Italy, Twentieth Century Fox in Germany, Les Films Ariane in France, Lauren Film in Spain and Village Roadshow Corporation Ltd. in Australia. These companies are not required to provide advances or guarantees. However, they acquire only theatrical rights to all Sovereign product, and we make every effort to offer them A-quality pictures.

The overseas video rights and television rights are held by Sovereign to exploit for the benefit of the producer/filmmaker and Sovereign. RCA/Columbia Home Video has acquired for worldwide video distribution outside the U.S. and Canada Sovereign's first two "packages" of films. We license television rights much like theatrical on a territory-by-territory basis, by various methods, sometimes directly to the end user of the product, the broadcasters themselves, such as BSB (British Sky Broadcasting) in the U.K. and Canal Plus in France, and other times to licensees, who in turn license them to the television stations or networks, such as Dan Valley Films, A.G. in Germany, Moonraker Films, A.V.V. in Italy and Tohokushinsha Film Co. in Japan.

Initial financing was provided to Sovereign by two equity investors in the company: Revcom/Les Films Ariane, based in France, and Nordisk, based in Denmark, and additional financing was provided by Credit Lyonnais bank Nederland (the Dutch subsidiary of Credit Lyonnais, France). The company is extremely cash-intensive, contrasted with sales agents who may be offered a project to represent without themselves putting up a guarantee. A sales agent would call up Rank Films, for example, to obtain a distribution commitment perhaps evidenced by a letter of credit for, say, $1.2-million for all distribution rights in the U.K. That process would need to be repeated with four or five other local distributors. The producer would then take six documents to a bank to borrow the money needed to make the film. Working with Sovereign, the producer takes one document from the company, pledging that portion of the negative cost attributable to overseas rights (in the form of a contract or other instrument, such as a letter of credit) so that the producer can borrow.

Some examples of Sovereign's pictures will indicate how the company functions. *My Left Foot* involved Granada Films, which owned rights in the U.K. and Ireland, and Miramax, which had acquired from Granada all other territories in the world. Miramax offered the picture to Sovereign at the inception of the company after others in the industry had turned it down. We screened the film and acquired it within twenty-four hours, to the dismay of others in the business who volunteered, prior to the film's opening, that the picture would destroy our fledgling company. The success of *My Left Foot* led to *The Field* from the same producer-director team of Noel Pearson and Jim Sheridan. Here Sovereign committed to the project at the screenplay stage. The picture was financed 50/50 by Sovereign and Granada, without a domestic deal and was cash-flowed by Granada against a letter of credit. Sovereign retained worldwide rights except for the U.S., Canada, the U.K. and Ireland. When the picture was completed, Granada made a domestic distribution agreement deal with Avenue Pictures.

On *Reversal of Fortune,* producer Ed Pressman had obtained from the Japanese distributor Shochiku Co., Ltd. the funds to acquire the book by Alan Dershowitz and to commission a screenplay by Nicholas Kazan. Shochiku acquired distribution rights to the picture in Japan and a worldwide equity position. Ed and Oliver Stone, as producers, along with the director, Barbet Schroeder, then solicited Sovereign for an investment to cover that portion of the negative cost attributable to the rest of the world absent the U.S., Canada and Japan. In fact Sovereign was the largest single investor in that picture. With that commitment the casting of Jeremy Irons, Glenn Close and Ron Silver was completed. The package was then offered to Warners, which made the domestic distribution deal, licensing all U.S. and Canadian rights in exchange for a distribution guarantee.

*Impromptu* and *Love Crimes* are rare examples of pictures Sovereign fully financed and produced. The "production" portion is unusual and something outside of our primary methods of operation. *Impromptu* was presented to Sovereign by the producers and ICM. The screenplay by Sarah Kernochan was outstanding. While Sovereign hoped a French-English coproduction would be possible, that did not prove feasible, so the company funded 100% of the film. When the picture was completed, Hemdale Film Corporation acquired the U.S. and Canadian distribution rights for a substantial advance against distribution.

*Love Crimes,* written and directed by Lizzie Borden, was originally developed by Miramax, which had released her *Working Girls.* Miramax offered Sovereign the international rights and subsequently requested Sovereign to cash-flow the film and supervise production. In return Miramax provided Sovereign with a bankable guarantee against U.S. and Canadian distribution rights. Sovereign secured a completion bond to cover production (see article by Norman Rudman, p. 216), which is usually a condition imposed by the lending institution.

Sovereign normally acquires no interest in the U.S. and Canadian rights unless the producers ask us to become involved. *The Comfort of Strangers* was financed jointly by Angelo Rizzoli's Erre Films and Sovereign with no regard to the domestic distribution market. For its half, Erre retained the U.S., Canada and Italy; Sovereign acquired all other territories. Erre then engaged Sovereign to obtain a domestic deal in exchange for a placement fee. The picture was placed with Skouras Pictures.

True to the business plan, Sovereign sees itself as a financier-distributor (in the traditional United Artists–Orion mold) rather than a producer. Instead of producing, as noted, Sovereign prefers working closely with established producers and helping to finance distribution of their pictures. It is equally unusual for Sovereign to acquire a completed picture, as with *Cinema Paradiso* and *My Left Foot.* In both cases extraordinary films were available. But these pictures are departures from Sovereign's business plan, which is worth repeating: to finance generally half the negative cost of a picture in exchange for international distribution rights in perpetuity.

All revenues from theatrical, television and video in the Sovereign territories are accounted for as follows:

- First, to Sovereign for its distribution fees (which are similar to the majors): 40% theatrical and 25%–30% for both television and video. All subdistributors' fees are included within these fees. (Note that sales agents add their fees—usually 15%—to these distribution fees, resulting in an effective fee of 55% to 60%.)
- Next, to Sovereign to recoup distribution expenses such as prints, advertising, other marketing costs and shipping. (Note that major studios apply a 10% surcharge on all advertising expenses.)
- Then Sovereign recoups its advance plus interest.
- The balance is normally divided 50/50 with the producers.

Although it is surprising in retrospect, *Cinema Paradiso* was completed without distribution agreements in place. Worldwide rights (outside of Italy and France) were still available. The producer, Franco Cristaldi, and the French coproducer, Les Films Ariane (one of Sovereign's investors), asked us to work as a sales representative only. There was no risk to Sovereign, other than an obligation to advance distribution costs. In this case Sovereign concluded agreements throughout the world, including the U.S. and Canada. Miramax licensed the film domestically. *Cinema Paradiso* led us to the cofinancing of director Giuseppe Tornatore's next project, *Everybody's Fine* (*Stanno Tutti Bene*).

International distribution differs from domestic in interesting ways. Sovereign works in fifty-two countries in thirty-seven different languages. Timing is a factor, in that a picture may be released in different countries at different times. The expenses and creative issues of dubbing and subtitling come into play. Dubbing is required in France, Germany, Italy and Spain for theatrical release, whereas subtitling is preferable everywhere else (all television requires a dubbed version). Censorship is another element, with different countries applying different standards. Deals between distributors and exhibitors are structured as a percentage of box-office gross, as in the U.S.

An overview of important territories and formats may be appropriate. For example, in theatrical there was a recent eroding of grosses in Western Europe, but now they seem to be on the upswing. However, there is a danger of American blockbusters squeezing out fringe product for exhibition time, such as pictures from the indigenous national industry and so-called independent English-language product. (This domination of exhibition by the majors happens in the U.S. as well.) In 1990 this occurred when three American studios dominated the global market: Fox's *Home Alone;* Disney's *Pretty Woman* distributed overseas through Warners; and Paramount's *Ghost* released as part of their partnership in UIP. This is a serious concern in all overseas territories. Although people are going to the movies in increasing numbers overseas, the question remains, how many movies are they seeing? How many admissions per title? The answers must be tracked and will have a major impact on the independent suppliers of product.

Home video is mature in most countries. But it is interesting to note that by and large the top-ten best-selling videos worldwide tend to be American pictures. If the future calls for movies on laserdisc, that will be another format through which to exploit motion picture rights. But sadly the stand-alone video distributors without a studio behind them are finding difficulty with B- and C-titles.

The exciting future lies in television, as demonstrated by Japan. This country is the biggest source of movie revenue outside the United

States, representing 18% of the international (9% of worldwide) theatrical movie revenue stream, and 30% of international (15% of worldwide) video revenue. But from television the percentage is very low since Japan has yet to develop pay- and free-television formats. The dramatic growth in television networks and stations, largely due to deregulation but also influenced by expanding technology, will prove to be a new revenue source.

For the European Economic Community (EEC) television has a different profile, characterized by the transfer of many national television networks from government control to commercial control. In the U.K. there are new broadcast channels emerging despite the temporary problems that caused the merger of British Sky Broadcasting into British Satellite. After a period of decline France has recovered theatrically, and in pay-TV, Canal Plus is the giant in Western Europe. Italy's Fininvest/Berlusconi is also a major player in television. Although pay-TV does not yet exist in Italy, it is inevitable. Sovereign feels it is particularly well positioned to deal with quota requirements in the European Community with a base in London and two European parents. (For more details on the global market, see the article by Peter Dekom, p. 417.)

Emerging formats, which are as fresh as today's headlines, are markets to monitor. Obviously the value of Sovereign's growing library is impacted: whether a new format will represent $5,000 or $500,000 or more per title is an evolving question, since these values reflect a developing consumer appetite. Certainly additional sources of revenue will inure to the copyright owner. Emerging markets in Eastern Europe, China and the Far East are also anticipated.

The movie business is now an integrated international business. Sovereign is working to offer the best of American and European sensibilities to reach a global market.

# XII

# THE FUTURE

# THE FUTURE OF TECHNOLOGY FOR MOTION PICTURES

*by* **MARTIN POLON,** *who forecasts technology in electronic entertainment for the financial industry. He has written over 350 articles for such publications as* Audio, Billboard, Broadcast Systems Engineering *(U.K.),* Computer Merchandising, Recording Engineer Producer, Studio Sound *(U.K.),* Television Broadcast, *and* Video *magazine. Mr. Polon has teaching appointments at the University of Massachusetts at Lowell and at the University of Colorado at Denver, after teaching at UCLA. A past governor and vice president of the Audio Engineering Society (AES), he has served as moderator and speaker for the Society of Motion Picture and Television Engineers (SMPTE), AES, the Aspen Institute, the U.S. Department of Labor and at conferences on video music, home computers, electronic mail and stereo TV. He consults widely on film-industry topics, has presented papers to forums such as the American Film Institute and is listed in* Who's Who in the World. *Polon Research International is based in Newton, Massachusetts.*

*The challenge to film posed by the new video hardware is the mounting pressure to replace 35mm film . . . with high-definition television (HDTV) and associated space-age electronics at all levels of the feature process. . . .*

For the decades since the advent of the sound motion picture, the technology of motion picture production and exhibition has been suspended in a genre of the industry's own making. It has been as though each camera, lens, light, microphone, sound-recording innovation, editing system, projector, and even the film itself has evolved to become the very best tool possible for the magic of movie-making. And the quality shows. Other forms of visual communications in the past have lacked the overall emotional impact of a properly exhibited theatrical motion picture, with its extraordinary resolution.

But the only constant in electronic entertainment and film technology is change. It is this change, as wave after wave of state-of-the-art computer and digital technology evolves in theatrical film applications, that has brought motion pictures and television to the onset of the twenty-first century. The challenge to film posed by the new video hardware is the mounting pressure to replace 35mm film and its attendant paraphernalia with high-definition television (HDTV) and associated space-age electronics at all levels of the feature process. The first time film was impugned by its brash upstart cousin was during the late 1940s, when television was a very unpolished system with crude resolution of picture on a ten-inch screen. This assault on the movie audience brought chaos to the film studios. Order returned only when the television industry had to turn to the studios for programming. The impact of that confrontation of technologies nearly broke the back of theatrical exhibition.

The relationship between film and video went through a different confrontation in the beginning of the 1970s, with the advent of electronic home-entertainment devices. Once home video was perfected, by the early 1980s, the availability of purchase or rental of movies for the home exploded as a new aspect of the movie business, eventually overtaking theatrical in revenue potential. The theatrical motion picture has retained primacy as the audience's first exposure to studio product, although at the expense of technical expansion.

The adoption of new systems and new technologies by the film business is inevitable, but the question of which electronic system and which technology remains a conundrum. Just as yesterday's science fiction is today's science fact, technology is changing so fast as to threaten logical decision making. The future is the present.

That future will come to electronic entertainment by evolution, given the changing pattern of worldwide exhibition. In years past a decision in Hollywood such as the adoption of anamorphic (widescreen) projection was followed quickly by all users of theatrical product. Today such a decision could as well be made in Tokyo or Osaka as in Hollywood. No purely domestic decisions can survive in the future marketplace of global entertainment technology.

Film will survive, either by being a part of a reasoned turn in direction to the new technologies or else it will be absorbed by a dynamic force that will reshape it without historic restraint. Colorization of black-and-white motion pictures is an example of this process. Change seems inevitable, and there is room for new technology in the creation and reproduction of the theatrical motion picture.

## CURRENT TECHNOLOGY

### In Production

In the past the making of a theatrical motion picture encompassed a good deal of modern technique, but more by adaption than invention. The industry has found it easier to borrow and adapt existing technology, with some notable exceptions. For example many production companies make extensive use of radio equipment. Radio provides the synchronism between camera and audiotape recorder. Wireless microphones are commonplace. In both studio and location production, two-way radios are used to cue all the interrelated elements instead of the old-style bullhorn. The technology to accomplish this came from military communications, television broadcasting and law enforcement. Filmmakers have adopted existing products.

Today technological innovation rules the roost in filmmaking, despite the fact that the spiral of inflation has not left a comfortable margin for research and development. The tools of the trade have progressed into better instruments to transfer a creative performance onto film. Cameras have become smaller and quieter, lenses have shrunk in size with less optical distortion, and all viewing is reflexed directly through the taking lens and frequently into a TV monitor. Sound recording is often digital and uses very few constraining cables, since the link to the camera and microphone is usually by radio, not wire. Tape recorders and mixers are significantly smaller than a comparable setup of twenty years ago; even analog equipment is quieter electronically and has further noise reduction built-in. The sound is recorded with a minimum of noise and distortion. Digital video equipment, such as night-vision lenses, has become the companion to film production. Lighting equipment is smaller and produces less heat with the use of new lamps containing exotic filaments and gases and new housings. Editing systems have shifted from the vertical to the horizontal plane, with wide-screen viewing.

Several innovations have been noticeable in releasing the act of filming from previous physical restrictions. The gyroscopic back-supported camera mount (Steadicam, etc.) with enhanced video viewfinder allows a cameraman to move about rapidly along with the action of a scene, while the camera captures the moment without any image vibration. The stabilizing system transforms any mode of transportation into an effective camera platform. The development of new film stocks has allowed more latitude of exposure within varying light conditions. Recycling of the silver-based emulsion after film processing has also been enhanced. Special effects have moved from the camera and optical lab to computer-assisted effects generation. Laboratory processing of exposed film offers more opportunities for creative development control when governed by computer systems.

Another improvement has been to reduce the size and weight of production equipment, allowing location shooting to be done from specially configured trucks and vans with a minimum of difficulty. More time is then available for creative aspects, since the equipment setup is simplified by the reduction in bulk. Current industry practice has brought theatrical film production technology to a point of polish; that gloss is sometimes allowed to rub off in exhibition.

**In Theatrical Exhibition**

Any discussion of theatrical exhibition must begin with the multiplex. Over years of development these multiscreen theatres have evolved to offering greater choice (from four to twenty or more screens), comfort (quality seating with cup-style armrests) and a sense of safety (patrolled theatre complexes and parking). But the reduced projection staff and increased automation using platter systems sometimes leads to projection-induced print mutilation. For the distributor the continuing spiral in petrochemical and silver costs has increased release-print lab and film expenses, and therefore damaged prints are not readily replaced. The problem is compounded when film exchanges transfer prints mangled by faulty equipment to exhibitors with quality projection systems. The exhibitor wanting a perfect picture may not be able to achieve that goal, short of demanding a new print. But automation is not the only culprit in print abuse. Any theatre with poorly maintained projectors or rewinds is involved. And with enhanced exhibition environments and prices, audience tolerance for scratched prints has evaporated.

There are many positive aspects of exhibition today. The troublesome arc lamps used in the past—burning rods of carbon with variable light output—have been replaced by xenon-gas lamps. This provides an extraordinary projection light source with no flicker or variance, requiring little or no maintenance. Screen efficiency in returning light to the audience continues to improve. To take advantage of this brighter picture, there have also been advancements in the wide-screen anamorphic process, both in cameras and taking lenses, as well as in projection. The whole range of Panavision and Panaflex cameras, for example, has evolved to the point where wide-screen production is no more difficult than using a normal aspect ratio.

The improvement of sound reproduction in theatres has seen a dramatic implementation noticed by patrons internationally. The most spectacular effect on sound has been the advancement of the Dolby system for film recording and reproduction. Improving upon developmental efforts at RCA and Eastman Kodak, the Dolby Laboratories took their system of reducing film noise and applied it to enhance the frequency response of movie sound and stereophonic effect in theatres. The Dolby system reduced high-frequency noise present in optical film sound by recording the noise-susceptible high frequencies with accentuated volume. In projection this accentuation is removed, leaving a clean signal up to at least 12,000 hertz. Dolby also enables a stereophonic sound track to be used on standard optical release prints. This has brought multichannel stereophonic surround sound with optimum fidelity to thousands of theatres around the world. The conversion equipment is simple to add to theatre projection systems and does not affect existing films that are not Dolby-encoded. One of the key connections of entertainment technology is that there must be enough entertainment software available for the hardware to succeed. The Dolby system has met this test, and Dolby-encoded prints are an industry standard. The Dolby SR system further improves film sound quality for exhibition with its near-digital realism.

What distinguishes the theatrical motion picture in exhibition from any other visual-communications medium is the quality of resolution it delivers, the equivalent of many thousands of lines of scanned information. The challenge is whether the film industry can use this quality as a base for further improvement or be forced to adopt a superior video system. Whether theatrical exhibition can continue to attract an audience will partly answer this challenge.

**In Home Viewing**

The world has standardized on three systems of video transmission. The U.S. system, known as NTSC (National Television Systems Committee) uses 525 scanning lines to create a single image. The NTSC system is used in North America, Mexico, Japan and over twenty other countries. The SECAM system (Sequential and Memory) originated in France and is now used in Eastern Europe, the former Soviet Union and in many other countries. The third system is PAL (Phase Alternation Line), used in nearly forty countries. Both SECAM and PAL scan at higher line rates, affording significantly better resolution than the 525-line U.S. system.

The home market for electronic entertainment has shown quantum expansion since 1980, when competitive home video was introduced. The arrival of home videocassette recorders (VCRs) in the entertainment marketplace took two incompatible forms in the early 1980s. The Beta system, introduced first by Sony Corporation, used omega-wrap tape cartridges (so named because the tape resembled the Greek letter Omega when it wrapped around the video head drum). Machines using this format were manufactured by or for Sanyo, Sony, Toshiba, Zenith and Sears. The Beta system is no longer being manufactured. The surviving system is known as the Video Home System or VHS. Originated by the Victor Company of Japan (JVC), the system uses an M-wrap cartridge and is being made by or for all of the consumer electronic makers today. VCR adoption in the United States has achieved the status of a universal appliance, in over 70% of television households.

In the high end of consumer electronics there is a third system, Sony's 8mm-size video format, which utilizes a smaller, lightweight camera, with self-contained playback system that plugs right into any TV monitor. In consumer audio, there is the DAT (digital audiotape) technology, which offers digital quality, the competing DCC (digital compact cassette) and the MD (digital mini disc).

Projection or large-screen systems for video viewing have increased their importance in the home. Several formats have been developed to produce the large pictures. The *picture tube system* uses a screen-type picture tube similar to that found in a conventional color TV. This tube has extra brightness that, when focused through at least one lens (usually fresnel) and reflected through a mirror (or electronically reversed to "read right" the reversed image), will project a picture to a screen. Another format uses *projection tubes*, most usually three tubes for the primary red, blue and green signals that make up a video color signal. The tubes are focused to converge the three colors on a screen, combining to form a complete color picture on the light-intensifying surface. The LCD (liquid crystal display) projection system uses color-emitting semiconductor surfaces that are picked up by high-intensity light sources and projected onto walls or screens. Although these and other consumer technologies for viewing theatrical product are very attractive in the home, adoption is still influenced by consumer attitudes and the outlay of discretionary income.

We have seen many technologies for home use fail initially to achieve significant acceptance. Eventually, however, the laser video disc, the interactive cable system, direct broadcast satellite and pay-cable have all become part of the fabric of home exhibition. The appetite for home video software has proven to be prodigious. The paradox of the emerging technologies in the movie business is that, if public interest in theatrically exhibited product wanes, some portions of the theatrical customer base will switch to or be absorbed by different markets already extant.

## FUTURE TECHNOLOGIES

### In Production

The feature industry seems committed to film as the visual vehicle for original production for the foreseeable future. That is a strong testimonial to the dogged determination of Eastman Kodak and other suppliers, such as Fuji and Agfa, to create new 35mm film emulsions capable of meeting the high demands of definition to the twenty-first century. The speed of original negative film stocks has been boosted without any loss of detail or of resolving power. The match of production color rendition from original negative to internegative to color positive for release prints has continued to improve to a point where "what you see is what you get." Release prints have longer-term color stability due to enhanced cyan dye dark-keeping characteristics and resistance to color fading. These advances have cured the problem of poor release prints not delivering the filmmakers' perceived rendition due to poor color match or color shift in aging prints. Camera makers Arriflex and Panavision have labored to create new and more flexible instruments to capitalize on the new film stocks. That film remains a successful production medium, by most measures, with its new visual qualities, is a given.

With the premium that quality movies hold in the home-video marketplace, studios are cleaning up their libraries so that subtle black-and-white movies and well-preserved color films, particularly in the original three-stripe Technicolor imbibition process, can be viewed in pristine condition. But this attention to print quality has also led to the concept of "colorization," whereby many black-and-white motion pictures and television series have been colorized, using computers to "paint" in colors, frame-by-frame. Although this technique has brought new revenue to the old product, it has also attracted extreme controversy to the colorizing of classic films, such that Congress has required disclaimers to appear on a certain number of designated classic films, whenever they are exhibited as colorized. However, colorization has been a lifesaver to those films produced between the 1950s and 1980s that have suffered color fading of the prints and negatives.

Whatever the direction of the public's taste for theatrical and home entertainment, the need for both film and video product is greater than at any other time in the history of the industry.

The use of energy-efficient systems is going to be one potential direction for the technology of film production to turn to. Solar-power assemblies will be used for location filming and remote video setups. This will begin to replace dependence on gas-powered electric generators. Solar-power packs are already in use to power military communications systems. The high level of reliability shown is well suited to location film production. The technology of solar cell assemblies and associated power systems is such that weather conditions too severe for filming would still allow energy storage.

Inflation touches all segments of the economy, and an industry as complex as the film business is buffeted by each ripple. The issue of high cost comes at a time when competitive technologies of video production offer substantial economies, allowing video to be interchanged with film in production and postproduction. As a production tool video assumes a role approaching unity since it can capture, balance and edit instantly. Postproduction, especially as video proceeds into the realm of digital electronics, will provide many effects not available or affordable on film. And with high-definition video, picture quality can equal that of film.

The drawback to the use of video, historically, has been poor resolution. At the beginning of the 1920s, 35mm film possessed such high image resolution that projection merely needed the arrival of a relatively reliable light source, the carbon arc, and improved lenses to fully capitalize on image quality. Television started out at that time, with 24 electronic lines scanned. World standards established via electronic improvements after World War II increased resolving ability to 500 lines or better. These postwar standards have remained in use because millions of television sets have been manufactured to operate on these standards. While there is no easy way to increase picture fidelity without casting aside all of the world's TV receivers, video production systems do not have to conform to any scanning limits under 1,000 lines.

High-definition video systems are now a reality, originally spun off from aerospace and military video research, where high resolution is vital to visual accuracy. In any situation where the final video image can be transferred, as in production, the image-producing system should resolve at the highest rate technically feasible. Scanning rates over 1,000 lines have proven reliable for tele-production systems. NHK, the government-owned Japanese Broadcasting Corporation, has been using entire video production systems operating above 1,100 lines for years. Since the resolving power of the human eye operates logarithmically, rather than in a linear mode, the difference between, say, 525 lines and 1,100+ lines scanned is better than the perceptive doubling in the clarity of image. Any system operating above 1,000 lines provides significant visual advantage in picture fidelity and definition. Production HDTV equipment offers a whole range of production video hardware in high-resolution systems. Although the evolving large-scale memory technology and microprocessor systems could allow the scan rate to eventually go higher still, HDTV systems in the range of 1,500 lines of resolution will most likely be standardized for studio production.

With such a high rate of visual definition, a production HDTV video system can be exchanged for a film camera, laboratory and editing facility. The rapid response via VTR playback allows confirmation of any given scene, eliminating the need for costly retakes upon viewing work prints or dailies. Most directors already refer to conventional video playbacks while shooting on film. Francis Coppola has been at the forefront in experimenting with applying the combination of computer and video technology in an effort to reduce costs and the lag time between first day of shooting and delivery of a completed picture. George Lucas has also invested in merging computer and video technologies, leading to the construction and marketing of such systems as the Editdroid for postproduction editing.

Productions today can choose between a film-based editing system such as Kem or Moviola, or video-based editing. A video system has the advantage of preview and of speed. Since video edits electronically by storing image information into a memory and then transferring the visual images, there is less danger of destroying original material inadvertently. Various shots can be viewed, stored and compared prior to the actual edit. The transitions can be done via electronics with cuts, fades, wipes and literally hundreds of other effects that would require lab processing if done on film. In addition, complex special effects can be generated directly by computer to videotape. The development of sophisticated multidimensional digital graphics programs for computer enhancement have proven vital to the successful effects in such pictures as *Backdraft* and *Terminator 2*.

The abiding issue for the feature-film industry and its existing technology is the way it handles the pressure to replace 35mm film and its attendant hardware with electronic high-definition television (HDTV) and associated electronics at all levels of the feature process: production, postproduction, distribution and exhibition. Most of the pressure on film technology is financial, and certain economies of video production could offset the rising costs of features. At present, film maintains its primacy over the latest video developments in projection and equals the best that video can offer with its new "high-definition" film stocks. Nevertheless no other technological event in the history of film, not even the coming of sound at the end of the 1920s, promises to be as revolutionary as the push to gradually "HDTV" the movie business.

The studios themselves are always looking at enhancing video systems for production. Paramount and Twentieth Century Fox, among others, have installed state-of-the-art electronic soundstages for video production on their lots. Other facilities could be converted to high-definition at the appropriate time.

In the editing of postproduction sound, digital audio work stations have become as much a part of the motion picture environment as they are on the recording studio scene. Outside as well as inside the film studio, audio and visual enhancement via personal computers and midi-based electronic musical systems has created an entire electronic cottage industry for postproduction. Composers working on film scores can produce all of the music at home, using a battery of keyboard synthesizers, mixing consoles and digital audio recorders.

The creation of sound effects has moved in two directions. The emergence of digitally recorded effects, distributed via the absolute digital purity of compact discs, has given the effects expert a new and more cost-effective tool. Similarly, computer sampling techniques allow an effect—from a CD effects disc, or recorded specifically, or recalled from a computer ROM (Read Only Memory) chip—to be enhanced or modified at the command of the effects designer sitting at a keyboard. The technology is also completely portable and can be accomplished in a home studio. This alternative to the Foley stage offers a vast array of tools, and many Foley houses now provide a hybrid of both technologies.

Visual effects, whether done on an affordable Amiga personal computer or via a much larger computer processing array, allow for the creation of any image a motion picture calls for. The full range of magic tricks from the film industry's past—in-camera effects including mattes and painted images, optically printed effects—has waned as computer animation and computer image creation advance. Needless to say, the larger the computer, the more realistic and spectacular the effects can be. Even smaller systems are providing a versatile array of functions at rock-bottom prices.

**In Distribution**

Currently, a major expense for studio-distributors in feature distribution remains the cost of striking 35mm prints from the master negative and transporting the prints to theatres. In the United States some distribution executives cite the cost savings from the elimination of film prints and film exchanges as justification for gradually shifting to video distribution for the domestic market. Despite the relatively low percentage represented by print cost to the total feature-film production and advertising costs, the millions spent on prints (in the range of $1,500 each) and print distribution per film receive increasing scrutiny in studio management circles.

The technical scenarios most often quoted to replace film use either satellite or laserdisc systems to feed theatre video projectors. Real or hypothesized savings would include the reduction of theatre technical staff. In order to transmit features via existing satellites at the lowest cost, off-peak hours would have to be used. However, this mode lacks one key element. The world electronics industry has not yet produced a reliable, high-quality recording format to store the features downloaded from the satellite ("bird") during nonpeak transmission hours. An alternative scenario would call for movies to be transmitted live via dedicated satellite to all theatres at listed showtimes continuously, obviating any need to record.

The cost to the movie industry of exclusively maintaining at least three and perhaps four birds for live, continuous distribution does not represent an appreciable savings over today's print-based system, at first glance. Upon closer inspection, however, the multiple costs of all film distribution (print runs and exchanges) for, say, 240 films a year (20 films a month times 12) multiplied by an average distribution cost of nearly $2-million per print run would total nearly $500-million. The costs of tariffs for equivalent, dedicated satellite time in the U.S. for all those films and all served theatre circuits might very well equal or be less than that figure. Over time, many expect satellite-distribution expenses to shrink, while film expenses will only grow. Add the potential savings from eliminating projectionists in most if not all theatres and the economics of satellite delivery become more cost-efficient than film. One could even envision one of the current satellite consortiums launching a new bird or birds for the exclusive use of the motion picture industry. Yet another scenario could use the multiplexing technology envisioned for direct broadcast satellites, in which case channel demands and corollary satellite expenses would be further reduced. All of this assumes, of course, the advent of acceptable large-screen video projection in theatres.

Piracy would have to be addressed if satellite transmission is used for theatrical distribution and exhibition. Currently movie thieves use the odd release print stolen from an exchange or while the print is in transit, or bribe copies from pay-television technicians or even from film studio staff. Elaborate and reliable electronic encryption will have to be used in transmitting movies to theatres, with the extra cost of expensive descrambler units at each theatre. Added to that is the technical problem of maintaining video waveform integrity throughout the scrambling process.

Laser video disc becomes another methodology to distribute and exhibit video feature films. Its high visual quality, eventual compatibility with high-definition video systems and ease of transportation (via overnight or mail) bode well for distribution to exhibitors. Since current recording time is limited to one hour per disc per side, codes recorded on each disc would have to trigger "changeovers" to a second or third theatre player to accommodate feature length. Other disc technology compression techniques could increase play time for a single disc. Add to that the antipiracy potential for the discs to self-destruct after the playdate, eliminating any need for return to the distributor. With players costing only about $1,000 for a robust industrial version, this becomes an intriguing option.

All of the ancillary markets of the movie business are serviced electronically—without film. Video (not yet high-definition) plays the integral part in transmitting motion pictures to markets that follow theatrical exhibition in the revenue stream. Features distributed to home video, airlines, pay-cable, network television, and for TV and cable syndication are all transferred to videotape, either at the studio or at the secondary distributor's place of business. Satellites are used to "flash" completed product to various facilities for syndicated distribution.

The film-vs.-video question is going to be tested in Third World and emerging Eastern Bloc countries. The goal in these countries is to standardize on very high quality video-projection theatres as a lower-cost alternative in the creation of a film exhibition and distribution infrastructure.

**In Theatrical Exhibition**

Film projection systems have continued to improve. Utilizing advances in release-print quality, picture brightness has been increased via faster projection lenses and improved light return from screens. Couple this with advances in film sound, and the synergy results in a viewing experience heretofore unrealized.

Projection systems become the weakest link in any proposed mass adoption of video for theatrical exhibition. Despite the advances made by such consumer electronic giants as Sony, Matsushita and Mitsubishi—plus the still impressive performance of the massive Swiss Eidofor systems—neither image quality nor affordability is on the side of theatrical video-projection planners as of this writing. Further, the cost of reequipping theatre projection rooms in every venue in America would be prohibitive.

Exhibitors have resisted spending money on technological innovations that might well help the studios more than themselves. Many are more eager to reconfigure armrests with cup-holders than invest in a high-end, sophisticated sound system. After all, as most exhibitors queried about these new developments observe, any dollar savings from the elimination of film distribution would not materially impact their own bottom line. They point out with some angst, for example, that the studios would expect theatres to pay their share of live satellite distribution costs, while gaining only the marginal benefit of possible projection automation. This perceived benefit is a dubious one. After all, moving parts and loadable features in laserdisc systems or early-morning satellite download-and-record schemes would mediate against the total replacement of theatre technical staff. In any multiplex complex, whether film or video, there will be plenty of equipment requiring trained personnel to be on hand to deal with failure of any component. The show must go on!

The final measure of the film-vs.-video-projection question will be found in its effect on audience attendance. It comes down, of course, to the issue of visual quality. Video systems capable of serving up to 150 people do provide acceptable alternatives to the projected film image for most viewers. But larger audience-capacity systems such as the aforementioned Eidofor are either prohibitively expensive (reaching into six figures) or provide an image unacceptable in relation to that of projected film.

To improve sound in the theatre, the film industry is turning to the digital audio technology that has had such a major impact on production sound recording and playback. Digital sound recording and reproduction can be likened to a perfect copy of a fine painting by an old master. If that reproduction is run over by a truck, it is seriously degraded. The film sound track likewise suffers from both the medium of recording and the physical movement of the film through the projector. To continue the artistic analogy with digital, in order to create a fine reproduction, the original painting would be copiously logged by computer so that each color and brush stroke would be recorded in memory as a series of numbers, allowing the painting to be reproduced at will, following the detailed instructions. So it is with digital sound on film, where the sound track is recorded as a series of binary numbers (*0*s and *1*s). These numbers are retrieved in playback, and the sound track is re-created from the detailed instructions. The use of digital technology allows existing methods of recording and projection to be used, since it is the sound signal itself that is encoded and decoded, totally unaffected by the flutter of the projection movement or the high-frequency noise of optical recording. The resulting digital sound is exceptional.

No example offers the potentials or pitfalls awaiting advances in feature-film technology better than the evolution of sound and sound systems in exhibition. The digitalization of audio will allow film sound finally to escape the frequency-response limitations of the optical film-projection system. The nonstable movement of the film through the sound gate of a projector has always limited sonic response due to low-frequency noise, while the high frequencies have been limited by the use of the optical sound process itself, which generates high-frequency noise. Until the Dolby system devised a technique for reducing the noise through electronic treatment, film sound was limited to a frequency response to 100 hertz (or cycles) to 6,500 hertz, with some sound reproduced as high as 8,000 hertz. The original Dolby system for film sound allows a high-end response of around 12,000 hertz, and Dolby SR extends that even further. The Dolby system has greatly improved film sound fidelity, and the promise of further enhanced sound from multiple channels with surround has finally matched the expectations of the audience.

A further jump in exhibition sound quality is expected with the addition of digital sound tracks to theatrical films. Digital optical audio was first pioneered by Eastman Kodak and Optical Radiation in a noncompatible system to existing sound tracks. The Cinema Digital Sound system was incorporated at some venues as part of the massive national release of *Terminator 2*. Digital competitor Dolby has perfected a truly compatible multitrack system, so that film distributors will be able to stock only one kind of print, capable of being reproduced monaurally, or in stereo, in analog Dolby or digital Dolby sound formats.

For some time the penetration of Dolby stereo-equipped theatres, some with high-quality HPS-4000 or THX sound systems, has been on the increase. But in multiplexes some theatre owners have been equipping only a percentage of their screens with stereo and an even smaller percentage yet with high-quality speaker arrays. In fact, speaker systems of reasonably high quality—new high-end units, less expensive new units, revamped older systems, enhanced existing systems—are estimated to exist in less that 25% of all sites in North America.

Occasionally purveyors of high-quality speaker systems have refused to install at a particular screen site because of acoustic deficiencies of early multiplex complexes or older theatres divided up into smaller ones. Add the complexity of installing surround and rear loudspeakers, and it is not always guaranteed that discernible surround stereo will be accurately provided to the audience. (Further complicating the issue of surround sound in exhibition has been the use of psychoacoustic adaptive systems, which attempt to introduce the "perception" of surround sound in a nonsurround-equipped stereo environment. These systems depend upon the use of delay and electronically altered phase and frequency response in the film's sound track.) So what the audience hears may have little or no resemblance to the magnificent sound track on the film, as mixed in the studio's superbly equipped environment.

These technical differentials are only a small pitfall when a consumer seeks out a print exhibited in high-quality stereo surround.

Theatre advertising in newspapers can thwart the search as well. The distributor will take a larger display ad, containing the campaign artwork and usually specifying that the picture is being presented in Dolby stereo (or Dolby SR) in certain theatres. But such stereo designation does not always appear in the theatre chain's own cluster ad. In a major East Coast city, for example, on a given Friday in April, of the twenty-one films featuring display ads, only fifteen indicate the presence of Dolby stereo "in selected theatres." Only two were identified as being in the SR process, whereas most if not all were produced in Dolby stereo. But the exhibitor ads were generally noncommittal as to whether or which specific screens offer a movie in stereo, even though most screens are stereo-equipped. Why? To allow the exhibitor the flexibility to move pictures to different screens within a multiplex depending on popularity. If a picture set for a large auditorium opens poorly, it is relegated to a smaller theatre, one less likely to be as well equipped. This is an example of how advanced technology can be thwarted at the point of sale. The moviegoer is wise to inquire at the box office, rather than relying on advertising, when seeking optimum sound.

A scenario for theatres of the future would rely heavily on state-of-the-art technology to provide the maximum entertainment experience for the audience. In this arena visuals could utilize the widest screen technology possible without the optical distortion at the edges of the picture that currently limits the dimensions of conventional projection. New lightweight lenses could allow extreme wide-screen projection onto wrap-around screens. The digital sound track of the film could contain two sets of computer information: one, a stereophonic multichannel surround-sound mix; a second cue track could contain information to operate various physical effects in the theatre. The cue track would instruct computers to provide these other effects: seating could be capable of sustained vibration by being mounted on a platform or shaker-table; air in the theatre could be climatized, controlling temperature and water vapor content; elements capable of producing scents by electrical current flow could provide trace aromas.

As "far out" as all this seems, some of this technology has already been used in certain touristic venues, such as "The New York Experience" multimedia presentation at the South Street Seaport in Manhattan, and at the many IMAX theatres around the country. The IMAX theatres are showcases for the largest projected image in existence. Each IMAX frame, at 24 frames-per-second, spans three conventional frames of 70mm film. The process was invented in the late 1960s and uses specifically designed cameras and projectors. The system moves horizontally through camera and projector (as did the Vista-Vision system of the 1950s), rather than vertically. Short films produced in this format, with commercial sponsorship, are shown at

IMAX theatres on flat screens as tall as five-story buildings and 70 feet wide. (Another IMAX process, OMNIMAX, projects onto a planetarium-style concave screen.) The IMAX experience is truly breathtaking, especially when transporting the viewer to the Grand Canyon, over mountains, or through the ocean. Another familiar format is Disney's Circle-Vision 360, found at Disney theme parks, whose nine-camera simultaneous photography from a central locus fans out over 360 degrees when projected onto nine adjacent screens that form a complete circle. The effect is a total suspension of disbelief. Other systems, such as Showscan, also use new technologies. Unfortunately it seems unlikely that all of the innovations capable of enhancing the moviegoing experience will be adopted for commercial theatrical exhibition due to cost and exhibitor reluctance.

**In Home Viewing**

The home will be the focal point for the continued development of technologies heretofore reserved for professional television and telecommunications. Video and computers will dominate the innovations to serve the home as systems for displaying information and as methods for creating, controlling and capturing entertainment, information retrieval and family activities.

The television set will change form and will exist on several levels. The range will span from pocket-sized color units to wall-size flat HDTV screens. Hanging on the wall, the home TV will have a 5-by-8 foot screen composed of liquid-crystal elements in a matrix, or may be composed of gas-discharge display materials. Within the gas display, fluorescent cells of red, blue and green materials will be arranged in a discreet pattern forming the picture.

The sound to accompany this improved television technology is already being enhanced. TV audio, digitally transmitted on satellite (or within the video signal) or in the analog domain, arrives in the home in a stereophonic-surround format. At the end of the 1970s the transmission of stereo television into the home was an accomplished fact in Japan; ten years later this was true in the United States. Using techniques developed for FM radio, the TV audio signal (which is FM) utilizes an MTS subcarrier and DBX processing to allow full stereo transmission into the home. In addition, movies viewed at home in Dolby stereo with surround can be reproduced by home components. When combined with large-screen video, a "home theatre" environment is created.

Recording and reproduction of video entertainment will continue to be enhanced. In the 1970s professional video recording used two-inch videotape. One-inch videotape then emerged as a new professional alternative, with better audio quality on the smaller machine. Today digital videotape recorders represent state-of-the-art for professional video recording and playback. The three-quarter-inch

U-matic video recorders are being replaced in service for the industrial and professional video market. The half-inch home VHS cassette, used as the standard for home recording, plus Sony's Beta system, have evolved as more compact, hand-held professional alternatives. The advantages of digital recording, that is, laying down and picking up video information from tape independent of the mechanical or electronic negatives inherent in the recorder, offer substantial quality improvements regardless of format. Once digitized, the signal-to-noise of the recording will generate flawless originals and subsequent copies. It is expected that, in the future, all half-inch video equipment will be in the digital domain, first for professional use and then in the home.

Spectacular growth for the reproduction of features in the home via laser video disc has not yet occurred. The continued involvement of Pioneer, Sony and Philips suggests the future potential for the laser video disc format.

The home computer, perhaps incorporated within the TV set, will continue to grow as a billion-dollar industry for home entertainment and information control, including home banking and shopping. In addition, the television signal could carry encoded data that could be retrieved as pages of information on the TV screen.

The VCR has achieved commodity status and near saturation. The laser video disc has had much more modest reach and penetration, though it still remains the high-quality home feature-viewing tool of choice. Cable-TV systems are commonplace and shall continue their presence on the American scene. Pay-TV service access, whether delivered by cable or satellite, promises to continue to add viewers beyond the already achieved sixtieth percentiles.

Pay-per-view systems, combining the technologies of the movie business, the cable industry and telephone-service providers, will continue to work at luring movie-watchers at home, while their success in offering one-time-only sporting and concert events will also continue. A measure of the potential can be appreciated as over 1-million subscribers recently paid $35 each to watch a live boxing championship. Certainly most urban American homes have access to current movies via broadcast, cable or satellite pay systems. But what of rural America? Companies such as TVN Entertainment (with backers including MCA and Paramount) and the SkyPix Corporation are positioning themselves to offer pay-per-view movies to this market. For example, after a rural homeowner invests at least $400 to as much as $3,000 in a satellite dish, and pays a service fee of perhaps $20 per month, movies will be available by ordering through an 800 phone number, at a cost of around $4 per title.

The home will be linked via video cable and/or telephone company fiber-optic line and/or direct-satellite broadcasting to the broadest base of entertainment and information sources imaginable. Some of the linkages will allow for interactive communications, as in addressable cable, so that an audience can be polled after a presidential speech, for example. (Several interactive systems have been tested in certain communities by various cable systems in the 1980s.)

Playback will take place on a sophisticated linkage of the home computer, video projector and disc player and the stereo sound system. The computer could be programmed to operate such accessories as audio delay. For example, a broadcast of the London Philharmonic playing from the Royal Festival Hall in London could be available live via satellite. The home computer could be instructed to present the program for viewing on the projector screen with the stereo TV sound enhanced via audio delay by the exact delay time one senses when sitting inside the hall itself. In the same operation the computer might also activate the audio unit to record the program while the video recorder is transcribing a screening of *Casablanca* from a local TV channel without the commercials (which the computer senses and removes). The gathering and interconnection of all the technology tools for home entertainment will yield the most powerful mechanism for consumer control and will continue to be a strong barrier facing exhibitors of theatrical films in the future.

The introduction of home high-definition television receivers is just cresting the horizon, as tests under the authority of the U.S. Federal Communications Commission work toward an eventual and timely U.S. domestic HDTV standard. The Japanese HDTV system is currently available by satellite on an experimental basis to certain households. The European Eureka system is forging ahead for eventual adoption within the European Community.

But in relation to key issues such as global compatibility, cost, and timing, there are dark and scary woods on the Yellow Brick Road of the advance of video high definition into the world marketplace. A common HDTV standard would allow coproduction and production exchange of programming over borders without any technical problems to be "standards converted." But this is not what is taking place. First, the technology under consideration by the FCC for the United States will be incompatible with the in-place Japanese system. Second, there is a tendency in other countries to adopt an analog standard now, in the short term, thus precluding a superior global digital standard in the long term. Third, the presence of the European Economic Community on an equal footing with the United States has fostered a political climate of technological dualism. Under a multinational technology project called Eureka, the Europeans (and by geography the Eastern European Bloc as well), have selected a systems standard for high-definition incompatible with anything proposed for North America.

Agreement on a single, global-compatible digital standard would allow for the development of home and studio high-definition equipment with complete accessibility for all. It would accommodate all of the regional differences in scanning standards and power frequencies, using software to define each region's features. The price that will be paid for the current disagreement over separate regional analog development of HDTV will be the higher costs for home and studio HDTV equipment, as well as international program incompatibility. Yet there are many who hope, at least for the production studio, that a common digital standard can still be adopted.

## CONCLUSIONS

The motion picture industry can dominate the very technologies that seem to menace it by supplying the software (programming) for the respective hardwares. But the film industry has to change on many levels. The making and exhibiting of motion pictures requires the latest equipment, even if that equipment is costly. And in theatres, studio-distributors must share that burden with exhibitors if exhibition is to survive.

Separately, the enormous potential of home entertainment will not be fulfilled solely by the theatrical motion picture, as demonstrated by the popularity of MTV, how-to videos, home-shopping networks and documentaries. The number of cable channels available, potentially up to eighty or more, will require new, creative production at reasonable prices. It is video that meets the software challenge, especially with high-resolution systems and rapid, enhanced edit and modification capability.

Perhaps one way to best consider the future of film is to look at the research and development (R & D) budgets of any major technology user, manufacturer or broadcaster. Communicators such as the telephone companies and computer companies also share the mantle of R & D. The amount spent at each company may approach 10% of gross profits or more. The motion picture industry, however, does not expend, in comparison, enough to sustain even a modest industry-wide research program.

One hopes that the film industry will meet its challenges. The vision of the future should provide motivation to enter a new era of integrated, cooperative efforts in production, distribution and exhibition. In the technological world of entertainment the future may well be decided by corporate and political discussions, by telecommunications multinationals and global standards, and by the timeless attraction an audience feels for entertainment. The industry can be entering a new age if it looks forward, not back.

# INDEX

# ABOUT THE EDITOR

*JASON E. SQUIRE is an independent producer and movie industry consultant who began his career in 1969. He has served as a creative executive in feature films in New York and Los Angeles for United Artists, Avco Embassy, Italian producer Alberto Grimaldi and in movies for television at Twentieth Century Fox.*

*Mr. Squire has acted as a consultant for producers, screenwriters, motion picture-related investment groups and for the USA Film Festival in Dallas. He has spoken on conference panels at the Aspen Institute, the American Film Institute, the Beverly Hills Bar Association and the Audio Engineering Society.*

*He is an adjunct professor at the University of Southern California School of Cinema-Television and in USC's Master of Professional Writing Program. A member of the Writers Guild, Mr. Squire wrote the screenplay for* Red Harvest, *an Italian-Spanish coproduction, and a stage play,* Waiting Room.

*In the recording industry he was executive producer of the soundtrack albums for* Conan the Barbarian *and* Conan the Destroyer.

*In 1972 he coedited the pioneering textbook* The Movie Business: American Film Industry Practice; *in 1983 he edited* The Movie Business Book, *which has enjoyed wide adoption at universities, a U.K. edition and a Japanese translation. His current work,* The Movie Business Book, Second Edition, *is a fully revised and updated volume that addresses the changing dynamics of the international movie business in the nineties.*

*An honors graduate of Syracuse University and UCLA, Mr. Squire is a member of the Pacific Southwest regional civil rights committee of the Anti-Defamation League. He is currently developing movie projects and is based in West Los Angeles.*